Cities and Gender

Men and women experience the city differently in relation to housing assets, use of transport, relative mobility, spheres of employment and a host of domestic and caring responsibilities. An analysis of urban and gender studies, as co-constitutive subjects, is long overdue.

Cities and Gender is a systematic treatment of urban and gender studies combined. It presents both a feminist critique of mainstream urban policy and planning and a gendered reorientation of key urban social, environmental and city-regional debates. It looks behind the 'headlines' on issues of transport, housing, uneven development, regeneration and social exclusion, for instance, to account for the 'hidden' infrastructure of everyday life. The three main sections: Approaching the City, Gender and the Built Environment and, finally, Representation and Regulation explore not only the changing environments, working practices and household structures evident in European and North American cities today, but also those of the global south. International case studies alert the reader to stark contrasts in gendered life-chances (differences between north and south as well as inequalities and diversity within these regions) while at the same time highlighting interdependencies which globally thread through the lives of women and men as the result of uneven development.

This book introduces the reader to previously neglected dimensions of gendered critical urban analysis. It sheds light, through competing theories and alternative explanations, on recent transformations of gender roles, state and personal politics and power relations; across intersecting spheres of home, work, the family, urban settlements and civil society. It takes a household perspective alongside close scrutiny of social networks, gender contracts, welfare regimes and local cultural milieu. In addition to providing the student with a solid conceptual grounding across broad structures of production, consumption and social reproduction the argument cultivates an interdisciplinary awareness of, and dialogue between, the everyday issues of urban dwellers in affluent and developing world cities. The format of the book means that included in each chapter are key definitions, 'boxed' concepts and case study evidence, along with specifically tailored learning activities and further reading. This is both a timely and trenchant discussion that has pertinence for students, scholars and researchers.

Helen Jarvis is senior lecturer in human geography at the University of Newcastle-upon-Tyne. She is sole author of *Work/Life City Limits* (2005, Palgrave) and co-author of *The Secret Life of Cities* (2001, Pearson).

Paula Kantor is Director of the Afghanistan Research and Evaluation Unit, based in Kabul. She was previously lecturer in social development at the University of East Anglia. Her research focuses on urban livelihoods and vulnerability, informal employment and gender relations – with a regional focus on South Asia.

Jonathan Cloke is Research Associate for the Global and World Cities Group within the Geography Department at Loughborough University. He has researched micro-credit and corruption and is undertaking new research into formal credit expansion in major Latin American cities.

Routledge critical introductions to urbanism and the city

Edited by Malcolm Miles, University of Plymouth, UK and John Rennie Short, University of Maryland, USA

The series is designed to allow undergraduate readers to make sense of, and find a critical way into, urbanism. It will:

- Cover a broad range of themes
- Introduce key ideas and sources
- Allow the author to articulate her/his own position
- Introduce complex arguments clearly and accessibly
- Bridge disciplines, and theory and practice
- Be affordable and well designed

The series covers social, political, economic, cultural and spatial concerns. It will appeal to students in architecture, cultural studies, geography, popular culture, sociology, urban studies, urban planning. It will be trans-disciplinary. Firmly situated in the present, it also introduces material from the cities of modernity and post-modernity.

Published:
Cities and Consumption, Mark Jayne
Cities and Cultures – Malcolm Miles
Cities and Nature – Lisa Benton-Short and John Rennie Short
Cities and Economies – Yeong-Hyun Kim and John Rennie Short
Cities and Cinema – Barbara Mennel
Cities and Gender – Helen Jarvis with Paula Kantor & Jonathan Cloke

Forthcoming:
Cities, Politics and Power – Simon Parker
Children Youth and the City – Kathrin Hörshelmann and Lorraine van Blerk

Cities and Gender

Helen Jarvis with Paula Kantor and Jonathan Cloke

LONDON AND NEW YORK

First published 2009
by Routledge
2 Park Square, Milton Park, Abingdon, Oxon, OX14 4RN

Simultaneously published in the USA and Canada
by Routledge
270 Madison Avenue, New York, NY 10016

Routledge is an imprint of the Taylor & Francis Group, an informa business

Typeset in Times New Roman by Saxon Graphics Ltd, Derby
Printed and bound in Great Britain by TJ International Ltd, Padstow, Cornwall

British Library Cataloguing in Publication Data
A catalogue record for this book is available from the British Library

Library of Congress Cataloguing in Publication Data
Jarvis, Helen, 1965-
Cities and gender / Helen Jarvis with Paula Kantor and Jonathan Cloke.
p. cm.
1. City planning. 2. Urban policy. 3. Feminist criticism. I. Kantor, Paula. II. Cloke, Jonathan. III. Title.
HT 166.J29 2009
307.1'216082–dc22
2008047605

ISBN: 10: 0-415-41569-1 (hbk)
ISBN: 10: 0-415-41570-5 (pbk)
ISBN: 10: 0-203-87806-X (ebk)

ISBN: 13: 978-0-415-41569-9 (hbk)
ISBN: 13: 978-0-415-41570-5 (pbk)
ISBN: 13: 978-0-203-87806-4 (ebk)

Contents

Boxes

Case studies

Figures

Acknowledgements

First and foremost, we are grateful to Malcolm Miles and John Rennie Short, editors of Critical Introductions to Urbanism and the City, for inviting us to contribute to this series on a perspective about which we each feel passionately. Our enthusiasm was tested over the three years it took to complete this book, especially given the logistics of collaborating from different institutional bases (typically across three continents), but it helped that we all firmly believed in the importance of critical analysis. For keeping us on track we wish to thank not only the series editors but also those who provided technical assistance and moral support in the production process, especially Antonia Edwards, Andrew Mould and Michael P. Jones. We also wish to acknowledge the two anonymous reviewers (and again Malcolm Miles) for providing important suggestions to improve the final version of this book. Any outstanding errors or omissions are our own.

In developing the proposal for this book Helen drew on many years of teaching on 'cities and social change', 'gender divisions at work' and more recently 'social geographies' at Newcastle University. This meant that many undergraduate students, along with postgraduates who contributed to these modules as teaching assistants, unwittingly helped to shape the content and learning style adopted here. Their collective enthusiasm and insights are duly acknowledged, with the hope that together we can keep gender on the agenda.

It is a harsh irony that writing about a more ethical and caring urban social future should have required the kind of workaholic excesses which completely undermine these ideals, but this is frequently the case now with academic writing. With this in mind, there are a number of people Helen wishes to acknowledge personally. Above all, Helen wishes to acknowledge the impact that writing this book has had on her daughter, Miriam, who now knows more than she should at the tender age of 6 of the pressures of missed deadlines and of a bad-tempered parent who never gets sufficient sleep. Heartfelt thanks go to the 'pit crew' of friends and colleagues who have pitched in with childcare and play-dates: Talya Leodari; Katie Brittain and David Peat; Caroline Austen and Ian Cowell; Sarah

Rees; Elaine and Brian Lilley; Chris Hagar. For conversations which go to the heart of a non-sexist city and transformative teaching and learning, thanks go to Alastair Bonnett, Seraphim Alvanides, Peter Hopkins (all at Newcastle), Rachel Pain (Durham), Marianna Pavlovskaya (CUNY) and Lia Karsten (Amsterdam), and to Duncan Fuller (Northumbria), whose premature death in the final days of writing cast a long shadow and painful reminder of the things that *really* matter.

Paula drew on data from her field work in India and Afghanistan in writing her contributions to the book. She thanks the respondents involved in the studies she drew from for their time and patience, and the hard work of the field teams involved in collecting the data. She also thanks Helen for involving her in the project and for Helen's unflagging energy in keeping it all going, across time zones and competing demands on time. It would not have come together without her leadership and dedication to the project.

Jon also wants to thank Helen for involving him and also for doing the lion's share of the work whilst fending off his more exuberant irrationalities. Thanks as well to his good buddy and long-time friend Ed Brown at Loughborough, to his patient and lovely partner, Bindee, and his dog, Jerry. Jon would also like to thank those people who agreed to have their photos taken for this book.

Abbreviations

ABC	Acceptable Behaviour Contract
AIA	American Institute of Architects
APE	Association for the Protection of the Environment
APG	Alternative Planning Group
ASBO	Anti-Social Behaviour Order
BME	Black and minority ethnic
CASE	Centre for the Analysis of Social Exclusion (UK)
DFID	Department for International Development (UK)
ECLAC	Economic Commission for Latin America and the Caribbean
GAD	gender and development
GDP	gross domestic product
GEMS	Grassroots Enterprise Management Services
GIC	global intelligence corps
GIS	Geographical Information System
HDI	Human Development Index
IBRD	International Bank for Reconstruction and Development
ICT	information and communication technology
IMD	Index of Multiple Deprivation
IMF	International Monetary Fund
JICA	Japan International Cooperation Agency
LAT	living apart together
LDC	less developed country
LTA	living together apart
MDG	Millennium Development Goal
MFA	Multi-Fibre Arrangement
MNC	multinational corporation
MO	Mass Observation
NAACP	National Association for the Advancement of Colored People
NGO	non-governmental organisation
NOW	National Organisation for Women
PGI	practical gender interest

PNUK	Planners Network UK
PREAL	UN programme for the Promotion of Eductional Reform in Latin America and the Caribbean
RIBA	Royal Institute of British Architects
SAP	Structural Adjustment Programme
SGI	specific gender interest
SLF	sustainable livelihoods framework
SPARC	Society for the Promotion of Area Resource Centres
UN	United Nations
USAID	United States Agency for International Development
USD	US dollars
UWEP	Urban Waste Expertise Programme
WAD	women and development
WID	women in development
WIEGO	Women in Informal Employment Globalising and Organising

General introduction

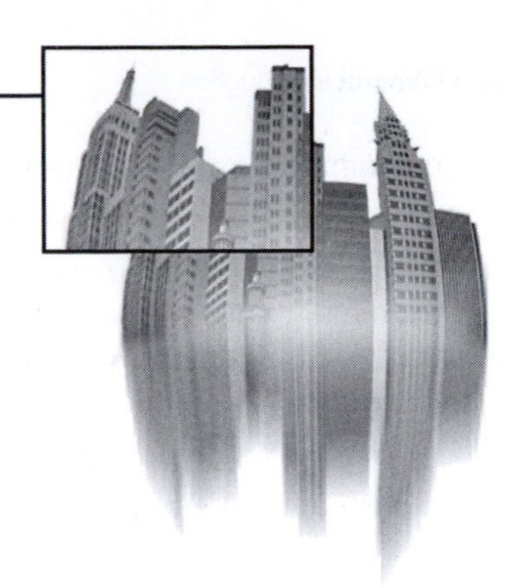

Rationale and scope

Cities function as key sites in the production, consumption and reproduction of gendered norms and identities. At the same time, cities are themselves shaped by the gendered embodiment and social reality of daily routines – at home, in public and on the move. This co-constitution is not yet widely acknowledged in the legions of books on cities, towns, urban social policy and development. This is because the concepts of 'urban' and 'gender', 'material' and 'social' are usually maintained as separate fields of study. While most introductory urban studies texts now recognise gender differentiation across a range of measures – in relation to housing, transport, employment and caring responsibilities, for instance – the systematic treatment of urban and gender studies, as co-constitutive subjects, remains long overdue. Our aim is that this book populates a more joined-up place on the library shelf, bringing these sometimes antagonistic disciplines into constructive dialogue.

This book has a three-fold rationale: to bring about the systematic intertwining of urban studies and gender studies; to expose persistent inequalities in the everyday lived realities of men and women in both the global north and the global south, through the analytic lens of gender; and to influence (and ultimately transform) the tone and substance of classroom debate as well as practitioner and civic engagement. We avoid making any claim to present a grand or total theory or method of cultivating inclusivity. Instead, we introduce the reader to a reflexive learning environment, combining critical engagement with competing schools of thought, 'boxed' key concepts and case study evidence with a series of learning activities, to charge the receptive student, teacher and practitioner with the responsibility to engage with an explicitly gendered urban theory in their own thought and practice. Readers are invited to explore new ways of seeing the world – and to reflect on their own and other people's taken for granted understanding of everyday environments, events, discourses and interactions.

Our approach draws on practical examples from everyday life to demonstrate the many ways that cities and 'the city' are gendered as well as classed, sexed and otherwise socially and culturally constructed. Throughout this book we encourage the reader to develop 'new ways of seeing' (Berger 1972) gender in the city. For this purpose we introduce skills of urban ethnography alongside a (re)conceptualisation of the ordinary moments of daily life. The urban ethnographer closely examines everyday social settings for hidden meanings and assumptions about masculine, feminine and heterosexual roles and identities, 'reading' such clues as are apparent in popular talk, texts and visual images. We draw explicitly on traditions of human geography alongside social theory to think about and view cities and gender through a space–time lens.

With each chapter we include wide-ranging illustrations of identity-moulding processes of gender 'at work' in different spaces of the city; in a variety of cities, across several continents. Our aim throughout is to move away from the impression that scholarly literature and theories of urban studies and gender studies reside exclusively within a narrow Anglo-American tradition, by highlighting similarities and differences in gendered lived realities from case studies drawn from around the world. The scope within which we are able to demonstrate global relevance is necessarily limited in a single text which also seeks ambitiously to intertwine two major disciplinary perspectives and cultivate new ways of looking at cities and urban social phenomena. We therefore pick up the thread of 'development' and the stark realities facing urban dwellers in so called third world cities in discrete sections and case studies. We also recommend that students and practitioners who wish to fully challenge the ethnocentrism of orthodox views of cities and gender supplement this text with others which are dedicated to the particular concerns of less developed countries. Suggestions for further reading are identified at the end of each chapter for this purpose.

A learning activity is included at the end of each chapter. The first activity, at the end of this chapter (see p. 4), asks you, the reader, to question everything (including your course syllabus) with a new alertness to the historical legacy of hundreds of years of gender (and class and race) discrimination. Like a small child on a long journey it asks the question: 'Are we there yet?' Have we arrived at a time and in a place where men and women have equal rights, opportunities and prospects? In subsequent learning activities you are asked more specifically to explore all that is deemed 'natural' in a given situation, what activities, modes of dress and behaviour are 'in place' or 'out of place'. Nudity, for example, would appear 'out of place' in the street. Minimal or provocative clothing would be 'out of place' in a place of worship and in places with a conservative or strict dress code – where this 'code' is regulated as much by peer pressure as by legal sanction. Ask yourself how you feel and in what situations you would think of yourself as being 'out of place' – removed from where you feel comfortable. What role does gender play in this? What would you find unsettling or threatening? Is this unease rooted in actual threat or experience of danger, or a lack of belonging or reference to that which is familiar and 'everyday'? Think about your sex; your sexual orientation; your class and upbringing; your physical characteristics (size, weight and appearance as well as skin colour). Being positioned behind one pair of eyes and one cognitive lens does not prevent the practice of empathy – stepping into the shoes of another. We shall return frequently to this notion of everyday as a dynamic concept through which to explore cities and gender.

Structure

The book is structured in three parts. Part I considers the plural disciplinary perspectives by which the city can be approached, notably the disciplinary lenses of 'urban studies' (Chapter 2) and 'gender studies' (Chapter 3) as separately conceived but also with respect to geopolitically uneven development and regional cultural specificity. Part I opens, in Chapter 1, with an introduction to the complexity of people–place interactions. This chapter provides signposts to core themes and concepts to be developed over the course of the book (such as bodies in/out of place), combining a short series of illustrative headline narratives with an introduction to the skills of urban ethnography which are needed to be able to 'read' the city with gender in sight. Chapter 4 rounds off our theoretical framework by reflecting on matters of scale, power and interdependence, notably the contradictory forces of the global in the local.

Part II (Chapters 5–7) takes a closer look at discrete aspects of the built environment and relates these to gender identities and practices. Chapter 5 (re)considers the notion of uneven urban infrastructures from a holistic, feminist, everyday life perspective. Chapter 6 looks at one of the most significant urban flows, the migration, movement and mobility of people within and between cities and outlying regions. It considers the material and social consequences of highly uneven and unequal access to mobility. Chapter 7 traces the complex flows, networks and interconnections binding and blurring the realms of home, work (paid jobs and unpaid reproductive work) and communities in space and time. Case studies illustrate the local and the multi-scalar implications of these intersecting flows. Comparisons are drawn between the direct exploitation of gendered migrant labour and the sometimes subtle exploitation of sex through emotion-work in the new economy.

Finally, Part III considers the way cities and gender are represented and regulated by the state and civil society and how government, non-government and informal initiatives respond to persistent inequalities and uneven development. Chapter 8 explores the way cities of the global north and south are planned and managed with specific reference to the social and cultural construction of gender systems and gender contracts. The role of resistance movements and collective action against the prevailing gender order are also examined. Chapter 9 focuses primarily on the global south to examine urban poverty and vulnerability and gendered experiences of livelihood insecurity and its consequences. Comparisons are made with 'new' expressions of urban poverty found in advanced neo-liberal and post-socialist cities. Chapter 10 draws together the intersecting conceptual strands of the book, emphasising the critical role of reflexive learning and transformative practitioner engagement.

Format

Each chapter has a similar format: key learning objectives are addressed through the introduction of substantive themes and boxed concepts (some including definitions) which are brought to life through selected case studies and accompanying illustrations and archive texts. The earlier chapters provide an integrated historical and theoretical framework drawing on a wide ranging literature and scholarship. The later chapters engage more specifically with empirical evidence and contemporary issues confronting planners, policy makers, commentators and civil society. We include here some of our own research from a number of projects conducted in Britain, the USA, Latin America, India and Afghanistan. At the end of each chapter we provide a brief summary of the topic covered in relation to the learning objectives. Suggestions are made for further reading with the aim of stimulating wider interest and critical thinking rather than to reinforce the topic covered. A complete bibliography and index appear at the end of the book.

In order to actively engage with new ideas and lines of questioning, we include with each chapter a suggested learning activity. Suggestions are made for ways of confronting the multiple and fluid realities of cities and gender through, for instance, self-guided walks, the 'deconstruction' of visual media, texts and archive research, as well as through on-line and off-line research. It is our hope that engagement with this book is active rather than passive, and that case studies, activities and suggested reading and websites are used to launch further independent and participatory study. We hope that teachers and students draw on the suggested learning activities (or develop others like them) to actively reflect, through participatory, transformative classroom activities, on the question of gender democracy and social justice.

We include a suggested learning activity with this general introduction. The suggestion is that you keep a 'reflexive learning diary' spanning the period of time over which you engage with different aspects of gendered urban studies – which may or may not follow the chapter structure of *Cities and Gender*. The diary should include five or six discrete instalments of about 300 words in which you reflect on your own experience of a particular issue (such as the barriers to movement that you yourself experience in certain parts of the city or at particular times of the day or night). This way you should build up a series of entries that reflect developments over time in your critical thinking on cities and gender. You should avoid making changes to earlier entries even if you subsequently become critical of what you wrote at an earlier stage in your learning. The idea is to be as self-aware and understanding of others as possible: put yourself in the picture (whether of sex segregation or fear) either by drawing on direct personal experience or by imagining yourself in different roles associated with a particular encounter. This activity seeks to develop skills of reflexivity as well as to practise a form of transformative learning where changes in your own behaviour may potentially result from critical awareness (see, for instance, McGuinness forthcoming; Haigh 2001; Cowan 1998).

References specific to bringing feminism into the classroom

Bohmer, S. and Briggs, J. L. (1991) 'Teaching privileged students about gender, race and class oppression'. *Teaching Sociology*, 19.2: 154–163.

Bondi, L. (2004) 'Power dynamics in feminist classrooms: making the most of inequalities?' *Women and Geography Study Group, Geography and Gender Reconsidered*. WGSG e-publication (CD): 175–182.

Fisher, K. (2003) 'Demystifying critical reflection: defining criteria for assessment'. *Higher Education Research and Development*, 22.3: 318–335.

Maddrell, A. (1994) 'A scheme for the effective use of role plays for an emancipatory geography'. *Journal of Geography in Higher Education*, 18.2: 155–162.

Maguire, S. (1998) 'Gender differences in attitudes to undergraduate fieldwork'. *Area*, 30.3: 207–214.

Merrett, C. D. (2004) 'Social justice: what is it? Why teach it?' *Journal of Geography*, 103.3: 93–101.

Robson, E. (2007) 'Listening to geographers from the global south'. *Journal of Geography in Higher Education*, 31.3: 345–352.

Skelton, T. (2001) 'Girls in the club: researching working class girls' lives'. *Ethics, Place and Environment*, 4.2: 167–173.

Thien, D. and Davidson, J. (forthcoming) *'Gender Interventions in Research, Teaching and Practice'* (special issue). *Journal of Geography in Higher Education*.

Valentine, G. (2005) 'Geography and ethics: moral geographies? Ethical commitment in research and teaching'. *Progress in Human Geography*, 29.4: 483–487.

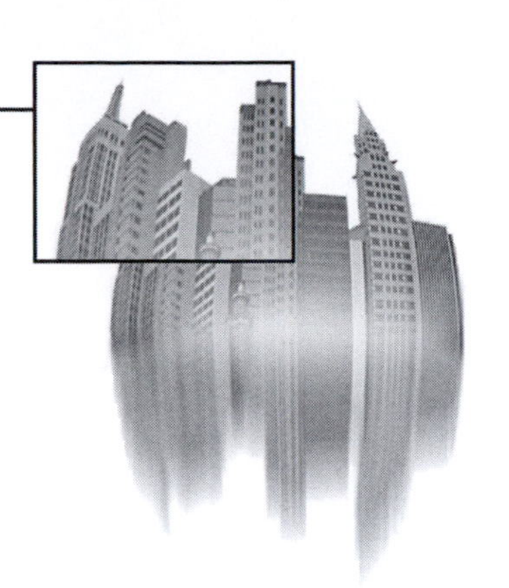

Part I

Approaching the City

1 From binaries to intersections

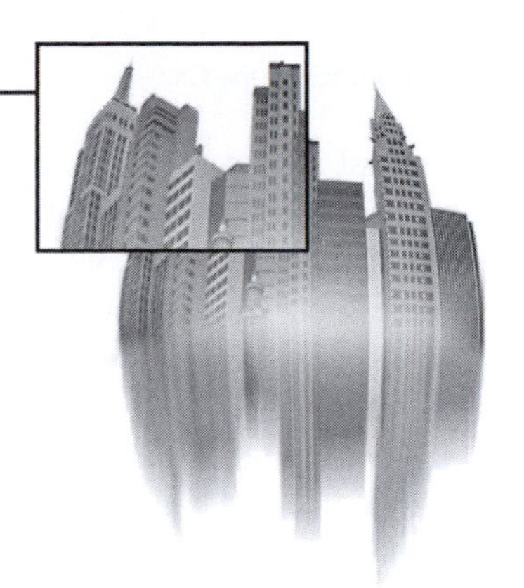

Learning objectives

- to appreciate the systematic intertwining of urban studies and gender studies as co-constitutive subjects
- to begin to recognise the city in 'ordinary' terms through an integrated and inter-disciplinary focus on everyday life
- to view urban space and gender identities beyond binary distinctions

Introduction

This chapter introduces the reader to new and different ways of seeing the city from a gender perspective. While we place gender in the foreground as a fundamental site of inequality, we acknowledge along with other feminist scholars that this constitutes just one side of a 'wicked triangle' of race, class and gender (Grünell and Saharso 1999). Travelling around the world, towns and cities are encountered which might not immediately appear to be 'gendered'; yet we argue that they are. The 'concrete' sense in which the city is constructed over many hundreds of years and at vast expense tells a different story from more ephemeral manifestations of popular culture such as fashion, television and advertising, for instance. Cities assume a semi-permanent spatial arrangement and material culture, filtered through the psychological architecture of belief systems in a constant state of flux. Over time these cultures sediment in the form of buildings, monuments, political and administrative systems, which in turn come to symbolise and reinforce powerful regulatory norms and stereotypes. A particular challenge for the systematic intertwining of cities and gender, indeed for gender mainstreaming in general, is to prevent gender being defined as 'women'. Meeting this challenge requires a comprehensive overview of the related disciplines as well as new tools and techniques of urban ethnographic analysis.

First among the skills of urban ethnography that we consider crucial to understanding the gendered nature of cities is standpoint awareness: from whose perspective, with whose voice, are cities and gender and power relations defined? This chapter begins by questioning the tendency for cities to be viewed in binary terms, from a narrow 'androcentric' and occidocentric (ethnocentric) standpoint. Then we explore the way that cities and gender are popularly represented, both by the media and in the public imagination (through newspapers, television, film, fiction, monuments and archives), and how these representations shape cultural norms and stereotypes and ways of 'doing' masculinity and femininity. This discussion in turn raises awareness of the many manifestations of gender and power at work in the city – beyond/behind the headlines and the picture postcard images which are themselves produced (and consumed) in segregated and exclusive ways. What captures the public imagination may be skewed by sensational journalism but it can also provide a telling insight into the tacit codes by which men and women, boys and girls define their gender and sexual identity in relation to powerful cultural norms. Over the course of this chapter we encourage the reader to differentiate a top-down 'gaze' from a grassroots or street-level 'lived experience', deploying skills of empathy, close observation and critical engagement.

Moving beyond binaries

An enduring fascination with real and imagined cities is captured in two recurring themes of introductory urban studies texts. The first is the quest for utopia; the vision of the well designed, well governed, 'good' or 'ideal' city. The second is the unhealthy, deviant or dysfunctional city. Historically the tendency has been to elaborate these themes of utopia and disorder through contrast theory and binary distinctions (see Chapter 2), such as between good and evil, urban and rural, civilised and primitive, modernity and post-modernity, collective and individual. Likewise, when we turn to gender we find persistent reference to a male/female binary defined by biological sex (see Chapter 3).

The first part of this book outlines and then dismantles these separate binaries (of urban studies, gender studies and global north/south). This journey takes us from an essentialised male/female binary of 'separate spheres' (public/private, outside/inside, economy/family, work/home, distance/intimacy, duty/love), to an awareness of the intersection of multiple identities, multiple economies and a blurring of space–time boundaries. This reflects the steady way in which urban studies has expanded to recognise the diversity of urban experiences, to the wider range of cities across the world, increasingly emphasising the 'flows and networks (money and migrants) that pass through cities rather than the territory of the city itself' (Robinson 2006: 93). In this book we continue this pursuit of a more integrated and nuanced gendered urban analysis, combining the lessons learned from historical developments in these separate disciplines and adding to these the skills of feminist urban ethnography and critical engagement with the similarities and differences of cities and gender encountered in a highly unequal world. Crucial to this integration are a number of 'bridging concepts', which we introduce (in boxes) in this and subsequent chapters. First among these bridging concepts is standpoint awareness, which is defined in Box 1.1.

Box 1.1: Standpoint

Significant among the contributions of feminist scholarship is the understanding that students, teachers, academics and members of the public do not gain their knowledge and understanding of the world as a 'god trick' 'as if from nowhere' (Haraway 1991). What individuals see, hear and believe is necessarily 'laden' with personal values, encounters, experiences and expectations; so a young black man growing up on the streets of a poor neighbourhood in Harlem is going to have a different 'position' or standpoint on crime, based on relative fear of crime, for instance, compared with a white middle-class mother travelling through that same New York neighbourhood on the subway with her children.

Recognising the existence of multiple 'lived realities' rather than a singular 'objective truth' alters both the subject and the manner of observation. Thus bell hooks (1982) urged white women to re-examine the female subject in their work, to theorise the complexities of 'raced' identities rather than to pronounce on a meaningless category of 'every woman' from the privileged experience of 'white woman'. On the one hand, standpoint awareness can lead to the view that 'there is nothing about being "female" that naturally binds women' (Haraway 1991: 155, cited in McDowell 1999: 22). On the other hand, it can be argued that standpoint is entirely compatible with the construction of multiple masculinities and femininities within a powerfully maintained norm of dichotomised male and female heterosexual bodies (Butler 1993, cited in McDowell 1999: 23). Here we largely take the latter view.

We encounter the legacy of binary distinctions in popular definitions of what is urban relative to a sense of the rural. Yet in reality we can identify a large number of settlement types that are neither rural nor urban in the conventional sense (see Chapter 2). We also confront the perennial binary of 'structure' versus 'agency' which has exercised theoretical debate for a hundred years or more. By highlighting both the mundane and the messy realities of urban daily routines, we seek to dismantle the extreme caricatures of people either as victims of circumstance or exercising full control over their own destiny. Linked to this, we move debates on urban livelihoods beyond the binary that typically defines a formal (cash) economy in opposition to an informal (barter or love) economy. Finally, moving beyond the male/female binary in the remainder of the book, we recognise sex-as-gender to be socially and culturally constructed, reflecting multiple and fluid ways of expressing masculinity and femininity; of 'doing' gender rather than 'being' essentially born into a fixed identity of boy or girl; as son/daughter, husband/wife, breadwinner/homemaker.

Androcentrism and occidocentrism

The discipline of urban studies which emerged from the late nineteenth century typifies a very narrow standpoint as it was without question a male domain. Not only has writing on the city been largely androcentric (or male conceived), but also it has derived from an occidocentric (or Western conceived) viewpoint that does little justice to the lived realities of men and women in either the global north or the global south. Many scholars argue that it remains so today.

> *Androcentrism* is defined as being centred or focused on men, often to the neglect or exclusion of women.
>
> (American Heritage Dictionary 2000)
>
> *Occidocentrism* (used synonymously with Eurocentrism) is defined as the projection of Western values (Occident meaning 'the west' of Europe, America, or both, as distinct from the Orient) and pervasive or under-examined notions of colonialism in relation to non-Western cultural contexts.
>
> *Ethnocentrism* is defined as the tendency for the author/observer to view the world from their own cultural standpoint and for them to believe that their own culture is superior to that of other groups.

Androcentric origins are evident not only in terms of who the urban scholars were but also in the subject matter and mode of analysis considered worthy of investigation. This is highlighted with respect to the 'top 10 ideas of urban studies' listed in a mainstream teaching text published by Blair Badcock (2002: 5) and distributed across the English-speaking world. Badcock's list begins in the 1880s with Charles Booth's survey of working-class living conditions in Victorian London. It then proceeds to emphasise 'the' Chicago school of Park, Burgess and McKenzie (1925), Alonso's (1964) economic modelling of urban land uses, the French architect Le Corbusier's 'cities in the sky' and the writings of Lewis Mumford, David Harvey, Manuel Castells, Mike Davis and William Cronin – all arguably 'macho', 'macro' grandstand views of the city. The list recognises just two female 'micro' (more qualitative) influences, both after the 1960s when the relatively new discipline of women's studies (later recast as gender studies) began to exert influence at the margins of urban debates by exposing discriminatory assumptions of life in a world 'made and managed' by men (Roberts 1991).

While the volume and sophistication of feminist interventions have increased over the intervening years, the puzzle remains that urban studies and gender studies continue to exist as separate disciplines whereby at best they co-exist. Again with few exceptions (but see Greed 1991; Matrix 1984; Roberts 1991), consideration of gender issues remains largely separate and subordinate within urban studies. Similarly, gender studies rarely engages systematically with the built environment (but see WGSG 1997; McDowell 1999; Booth *et al.* 1996; Domosh and Seager 2001), tending instead to focus on the social relations and economic implications of gender segregation at home and in the workplace. What is missing then is a disciplinary intertwining of cities and gender through the engagement of students and practitioners of urban planning and urban social policy in an integrated theory and practice.

Second among the skills of urban ethnography we wish to cultivate in the course of this book is an appreciation of the empirical richness of everyday life. Much as Jenny Robinson (2006 : 1) argues that all cities are best understood as 'ordinary' to resist the ethnocentric labelling of certain cities as Western, Third World, Developed, Developing, World or Global (but also see Chapter 5), we argue that urban ethnographic analysis is best conducted through closer scrutiny of the 'ordinariness' of everyday life. Recognising the mundane moments of daily life attests to the diversity and complexity of city dwellers – from those who sweep the trading floors to the traders themselves. This is why everyday life is introduced here as another of our bridging concepts. Not only has this concept been similarly mobilised by the separate disciplines of urban studies and gender studies, but it also combines with urban ethnography (and mass observation) in potentially transformative ways. The bridging concept of everyday life is introduced in Box 1.2, while the transformative potential of an explicitly feminist 'everyday' is discussed further in Chapter 5.

Box 1.2: Everyday life

As a basic definition, the term 'everyday life' refers to those ordinary, taken for granted, habitual thoughts, activities and settings that are close and familiar to all of us but which are rarely measured by governments or scholars or endowed any particular significance. Henri Lefebvre used the metaphor that everyday life is like fertiliser: it functions as a source of life-giving power but it largely goes unnoticed as it is tramped underfoot:

> A landscape without flowers or magnificent woods may be depressing for the passer-by but flowers and trees should not make us forget the earth beneath, which has a secret life and a richness of its own.
>
> (Lefebvre 1991: 87)

Henri Lefebvre also famously coined the phrase 'the familiar is not necessarily known' to capture the understanding that while an activity such as shopping or walking may be ubiquitous it lacks meaning if it is not recognised – by naming, counting, researching and assigning value (Jarvis 2009).

The concept and language of 'everyday life' have their roots in the 'transformation of intimacy' or 'new emotional style' that infused the social sciences (sociology, psychology, geography, anthropology), stimulated by Freudian psychoanalysis in the 1920s (Giddens 1993; Illouz 2007). Further momentum can be attributed in the 1960s to Erving Goffman's symbolic interaction and dramaturgical approach elucidated in his seminal 1959 publication *The Presentation of Self in Everyday Life*. Illouz (2007) observes that 'emotional behaviour' (and face-to-face interaction) has become so central to economic behaviour since this time, especially through the transition from manufacturing to service sector 'post-industrial' economic expansion, that popular classifications of emotional intelligence and self-help guides are now deeply embedded in corporate business culture. Notions of 'emotional competence' as the 'ability to monitor one's own and others' emotions, to discriminate among them, and to use the information to guide one's thinking and actions' were formalised by Daniel Goleman in his 1995 international best-seller *Emotional Intelligence* (Illouz 2007: 64). Alongside this 'cultural turn to therapy' can be observed a second aspect of the 'modern emotional style' which locates the relations of the self in a realm dubbed 'everyday life' by Stanley Cavell (1996). From this we see a shift in research interest to the ordinary and the mundane.

There are a number of urban social theorists whose work is intimately bound up with conceptions of everyday life. Of particular note are Michel de Certeau (1984), who developed a 'grammar' of everyday practice; Henri Lefebvre (1984) and his philosophy of the ordinary; and Dorothy E. Smith (1989), who contributed a feminist understanding of the asymmetry of power between men and women. We expand on the contributions of these scholars in subsequent chapters (for an overview and critique of these key thinkers, see Gardiner 2000).

It is important to note that notions of 'ordinariness' have been mobilised far more literally by feminist scholars and in household research with the more progressive intention of systematically overhauling, as inclusive and enabling, the material fabric and institutions of the city.

Renewed interest in 'everyday geographies' also forms part of a shift in the scale of the analytic lens (from macro to micro) and the favoured choice of methods (promoting ethnographic applications of empathy and participation). According to Rigg an everyday perspective recognises that 'the social, environmental, political and economic micro-processes and micro-dynamics – the micro-geographies – often provide not just a more finely detailed understanding of change, but a *different* view' (2007: 8). Thus it is grounded in the micro-level 'to ask critically important "why" questions' (Rigg 2007: 8).

Ordinariness and mundane observations

We are not claiming anything especially new about interest in the subject and methods of mundane observations, although arguably the term 'everyday life' has taken on renewed significance in the twenty-first century, as indicated above, following a turn to post-structural epistemology. Back in 1937, Mass Observation (MO) was established in Britain as 'an anthropology of ourselves'. Still in existence today, its primary aim remains to study the everyday lives of ordinary people. MO encompasses numerous projects focusing on different and diverse aspects and views of daily life; including the Bolton Worktown Project, the National Panel of Diarists, Pub Conversations, the Cooperative Correspondence Club of young mothers (Bailey 2007) and, more recently, an email communications archive (Mass Observation 2007, available on-line http://www.sussex.ac.uk/press_office/bulletin/01dec06/article8.shtml, accessed 16/07/07).

There are other fascinating applications of mass and participant observation including those used for commercial product development. The technology firm Intel routinely employs a team of 'design ethnographers' to study the way people use products such as mobile phones and handheld computers in their daily lives. A typical set of observations are made by a design ethnographer riding around on a bus, participating as a regular passenger, watching and noting the context in which particular products are used (*Electronic Times* 2001).

While scholars express renewed interest in everyday life there is little evidence of an equivalent interest in 'ordinariness' in the news media. Looking at how cities and gender are popularly represented, we sense that it is fear rather than boredom which typically defines urban life in the urban imagination. If a news headline does not project fear, anxiety, violence or deviance, then cities are presented in terms of glamour, distinction and decadence. Domosh and Seager suggest that cities have always been portrayed as 'risqué and exciting' to reflect male pleasures – as typified by 'the lone man (the striding flaneur) wandering the streets of nineteenth century Paris or New York' (2001: 67).

The sensational bias that determines whether an experience, place or event is newsworthy or not also skews the way a seemingly monolithic identity is read off from human behaviour, as a simple matter of 'choice' or voluntary agency. This is illustrated in the language used by journalists describing high-powered working mothers as 'selfish' or 'uncaring' for leaving their children with 'foreign nannies'; young Muslim women as 'submissive' and at the mercy of 'honour killings'; the inner city school and neighbourhood written off as 'a place gone bad', from where those who can do so make their escape; and it is said that 'boys will be boys' with little expectation that they can change the path of biological destiny. One British headline asked, for instance, 'are working mums as bad as junk food?' (Johnson 2006).

Setting aside for one moment the issue of what constitutes 'news', how realistic is it that any individual or place is reduced to a single identity, motive or label? Would you define yourself by a single identity? As a student? A town planner? Girlfriend/boyfriend/spouse/parent? Working class? Black or minority ethnic (BME)? Or would you define yourself by a combination of intersecting identities, beliefs, cultural practices and experiences which you recognise as being embedded in locally specific but continually changing environments and milieux? Consider, for example, how concern with the rise of youth gun and knife crime has been pathologised in the British press as a 'problem of boys' and 'bad neighbourhoods', as illustrated in the next section (see also Nayak 2006).

The problem with boys and the hood

On 14 February 2007 British media headlines stimulated widespread public alarm and debate at the news that a third schoolboy had been murdered in the same South London neighbourhood in as many weeks. All three murders were boy-on-boy (two of them black-on-black) fatal shootings. The news media highlighted the rising incidence of youth violence in sensational terms, holding a mirror up to popular anxieties in search of explanations in the 'breakdown of the family'; lone mothers, absent fathers, lack of parental discipline and a loss of respect for family and community. Early news reports of the third murder described the scene of a boy shot dead 'home alone' after school on the doorstep of a run-down inner-city council flat. Over the following days, as more details emerged of the victim and his family, it transpired that Billy Cox did not neatly fit the stereotype image initially presented. Billy had a loving mother and a father residing together in the family home; he was discovered by his sister minutes after the end of a normal school day; both were 'latchkey' kids for less than an hour before their mother returned from work, as she always did, to make their evening meal. Yet neither was this murder entirely unexpected nor a freak accident. Violence, guns, gangs, a culture of graffiti 'tagging' and 'street reputation' were an apparently normal part of Billy's life at school and in the surrounding neighbourhood.

In the same week the British national press reported significant gender and ethnic divisions in educational achievement at age 16. The reported statistics presented a now familiar trend of girls significantly outperforming boys; Chinese and Indian girls and boys outperforming white children (especially white working-class boys) and Black Afro-Caribbean boys at the bottom of the heap of educational prospects in British state schools. It is a picture of 'redundant' or deviant masculinities which is painted repeatedly with respect to male unemployment, yob culture and absent fathers. Billy Cox was mixed-race White-British-Asian and he attended a 'failing' state school. But metres from his home was gentrified Clapham. What would Billy's prospects have been had he and his family had the resources to set up home in a 'safe' middle-class community (Butler and Robson 2003)? We suggest that this level of questioning needs to be applied to a wider range of urban social and environmental concerns than is generally the case in the media.

New man, redundant man and the ladettes

Notwithstanding the sensational bias and partisan politics of print journalism in general, it can be constructive, as a learning activity, to scrutinise news headlines and their content for evidence of continuity and change in the way cities and gender are popularly represented. Doing so captures an important theme of this book in the tension between powerfully persistent social divisions, norms and roles, alongside shifting fashions, technologies and transgressions. The invisibility of 'wifework' comes into sharp relief, for instance, when childcare and housework are undertaken by a man: where 'Mr Mom' is a novelty worthy of headline news. Likewise, unruly behaviour is less noteworthy for schoolboys (although obsessions with 'hoodies', 'gang culture' and 'street-corner thugs' can still be relied upon to sell newspapers), but when schoolgirls are observed to become more assertive, 'drinking, smoking and being disruptive in class', it makes headline news as the problem of 'girl-power' and a disturbing 'ladette culture'. Girls appear to enter the picture only when they over-step the boundaries of 'acceptable' behaviour. Mary Thomas (2005) observes, for example, the way girls 'hanging out' in public experience greater control and criticism of their behaviour with respect to 'ideals of femininity' and spatial constraints imposed by adults and older youth. She concludes that 'hanging out' is a fraught activity for girls in the public urban sphere (Thomas 2005: 599; Blackman 1998: 216). Questions also have to be asked about the parochialism and bias of such media attention. Why are some subjects (murders in London, but not in Mumbai) considered newsworthy and likely to elicit an emotional response in the audience?

Bodies in and out of place

People mark out and colonise space and territory in numerous, powerful ways – from the body, by their posture and dress, through occupation of motorised vehicles, to the defence of the home space, surveillance and fortified communities. Again, these are gendered practices – but not in the simplistic sense that *all* men colonise space aggressively, or that women are 'naturally' victims (Pain 1999).

All around us are examples of gendered bodies in and out of place – where 'belonging' or 'fitting in' reflects the intersection of multiple and fluid identities and identifications; in the use of aggressive body language; youths patrolling the streets in large 'packs'; young men cruising the streets in big-wheel cars with loud engines. In such examples we can see how the body, dress, language, movement and technology are routinely used to powerful effect. Yet, simultaneously, these same attributes can be turned against the less powerful, in situations of minority status, in the sense of being 'out of place' or in the wrong place at the wrong time. The classic example is the way newspaper reportage appears to blame female victims of sexual assault for provoking their attacker, 'asking for it', simply by wearing tight or revealing clothing in public.

Being 'out of place' assumes many forms. It can mean, quite literally, being locked outside, barred from entry; or it can mean that all eyes are on the stranger. In her book on Britain's inner city 'badlands', *Goliath*, Beatrix Campbell (1993) vividly portrays the crisis wrought by 'angry young men' (typically white and working class) abandoned by de-industrialisation, no longer valued economically for their physical strength and stamina, who are ill equipped to blend in or adapt to the subtle performance of emotion-work required in the service sector such as decorative jobs in retail. Illustrating this point, a recent recruitment drive for an international clothing retailer specified that the new store would be employing 'store models' rather than shop assistants. The employment ad was clear, if not spelled out in so many words: only young, slender, bright, beautiful people need apply (Keating 2007). We return to this theme of gendered economic restructuring in Chapter 2.

No longer identified by the role of male breadwinner or, as more than one journalist has described, a 'walking wallet', the young men of Campbell's (1999) 'white ghettos' find themselves increasingly redundant at home, and with teenage mothers more likely to manage alone in situations of poverty and poor housing, young fathers are increasingly absent and thus vilified as random sperm donors (Thomas 1999; see also McDowell 2003). Of course, peppering this account with sensational phrases only serves to perpetuate the problem of urban myth and caricature. It is the purpose of this book to peer behind such headline trends to explore the mundane moments beneath the media radar, focusing on the lived realities of everyday life. At the same time, these media representations offer a provocative route into our exploration of gendered geographies and histories.

Gendered space, time and belonging

Men and women are rarely confined to specific same-sex quarters in any strict sense, although a legacy of single-sex schools, colleges, gentlemen's clubs and meeting places has been surprisingly enduring (see Case 1.1). Moreover, 'there is disagreement whether women-only provision and spaces empower women or trap them in a ghetto of special needs' (McDowell 1999: 119). While some women assert the right to establish women-only spaces and organisations, as a sanctuary from male violence against women (Vancouver Rape Relief Society 2007), a 'more interesting project' for gendered urban theory is 'to think through what a non-sexist or non-gender specific (non-androcentric) urban environment might look like' (McDowell 1999: 120).

Although we are familiar with the phenomenon of urban space being highly segregated by income, class and race, urban spaces are rarely demarcated according to explicit sex separation, with the exception of toilets and changing rooms. Nevertheless, we are all familiar with the way that popular definitions of 'masculinity' and 'femininity' persist in statements such as 'he's really masculine' (strong, silent, commanding) or 'she's a girly girl' (pretty, chatty, nurturing). This routine normalisation of a hegemonic masculinity and homophobia has a profound bearing on urban experience, life-chances and well-being. The resulting inequalities are wide ranging – from legal barriers to owning property, through real or perceived threats of violence, to subtle but insidious labelling by mode of dress and appearance, and the sense of 'fitting in' or 'passing' (as 'normal') or being 'out of place'.

Even more readily apparent are the stories of power and subordination, domination and resistance, segregation and stratification that can be 'read off' the form and function of certain buildings, monuments, streets and signage. As Knox and Pinch observe:

> One well worn theme in architectural theory has been the manifestation of 'masculine' and 'feminine' elements of design. For the most part, this has involved a crude anatomical referencing: phallic towers and breast-like domes. Skyscrapers, for example, can be seen to embody the masculine character of capital.
>
> (Knox and Pinch 2006: 139)

Case 1.1: Archives as 'peep holes' into gendered urban development

This atlas suggests that the public face of London in 1929 was an all-male preserve. A description of the Metropolitan Police Force and London County Council Fire Brigade lists the facts and figures relating to 'officers and men of all ranks' employed to safeguard and 'protect' an area of some 700 square miles.

The 60 or so 'principle public buildings and sights' identified in alphabetical order focus on Church and Government buildings, Royal Palaces, War Memorials, citadels of commerce and tunnels (designed by and portraying men). The list also includes 'Clubs' (all of them restricted to men) – a list of some 68 all-male literary, sporting, political and drinking societies – including the Garrick, Marylebone Cricket Club (MCC) and the Reform.

The Twickenham Rugby Ground first admitted women in 1995 and the MCC did not admit women as club supporters until 1998.

The Royal Geographical Society, South Kensington, excluded women as 'fellows' until 1913.

Interestingly, the first ever London A–Z street atlas was compiled by a Bohemian woman artist and travel writer who would have felt thoroughly excluded by Bacon's 1929 guide. In 1935 Phyllis Pearsall (1906–1996) conceived the idea of mapping London following a conversation she had at a party, having got lost while making her way there on foot, following the best (clearly inadequate) ordnance survey map of the time. Pearsall's mapping project involved her walking the 3,000 miles of the 23,000 streets of London, all the while noting the names and drawing up the streets and landmarks which made up her road atlas. This story is recounted in a vivid biography which draws on diaries and business accounts for the Geographers' Map Company that Phyllis founded to produce the iconic A–Z (see also Pearsall 1990).

Source: Bacon's *Up to Date Atlas and Guide to London* (1929 edition) (Price 2/6). London: G.W. Bacon & Co. (.3–25).

Much has been made in the architectural press, for instance, of the phallic 'fat cigar' silhouette of the Swiss Re London HQ at 30 St Mary Axe, designed by Sir Norman Foster and Partners. Built in 2004, this building, nicknamed 'the gherkin' (Figure 1.1), has become one of the most iconic buildings of the London landscape. Yet, as Knox and Pinch go on to observe, 'feminist interpretations of architectural history

Figure 1.1 ***Profile of the 'gherkin' against the London skyline***
Source: Helen Jarvis

have shown [that] the silences of architecture can be more revealing than crude anatomical metaphors' (Knox and Pinch 2006: 139). These silences (the taken for granted infrastructures of daily life; of fetching, carrying and caring) are the focus of our attention in subsequent case studies. These are the kinds of observations which offer rich material for the reflexive learning diary introduced in the general introduction (p. 4) and in the learning activity at the end of Chapter 4.

This gendering of space is subject to continual transformation alongside changes in social conventions and formal legislation. Take, for example, the incremental shift away from women-only access to baby changing facilities in the public spaces of Western cities, in favour of shared unisex facilities which allow fathers as well as mothers to care for their young children. Changes like these are very gradual because social attitudes and conventions do not change overnight and it takes time and money to remodel buildings in response to changing tastes and expectations. On a day-to-day basis locally constructed norms of behaviour are taken for granted, but over the longer term they appear to be 'enduring rather than eternal', evolving over time (Bourdieu and Wacquant 1992: 133). Looking back in history, for example, examples of sexist, racist and socially elitist language and behaviour can seem glaringly obvious and quite outrageous. Some insight into this can be gained from exploring historic archives such as maps, diaries and letters, as suggested in Case 1.1. Further insights can be gained from images and observations of the fabric of the city itself; the interior and exterior of buildings, public monuments, street names and evidence of old and new industries. Consider, for example, the contrasting gendered interpretation of civic pride represented by the subject, scale and style of each of the two statues in Figures 1.2 and 1.3.

Figure 1.2 shows the statue of King Dom Jose I located in the centre of Lisbon's imposing Commercial Square. Cast in bronze in 1774 by the renowned sculptor Machado de Castro, the king is sat astride his horse in full military regalia, high on a monumental plinth, surrounded by allegorical figures. The location, scale and figurative arrangement are skilfully employed to convey the domination of domestic leadership (patriarchy) and overseas colonial expansion (empire).

Figure 1.3 shows the statue 'Meeting Place' located in the newly extended St Pancras International Station in London. Cast in 2006 by Paul Day, this nine metre high bronze depicts a young couple in an intimate embrace beneath a large clock-face – evoking a timeless romantic narrative of separation and reunion. While the subject and style of the second statue reflect a thoroughly contemporary (albeit Western) expectation of 'democratic' social relations

Figure 1.2 ***Statue of King Dom Jose, Lisbon***
Source: Helen Jarvis

Figure 1.3 ***Statue 'Meeting Place', London***
Source: Helen Jarvis

(this could be any of the couples milling about the station – other than that they are many times larger than life), the man and woman otherwise conform to social conventions of 'masculinity', 'femininity' and heterosexuality. This is an idealised and arguably traditional, highly conservative image of contemporary social mobility.

Blurring separate spheres

The most enduring argument for why cities are gendered in their construction and spatial arrangement resides with the (false) separation of discrete functions of production, consumption and social reproduction. Production entails not only the 'making' of things but the whole chain of commodities and services bought and sold in a capital economy. Social reproduction covers all of those activities that are fundamental to the continued maintenance of human life (subsequent generations) and social existence. In the urban survival economies of the poor in the global south the household itself is a vital source of production and here there is a complete erosion of the boundaries between livelihood and household, spatially, physically and economically. Thus, what it actually takes to reproduce cities and daily life for urban poor and rich alike goes far beyond biological reproduction to encompass all of the mundane, typically unpaid (and disregarded) activities of feeding, clothing, sheltering and caring for current and future wage workers and all those whom the wage economy no longer has or never had any intention of rewarding in exchange for productive labour.

Setting aside for the moment the activities of consumption (where this can be viewed as both an aspect of social reproduction and a profitable service sector industry), we are once again confronted with another binary. In Western post-industrial cities the distinction is popularly made between a masculine 'core' ('the city') and a feminine 'periphery' (the suburbs). Moreover, classic feminist texts long ago raised awareness of the status distinction popularly drawn between 'high-status' social relations of paid work conducted in the 'public' eye and 'low-status' 'hidden' social relations and activities of unpaid domestic work in the home and community. This binary (or dualism) has been criticised by contemporary 'post-structural' feminists, as we shall see in Chapter 3, not least because it ignores the significant proportion of women around the world who have for generations combined social reproduction work with long hours committed to wage employment and vital self-provisioning. Nevertheless, it is necessary to understand the historic origins and enduring popular appeal of these binaries in order that we can dismantle them. The blurring of these spheres (of formal and informal economies, public and private space, core and periphery) is articulated especially clearly in research and analysis which focuses on the household.

Cities of households and multiple economies

Focusing on the dynamic interactions of the household allows a holistic (or pluralist) approach to urban flows and networks – recognising formal and informal (cash in hand) work, remittance economies, domestic food production and self-provisioning, reciprocity and the 'economy of regard', state benefits and redistributive economies and, crucially, care work (Jarvis 1997). This complexity is usefully understood with reference to 'multiple economies' and we introduce this as one of our 'bridging concepts' in the intertwining of urban studies and gender studies in Box 1.3. We argue that the household is fundamental to understanding urban lived experience because it is a major welfare provider and the foremost realm of non-capitalist economic production (Gibson-Graham 1996; Jarvis 2006: 351). Focusing on the household highlights the role of social (re)production as an expression of 'the diverse conditions and organizational relations which allow human beings to survive in various social contexts and groups'. In this sense it is synonymous with 'livelihood' or 'survival strategies' (Mingione 1991: 124) (see also Chapter 9). We return to consider the widespread shift from viewing the household as a unit of analysis to a dynamic set of processes or 'lens' through which to view the city in Chapter 6.

Box 1.3: Multiple economies

Feminist scholars differentiate between a holistic understanding of multiple economies and a partial and narrow view of material and monetary economies of income and spending. A 'whole' economy comprises multiple interdependent assets and capabilities which have to be recorded across a spectrum of paid, unpaid, informal and non-financial activities, where social reproduction work is acknowledged as pivotal to the capacity of individuals and households to 'go on' from one day to the next. A 'whole' economy also takes into account the fluid and contingent quality of action-spaces, time (and timing) and the material world.

Household members engage in 'multiple economies' in order to secure income, domestic reproduction, raise children, and care for the disabled, frail and sick (Pavlovskaya 2004). They work formally and informally, in many economic spaces outside and inside their homes, do paid and unpaid work, produce and exchange goods, services and emotional care. We see the operation of multiple economies in the diverse range of goods and services (including sex) bought and sold from different sites: on the internet, through small ads and even from a shopping centre car park, as illustrated in Case 1.2.

A number of attempts have been made to fully account for privately exploited 'free goods' (including love, care and biodiversity). Hazel Henderson (1982, 1990) characterises what she calls the 'total productive system' of an industrial economy as a 'three-layer cake with icing'. Following this metaphor, gross domestic product (GDP) represents just one layer of the economy, and private market 'production and consumption activities' are merely the icing on top of the cake. The majority of the 'economic cake' is in fact made up of a non-monetised economy of unpaid labour; a 'love economy' which takes care of dependants, involves sweat equity, volunteering and home-based production. More recently, Miriam Glucksmann (1995) charts the gendered divisions of paid employment, unpaid work and care as multiple, interdependent parts of the 'total social organisation of labour'. Other attempts have been made to develop more inclusive and holistic systems of national accounting, such as with the Measurement of Economic Welfare by Torbin and Nordhaus (1972), which ascribes a monetary value to household work and deducts costs (negative externalities) for aspects of urbanisation such as commuting to work. This recognition of multiple economies builds on the general concept of 'social capital' first used in relation to urban neighbourhoods by Jane Jacobs (1961) as an expression of the value of informal associations.

Case 1.2: Underground economy taps into illegal workers

Every day thousands of people in Arizona buy and sell goods and services in what is best described as a 'word of mouth' economy. It's a system that plays out every time a homeowner pulls into a shopping centre parking lot and hires someone to do yard work; or when a cement contractor enlists the help of a worker to lay sidewalk for a day. It can even be high-tech, such as when a person logs on to the Internet and places a classified ad for painting services or a baby-sitter. This underground economy is here to stay, regardless of pending legislation that aims to crack down on businesses that hire undocumented workers: it thrives on everyday citizens who pay 'cash in hand' for casual workers to retile their roofs, clean their homes or mow their lawns. Eliminating the day-labour market is particularly difficult because of the high demand that exists for outside housework for affluent and busy working families. Indeed, though considered 'underground', this economy is very much visible. The most recognisable examples of day-labour markets are groups of workers who assemble at informal sites such as home improvement store parking lots, gas stations, parks and high-traffic intersections. Most are tapped for physically arduous jobs, and the work opportunities can be sporadic.

These are not the only places this market is playing out. Homeowners, renters and the businesses that aim to serve them have word-of-mouth referral networks such as community centre bulletin boards and web sites (such as Craigslist.org) where people can buy and sell goods and services. Few of those who regularly use these sources to find help for odd jobs claim to be concerned about the legal status of those they hire.

(Andrew Johnson, *The Arizona Republic*, 1 July 2007,
available at http://www.azcentral.com/arizonarepublic/news/
articles/0701biz-workers0701.html)

US law enforcement officials are focusing new sting operations on Craigslist.org because it is increasingly used to trade sex for money. In July raids, the Sheriff of Cook County Illinois rounded up 43 women working on the streets – and 60 who advertised on Craigslist. In Seattle, a covert ad on Craigslist resulted in the arrest of 71 men, including a bank officer, a construction worker and a surgeon.

Bruce Lambert, *The New York Times*, 5 September 2007,
'Law enforcement targets prostitution on Craigslist',
available at http://www.nytimes.com/2007/09/05/nyregion/
05craigslist.html?_r=1&hp&oref=slogin

Comment

This case study highlights the intricate 'chains of exploitation' binding some individuals in servitude while at the same time linking them indirectly to more affluent households (especially working parents) who themselves seek out solutions to time pressures and the squeeze on social reproduction work. There is a growing academic literature which critically examines both the motivations for, and human consequences of, households bound together in frequently exploitative multiple economies such as those identified above. Motivations and driving forces behind the 'out-sourcing' of social reproduction work, where one household directly employs 'help' in the home, or purchases labour-saving 'concierge' services, can be attributed to the rise in women's paid employment, such as with the normalisation of a dual-income household, as well as the 'speeding up' (and stretching out) of daily life through long-distance commuting and a proliferation of information and communications technology. One of the first scholars to make these connections, though without making any direct reference to the material context (of the city), was the eminent sociologist Arlie Russell Hochschild. In her book *The Time Bind*, Hochschild (1997) explored the repercussions of a work-driven lifestyle, highlighting the marital tensions in dual-income couples in which parents worked longer hours, husbands rarely contributed their share of the housework, all in a climate of rising cultural expectations of intensive parenting, labour-intensive home decoration and high standards of cleaning and garden maintenance.

A number of other 'demand-side' analyses of intra-household negotiations and work/life imbalances followed (including Nelson and Smith 1999; Hardill 2002; Presser 2003; Jarvis 2005a). Hochschild later shifted her attention to the supply side of this treadmill and the 'deficit' in social reproduction work in affluent dual-career households, typically plugged by migrant female workers – the nannies, maids and sex-workers – in *Global Woman* (Ehrenreich and Hochschild 2003). It is to the 'supply-side' issues of migration, low-pay, human trafficking and bonded labour that the greatest attention has now turned. Key readings in this growing field include Geraldine Pratt's (2004) study of Filipino domestic workers caught up in a global chain of feminised and racialised exploitation in Vancouver, in her book *Working Feminism*; Bakker and Gill's (2003) edited collection *Power, Production and Social Reproduction*; Ehrenreich's searing analysis of poverty wages in the USA in *Nickel and Dimed* (2001); and Polly Toynbee's (2003) equally vivid portrait of life in low-pay Britain. (See also Chapters 4 and 9.)

Urban imaginations

Finally, by way of introduction, it is important not to neglect the *imagined* subject of cities, even in a book which prides itself on drawing on empirical evidence of everyday life from across the globe. This is because so much of what is socially constructed (or taken for granted) as 'normal' in the way cities work and how people behave in particular local contexts rests with a medley of fiction, myth, legacy and law. Also it is because, rich and poor alike, those who come to live in the city interweave their realities with their dreams, to create their own urban magic realism. Who knows, for example, how cities came to be associated with the promise of riches and a land of opportunity (Lees 2004: 3)? Certainly by the fourteenth century a classic British children's story told how Dick Whittington, a boy from a poor Gloucestershire village, journeyed to London to make his fortune, believing the streets to be paved with gold. While the mean streets Dick found himself walking were certainly not paved with gold, this legend sees him become rich and influential as 'three times Lord Mayor' of London. Employed as a kitchen hand, Dick strikes it lucky in a number of business transactions and, by marrying his master's daughter Alice, he inherits a fortune.

A latter-day 'streets are paved with gold' belief can be witnessed in the enduring attraction of major 'world cities' such as London, New York and Paris to would-be migrants seeking better economic and social opportunities (McKenzie *et al.* 2007). Many thousands of men and women risk their lives and accumulate impossible debts, making the difficult journey from so-called 'third world' countries to so-called 'first world' countries. Most are attracted to a few 'honey-pot' cities through the myth of anticipated opportunities and riches; the reality is more likely to be a precarious existence of servitude. This movement is also evident within less developed countries (LDCs) as economic globalisation is increasing the tide of rural to urban migration, linked to expectations of economic opportunities, greater independence and the 'bright lights' of the city.

In Dhaka, for example, rising international investment in the garment export industry has meant increased jobs in export processing zones – largely for women. Many women move from impoverished rural locations, where economic stagnation and landlessness are breaking the patriarchal bargain of male support for female dependants, to take up these opportunities; the women are more willing to risk more to earn their own income under changing social and economic conditions (Ahmed and Bould 2004). Outcomes for many have been positive, but the sustainability of the changes brought by women's increased economic independence is questionable, given the footloose nature of global capital compared to less mobile labour (Kabeer 2000). In Chapter 6 we explore the gendered push and pull factors of transnational migration and in Chapter 9 the diversity of outcomes of export employment for women workers. For now it is sufficient to observe the inevitability that this legend has Dick rather than Alice as the central protagonist; neither is it a relic of the past that property might pass to a son-in-law in preference to a daughter[1].

Summary

This chapter has clarified the purpose of the book, namely the 'marriage' of urban studies and gender studies through an integrated and inclusive approach to the structures, materialities, practices and discourses of cities and gender around the world. We have suggested ways that the systematic intertwining of urban studies and gender studies might alter the subject and method by which people–place urban interactions are viewed by critically examining popular representations of cities and gender in the media. The bridging concepts of 'standpoint', 'everyday life' and 'multiple economies' were introduced alongside techniques of urban ethnography such as the close scrutiny of archives, maps, diaries, statues and monuments. Efforts to promote 'new ways of seeing' and participatory or action research were related to the capacity for gender mainstreaming in the classroom and in professional practice to be fundamentally transformative.

We began by questioning the tendency for cities to be viewed in binary terms from a narrow androcentric and occidocentric standpoint. We stressed that it is simplistic to reduce the social world to binary categories of urban/rural and man/woman. These tendencies were brought to life through everyday examples of the way urban and social 'problems' are labelled and represented in the media and in urban social and material landscapes. Together, these bridging concepts and illustrations of their application in urban ethnographic methods raised awareness of the myriad ways that masculinity and femininity are framed by powerful cultural norms.

Suggested learning activity

Search the internet for two or more examples of college-level urban studies syllabi. Try to identify syllabi for different institutions where the module or course selected focuses on urban topics/the city/cities. Critically examine the aims and objectives of the course as well as the range of topics and suggested reading material. Is the gendered nature of urban daily life made explicit, or is it implied? Is there enduring evidence or subtle examples of androcentrism, or occidocentrism? Which 'key thinkers' and 'key concepts' are identified? How far does the content of the urban studies syllabi you have identified differ from the content (and general introduction) of this book?

Note

1. It was not until 1922 that in Britain the Law of Property Act enabled both husband and wife to inherit property equally and not until 1964 that in Britain the Married Women's Property Act entitled a woman to keep half of any savings she had made from the allowance she was given by her husband. Property law varies from country to country around the world and there are states where women are still barred from holding or inheriting land, property or capital assets.

Further reading

Fuss, D. (1989) *Essentially Speaking: Feminism, Nature and Difference*. New York: Routledge.

Goffman, E. (1959) *The Presentation of Self in Everyday Life*. Edinburgh: Anchor Books.

Hayden, D. (1976) *Seven American Utopias: The Architecture of Communitarian Socialism 1790–1975*. Cambridge, MA: MIT Press.

Mohanty, C.T. (1988) 'Under western eyes: feminist scholarship and colonial discourses'. *Feminist Review*, 30: 61–88.

Spain, D. (1993) *Gendered Spaces*. Raleigh, NC: UNC Press. Focusing on non-industrial societies, this book offers fascinating insight and historical analysis by which to explore the spatial construction of gender cross-culturally.

Web resource

Harvard Library – Working Women Research, http://ocp.hul.harvard.edu/ww/. An excellent resource for access to historical documents, statistics and images of women working from the nineteenth through the twentieth centuries.

2 Historical trends in cities and urban studies

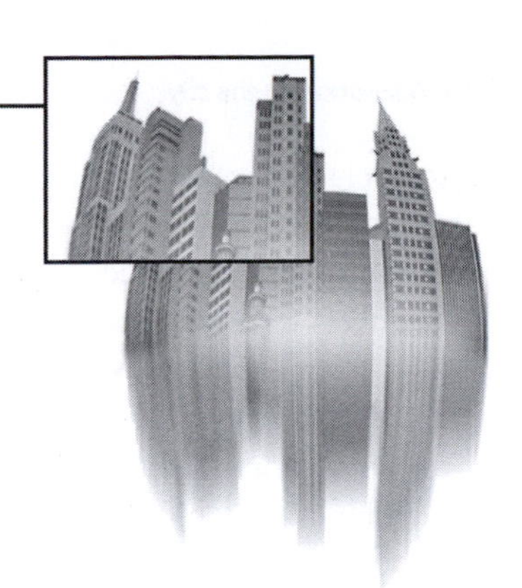

Learning objectives

- to gain familiarity with historical developments in trends of urbanisation alongside those in the discipline of urban studies
- to make connections between this historical legacy and contemporary urban issues
- to understand the origins and significance of competing schools of thought
- to consider how notions of 'development' and 'modernisation' intersect with urban studies and social theory

Introduction

This chapter charts some of the key historical trends in the development of cities and urban studies, with two aims in mind. The first aim is to understand the significance of cities for the social and spatial construction of gender. Evidence of the variety and complexity of urban settlement types and the multiple processes of urbanisation helps explain the rationale and resilience of a discipline called 'urban studies' and why opinions and theories compete to explain the complex patterns of urbanisation observed. For this purpose we introduce in the first part of the chapter an overview of urban transformations spanning the nineteenth and twentieth centuries. This represents a partial selection from a whole continuum of non-rural settlement types found across the globe.

Against this backdrop of urban diversity the second aim is to identify, from among the competing explanations and theories which have emerged over this timeframe, key paradigms and epochs which have shaped the largely androcentric and ethnocentric trajectory of urban social inquiry. This partial introduction to urban social theory attests both to the divergent ways that cities and societies are understood, according to the assumptions and standpoint adopted (see Chapter 1), and to the persistent neglect of gendered subjects (and subjectivities, feelings and experiences) within orthodox urban studies.

Also supporting these two aims, we consider a third strand of influence which we believe must be understood as a prelude to cultivating an explicitly gendered urban theory. At key points in the discussion we turn to the wider geo-political context and the rising application of urban social theory to the 'development project'. In brief, we focus on the coining of the term and creation of the discipline of development studies; the quest to ameliorate the risks, insecurities and inequalities creating and sustaining high levels of poverty found concentrated in the global south, alongside the cultural bias and standpoint assumed by this quest over time. We consider both notions of 'development' and 'modernity' alongside trends in the field of development studies itself.

In this chapter and Chapter 3 we generally adopt a chronological framework as a convenient organising principle. Yet it has to be recognised that competing explanations and alternative ideals overlap in the frequent revival (or reworking) of earlier paradigms. A good example of this reworking is the 'new urbanism' movement which is prevalent in twenty-first-century urban studies and town planning practice – reviving many of the ideas of Jane Jacobs[1] and Kevin Lynch from the 1960s and, before this, Ebeneezer Howard's Garden City Movement from the 1890s.

Notwithstanding the contemporary nature of urban scholarship, writing on the city dates back many centuries and we can learn much today from re-reading what nineteenth-century commentators interpreted as the adventure, corruption and danger of urban daily life. The history of urban studies is after all a history of ideas which advance as cities themselves grow and change. Because of this fluidity it can be misleading to fix on certain epochs. Nevertheless, for the purposes of intertwining separate disciplines in this book we believe it is useful for readers, while engaging with the content of Chapters 2 and 3, to have in mind a simplified chronological overview (included as an Appendix, pp. 298–300) of urban studies and gender studies together with a shared social and political context. Viewed in parallel, significant disjuncture is apparent between trends in thinking on cities and gender in academic scholarship.

The significance of cities

In Chapter 1 we stressed that it is simplistic to reduce the world to binary categories of urban and rural. Yet a common theme of writing on the city over the nineteenth and early twentieth centuries is the tension between urban and rural ideals, between utopian and dystopian visions and ultimately with notions of pre-modern 'under-development' and modernity. The language used to distinguish urban from rural has not only tended to assert a determinable binary division, as noted on p. 10 with respect to contrast theory, but also typically adopted metaphors which present the city as a corruption of nature. Of course, the opportunities and constraints confronting a *campesino*'s/farmer's wife living self-sufficiently many miles from shops and schools are going to be very different from those of a factory employee living in an overcrowded apartment block in the city; however, the moral distinction is dubious and counter-productive given the rapidity of contemporary urbanisation. Moreover, contrasts and evocations function as the product of time, place and standpoint; whether that of a 'rural arcadia' or a 'modern' city (where modern and urban are typically considered 'good' in the context of 'under-development').

Definitions of cities and urbanisation are problematic. While traditionally the city has been defined by its size and status, urban/rural distinctions today tend to be more experiential, based on density, intensity, economies of scale, diversity and cosmopolitanism. As we suggested in the general introduction to this book, there has been a steady shift away from distinct definitions, categories and hierarchies, to recognise diverse views of the world and alternative models of what cities of the future might, or should, be like. Rather than differentiating north from south or large from small, there is increasing awareness of the function of complex gender, class and ethnic conflicts which play out within and between households, in communities and against the state and in opposition to multinational interests.

In this section we trace the main trends and patterns of urban transformation with respect to urbanisation, suburbanisation, mega-cities, gentrification and 'gated' communities. We have chosen to begin our (largely chronological) series with the growth of the industrial city, with the challenge of accommodating the mass migration of people from rural villages to the cities of early trading posts and the ongoing need to build new, or adapt existing, buildings and regulatory frameworks. This is because we shall see on p. 35 that practical concern for population density and housing (and indeed public health) initially prompted what was to become an enduring fascination with all things urban. It is nevertheless important to recognise these to be highly selective, partial and in reality overlapping frames of reference, whereby new settlement patterns and living conditions never entirely replace earlier patterns but rather co-exist as a patchwork or mosaic. It is important to keep in mind this layering of new onto old when studying cities. Rarely is it possible to accommodate new technologies or social developments in purpose-built premises. Wholesale renewal (clear site demolition and reconstruction) is rare and most social and economic transformations sit awkwardly or anachronistically on top of a land use and infrastructure designed for a previous epoch. A good example of this is to compare the narrow roads of central London, originally built for horse-drawn carriages, with the imposition of wide boulevards onto medieval Paris by Haussmann, to facilitate the movement of troops.

Urbanisation

Urbanisation is typically associated with industrialisation and the creation of a market-based (wealth accumulating) economic system. It essentially represents the migration of people and jobs from rural areas and small towns to higher density metropolitan areas. This process acknowledges cities to be 'where most of the world's population will live and work; where most of the economic activity will take place; where the most pollution will be generated; and where the most natural resources will be consumed' (Badshah 1996: 1). It was through the rapid growth of the industrial city alongside new patterns of shift-work factory production that sites of industry were increasingly separated from over-crowded 'slum' residential districts. It is therefore always important to remember that the material transformation of Euro-American cities coincided with similarly significant changes to family life and social organisation (see Chapter 3).

The pattern of human settlement is changing rapidly around the world and for some time now we have known from United Nations (UN) statistics that more than half of the world's population lives in areas defined as 'urban' by their settlement density. Moreover, it has been predicted that twice as many people again will be urban dwellers by 2030. According to the *World Development Report* (2003), two hundred years ago there were only two cities of one million population (London and Beijing), but by 1990 there were 293. Whereas in 1800 less than 5 per cent of the 980 million world population at that time lived in cities, by 1950 this had more than tripled, to 16 per cent, and the 50 per cent mark has now been passed, probably in 2004. This global trend of urbanisation began with the first industrial cities, notably Manchester, Chicago, Detroit, Pittsburgh, which grew rapidly in the late nineteenth century to compete for the first time in scale with historically dominant trading cities such as London and New York.

It is tempting when making these connections between urbanisation, industrialisation and economic reorganisation (on which we say more in Chapter 3) to imagine that Europe and North America always led the way in the growth of the largest cities. Yet since the 1950s the growth of urbanisation has been far more rapid in Asia, Africa and South America. In Latin America as a whole, for example, in 1950 over 59 per cent of the population lived in rural areas, whereas by 2000 over 75 per cent of the population lived in urban areas (UN 2000, cited in Cerrutti and Bertoncello 2003); the pace of urbanisation in that region has been roughly twice as fast as the process which took place in Europe and the USA in the nineteenth century. The pace and extent of that urbanisation plainly vary considerably from country to country (with Chile leading the way) but the problem with that pace of change and the headline statistics it elicits is that complex socio-economic ebbs and flows, not to mention the real-life agency of the individual, household or social movement, can be hidden in the sheer bulk of data. This again points to the contested terrain of urban studies where competing theories are continually advanced which seek to explain patterns and processes of inequality and uneven development.

Suburbanisation

In the global north, rapid urban growth has been accommodated over time through a combination of upward and outward expansion where growth management has been shaped to a large extent by engineering and technological innovations (elevators enabling skyscrapers, motorways and electric rail extending the reach of the commuter belt). The construction costs of building tall 'cities in the sky' (to coin Le Corbusier's term) effectively confine these to premium land at the city core where multinational corporations see a good return on their investment. By contrast, low density outward expansion, creating suburbs, bedroom settlements and 'sprawl', tends to be associated with widespread car ownership and the life-course effects of residential mobility and housing choice. In turn, periodic regeneration feeds into the competitive spatial differentiation of residential neighbourhoods. For example, a 'move to the suburbs' is strongly associated with family formation, especially the quest for detached family housing, a yard or garden and family amenities, such as play spaces and well-equipped schools (Germain and Rose 2000).

The term 'suburbia' typically evokes in the popular imagination anachronistic stereotypes of the idealised 1950s symmetrical (but undemocratic) family (see, for instance, Hayden 2003). A widely held perception is that cities 'emancipate' people (offering freedom to be oneself sexually, politically, intellectually) while suburbs oppress or suppress these desires (Lees 2004: 9). This is revealed, for instance, in the way suburbia and oppressive gender role identities are satirised in the film *The Truman Show* (see Pratt and San Juan 2004). Notwithstanding a long tradition of feminist writing on the apparent 'entrapment' of women in suburbs and the androcentrism in town planning which underpins suburbanisation, recent feminist critiques have sought to dismantle this city/suburb binary (see Spigel 1992, in contrast to Wilson 1991).

Although it is widely perceived that suburbs are 'bastions of class and racial exclusivity' (Domosh and Seager 2001: 104), over time they have evolved and diversified with respect to the lifestyle aspirations and residential choice of different ethnic groups making up cosmopolitan metropolitan areas. Li (1998) describes what are referred to as 'ethnoburbs', where distinct ethnic communities emulate the settlement patterns of their suburb-pioneering predecessors. Ethnoburbs are effectively suburban ethnic clusters of residential areas and business districts where vibrant ethnic economies depend on large numbers of local ethnic minority consumers[2]. These neighbourhoods have strong ties to the global economy (through, for instance, travel agencies arranging transport to and from the main countries of ethnic origin of the resident communities, as well as money-transfer services back to the same places) reflecting their role as outposts in the international economic system. They are multiethnic communities in which no one minority community necessarily comprises a majority.

Suburban expansion and uncontrolled growth tend to be negatively associated with urban sprawl, or a lack or urban containment (both defined on pp. 36–7) and pejorative descriptions of sterile, monotonous landscapes. In affluent cities the problems of urban sprawl and a general shift in taste and lifestyle in favour of more cosmopolitan and 'liveable' mixed land use arrangements has prompted a growing trend for older suburbs to 'urbanise' (Domosh and Seager 2001). This typically involves the incorporation of new nodes of economic activity, retail parks and 'village-like' neighbourhood centres within the original footprint of the suburb. This makes an important point that the suburban hinterland is no less dynamic and varying than the cities from which they spring. Arguably it also reflects changes in family life and social organisation, notably changing gender roles and power relations.

In affluent areas where basic needs have been met urban planners increasingly seek to deliver on 'post-material' 'quality of life' measures of development. The question becomes one of whether a place is 'liveable' inasmuch as individuals and households are able to successfully juggle the time-pressures and multiple activities in their daily life, especially those associated with the rising number of dual-breadwinner households. These trends also reflect a shift in thinking among Anglo-American planning practitioners (notably reflected in Smart Growth and the new urbanism), as well as the spatial economic effects of the high-tech and dot-com industries. The definitions provided below contrast 'live-ability' and urban containment with urban sprawl.

> *Urban sprawl* is typically defined as the unplanned, uncontrolled spread of urban development where this tends to lead to a monotonous landscape of standardized buildings and significant distances required to travel between homes, jobs, schools and shops. The sterile appearance of sprawl attracts the pejorative term 'subtopia' (loss of place).

Urban containment is the policy of limiting urban sprawl by imposing restrictions on edge of town and out of town development (such as shopping malls). Urban containment (zoning and/or green belt or growth boundaries) vary geographically and over time according to changing fashions in land use planning and a willingness and ability to impose such regulations.

Liveable cities/neighbourhoods are basically those that are 'habitable' or suitable to live in. Vukan Vuchic provides a good definition of the 'qualitative' indicators of liveability. These broadly encompass 'those elements of home, neighborhood, and metropolitan area that contribute to safety, economic opportunities and welfare, health, convenience, mobility, and recreation' (Vuchic 1997: 7). Within the overall goal of achieving a liveable city, he identifies three sets of characteristics:

- *human-oriented and environmentally friendly*, with features that make it attractive and make living in it convenient, safe and pleasant – a high degree of sustainability is a component of this characteristic;
- *economically viable and efficient*;
- *socially sound*, that is, without social, economic or ethnic barriers, or wide variations in income, crime and unrest – there should be a sense of togetherness and pride in the city and region.

It is important to note that the problems of unplanned or unconstrained growth are not limited to affluent Euro-American cities. Uncontrolled and illegal 'squatter' settlements are a significant feature of the 'mega-cities' we go on to describe on pp. 37–8. Figure 2.1 shows, for example, the mass of irregular housing and poor living conditions in the favela La Rocinha on the outskirts of Rio de Janeiro.

Mega-cities

As the statistics above have indicated, the fastest growing cities are those of rapidly developing and less developed/industrialised countries. Rio de Janeiro and Sao Paulo are prime examples in Latin America, while Mumbai is the quintessential Indian mega-city. Here urban quality of life rests with the fulfilment of basic human needs and access to fundamental infrastructures: drinking water, sanitation, transport, communications, shelter and the like. Efforts to 'develop' these basic infrastructures (typically following a top-down Western template) can have unintended consequences for women as compared with men, or particular family structures, as witnessed in changes to household decision making and strategies of management and survival. In turn, gendered coping strategies and livelihoods are implicated in some of the most pressing debates of the twenty-first century. This is why there is a strong case for gender mainstreaming in urban studies – both in 'advanced' and 'developing' world contexts. We expand upon this theme in Chapter 8.

Figure 2.1 ***The mega-city sprawl of Favela La Rocintha, Rio de Janeiro***
Source: Jonathan Cloke

Much of what has been written on suburbanisation in mainstream texts can be said to be ethnocentric, taking an Anglo-American standpoint and assuming this cultural perspective to represent 'progress'. Orthodox understanding of technology-led urban growth and the life-course effects of residential mobility does not translate to the conditions of poor settlers on the hinterlands of the mega-cities. As we see in the case of Guinean students studying by streetlight in Case 2.1, for those around the world without reliable access to electricity and clean drinking water (at least that they can afford), the lived reality of the fracturing and splintering of mass rural to urban migration is very different from that conveyed by textbook models of suburban development and poly-centric dispersal.

Gentrification

Accompanying the growing diversity of urban, suburban and indeed rural land use and lifestyle arrangements there is evidence of a selective 'return to the city' movement popularly identified as 'gentrification'. This trend is associated in affluent economies with the rise of a 'post-industrial' society dominated by service sector employment and a growing wealth differential between those employed in professional, high-tech and 'creative' sectors and those in poorly paid, feminised,

Case 2.1: When the lights go out, students take off to the airport

It's exam season in Guinea, ranked 160th out of 177 countries on the United Nations' development index, and students flock to the airport every night because it is among the only places where they can count on finding the lights on. Groups begin heading to the airport at dusk, hoping to reserve a coveted spot under one of a dozen lampposts in the parking lot. Some come from over an hour's walk away. Only about a fifth of Guinea's 10 million people have access to electricity and even then the reality is of frequent power cuts.

With few families able to afford generators, students long ago discovered the airport. Parents require girls to be chaperoned to the airport by an older brother or a trusted male friend and even young children are allowed to stay out late so long as they return in groups. They sit by age group. Few cars disturb their studies. The students at the airport consider themselves lucky. Those living farther away study at petrol stations. Others sit outside the homes of affluent families, picking up the crumbs of light falling from their illuminated living rooms.

Comment

This report vividly conveys the systematic exclusion from basic services of poor residents in rapidly developing urban areas. The lived realities of this exclusion at the domestic level are increasingly subject to critical scholarly analysis, highlighting the widening gap between rich and poor within and between regional economies in a highly unequal world. This is particularly true with respect to water privatisation and the effective rationing of basic needs. The following extracts from the book *Social Power and the Urbanizaton of Water* by Erik Swyngedouw provide a compelling argument for this line of analysis:

> In Mexico City, 60% of all urban potable water is distributed to 3% of the households, whereas 50% make do on 5%. In Guayaquil, 65% of the urban dwellers receive 3% of the produced potable water at a price that is at least two hundred times higher (20,000%) than that paid by the low-volume consumer connected to the piped urban water network. The mechanisms of exclusion from and access to water manifest the power relationships (gender, race, class) through which the geography of cities is shaped and transformed.
> (Swyngedouw 2004: 30)

Source: Callimachi (2007).

personal service occupations, who are to a large extent priced out of the city. We explore these labour market transformations further in Chapters 3 and 7.

The hallmark characteristics of gentrification, notably the upgrading of run-down vintage property and the proliferation of coffee shops, wine bars and pricey independent boutiques, attest to the enduring cachet of a distinctly urban lifestyle (Pacione 2001: 7). The term 'gentrification' immediately signals the process most commonly believed to drive this transformation, namely the replacement of working-class by middle-class residents (McDowell 1999: 104). Use of this class-based term suggests a straightforward process of succession whereby the superior purchasing power of one social and economic class overwhelms and displaces another, where resettlement of previously abandoned central city locations generates fresh competition for space. Yet for several decades social scientists have engaged in heated debates concerning the motivations for this material and demographic transformation (for a good overview of this, see Lees 2000). Attention has focused in particular on the influence of shifting gender relations and the formation of new household employment structures, the emergence of 'gentrifiers', including 'yuppies' (young urban professional), 'dinkys' (dual income no kids yet) and dual-career families (Rose 1989; Warde 1991). Attention has also turned to the impact of changing consumer tastes and urban lifestyle identities (Mills 1993; Bondi 1998) and demand by city professionals to cut the cost (in time and money) of commuting to outer suburbs. A 'return to the city' trend of gentrification has to be understood alongside a trend of 'counter-urbanisation', whereby the absolute decline in resident populations of inner city areas (disproportionately reflected in an exodus of families with children) can be traced to significant residential growth in the lowest density suburbs as well as commuter villages beyond the urban fringe (Champion 2001).

Demographic trends suggest that a return to city living is dominated by growth in the number of single person and childless couple households; however, we find evidence on the ground of a more generalised splintering and fragmentation of household types. Hall and Pfeiffer (2000: 96) suggest that families with children are the new urban minority. Failure to attract and retain a representative cross-section of the population suggests that otherwise ostensibly 'successful' (economically competitive) cities gain competitive advantage by creating an environment which is hostile to dependent populations such as children, the poor, the sick and the elderly (Jarvis 2005a).

Evidence in the trend of gentrification of a renaissance of urban living has excited considerable interest among feminist scholars seeking to advance a more nuanced account of gendered urban structures and residential preferences. For example, Wekerle (1984) endorses the 'placing' of women in the city on the simple grounds of geographic proximity, by noting the benefits to women of a greater concentration of jobs, provision of public services, social support networks and a more liberal politics and tolerant climate associated with urban areas. At the same time she recognises that cities can be dangerous or unsettling – as sites of sexual violence, poverty, anonymity and alienation. Similarly, Lia Karsten (2003) seeks to counter the enduring belief that 'cities are not for families'. She traces the motives and coping strategies of Dutch families with children who choose to settle in the city of Amsterdam. She identifies two discrete types of family gentrifier: immigrant families and a new urban middle class. Despite many positive benefits of higher density urban living, she concludes that in the absence of active intervention to realise family friendly cities, discourses and practices of urban planning will continue to assume that suburban neighbourhoods best suit families with children (Karsten 2003; 2005).

While noting the realities of a continuum of non-rural settlement types, beyond the traditional/emancipatory, suburb/city binary, it is widely recognised that cosmopolitan and 'creative' cities and neighbourhoods explicitly thrive on a reputation of gender democracy and freedom of sexual expression – in a manner hard to imagine of more parochial places (Florida 2002). The sense in which particular places embrace difference and diversity is suggested by New York's annual gay pride march, illustrated in Figure 2.2.

***Figure 2.2** 'Married but not monogamous'; Gay Pride march, New York City*
***Source:** Helen Jarvis*

Larry Knopp (1998) observes a clear association between sexuality and urban space, noting for instance the explicit gay male identity politics (and later lesbian migration) underpinning San Francisco's Castro district, which has been a potent symbol and concentration of gay liberation since the 1960s. Paradoxically, while much has been written about the Castro district as an example of 'gay gentrification', this neighbourhood has more recently seen an erosion of gay identity as a consequence of the rising value of housing and retail premises. In a market fuelled by property speculation, independent retailers are losing out to chain stores and at the same time gay and lesbian residents exhibit growing confidence that they can live outside the 'safe space' of the Castro without fear of discrimination (Rubin 2000; see also Bailey 2001 and Kenney 2001 on similar trends in Los Angeles).

It is important not to overplay the relationships between gentrification and a gay and lesbian spatial politics. Brown and Knopp (2006) suggest that census surveys of same-sex households inaccurately reflect the diffusion of gay households across urban space. More particularly, Julie Podmore (2006: 595) notes that while gay men may have been known to produce highly visible enclaves in major post-industrial cities (such as San Francisco and Los Angeles), lesbian forms of territoriality on the urban scale have been relatively 'invisible'. She argues that this is because lesbian communities tend to be constituted through diffuse social networks rather than fixed commercial sites (Podmore 2006; see also Valentine 1993).

Gated communities

A noted consequence of the post-industrial gentrification described above is the way property speculation has priced low and even average income households out of previously accessible inner urban locations. More extreme still is the proliferation of security-controlled or gated communities which effectively exclude the general public from previously accessible sites by limiting entry to key-carrying or fee-paying community members. Also evident in this international trend of gated and common interest developments is the relatively new phenomenon of adult-only communities, where residents are required to be childless and with no intention of having children (even in some cases required to sign a contract stipulating that they will have no children as visitors) (Townshend 2006). Figures 2.3 and 2.4 show how individual buildings, streets and communities are designed for exclusive access, using private security firms, gates, cameras, signage (privatised streets) and intimidating architectural design to prevent passers-by from gaining access or peering inside (see Atkinson and Blandy 2006 for an overview). The private policing and surveillance of shopping malls is another manifestation of this.

These trends of urban exclusion and privatisation are not restricted to the affluent world. Most recently, the sharpest growth in gated community development has

Figure 2.3 ***Gated community, Spain***
Source: Helen Jarvis

been witnessed in Latin American, South African and Indonesian cities (Borsdorf *et al.* 2007; Leisch 2002; Landman and Shönteich 2002). The socio-spatial implications of this are raised in Case 2.2.

Transformations in development and modernity

So far in this chapter we have outlined key patterns and trends of urban settlement which sit squarely under the urban studies rubric. Yet recent years have witnessed increasing frustrations with the narrowness of individual disciplinary perspectives, especially with respect to complex and persistent problems of inequality and uneven development which arguably call for inter-disciplinary collaboration and pluralistic vision. Jenny Robinson (2006: 5), for instance, traces the influence both of transformations in thinking on notions of development (or developmentalism) and of modernity in contemporary urban studies. This reflects not only the contributions of a discrete field of 'development studies' but also the significance of 'epoch' in all aspects of thinking on cities and gender. By the term epoch we mean evidence of increasing rupture with the past. We can perhaps best illustrate this

Figure 2.4 ***Gated community, UK***
Source: Helen Jarvis

with reference to definitions of colonialism and post-colonialism and scholarly efforts within the latter to sever ties and make amends with the tarnished legacy of the former. In the following discussion we trace a number of shifts in thinking on 'urbanism', 'development' and the active pursuit of 'modernity' (Dennis 2008) which are best understood in relation to a standpoint legacy of colonialism.

> *Colonialism*: control by one national/sovereign power over a dependent area or people, imposing European 'civilisation' on other 'primitive' societies. Colonialism is associated with racial and cultural inequalities between ruling and subject people – with clear parallels between this and patriarchy (the rule of the father). Colonialism typically serves to maintain persistent inequalities and allow the exploitation of under-development.
>
> *Post-colonialism*: the time following the establishment of independence in a colony and the social, political, economic and cultural practices which arise through resistance to colonialism. Post-colonial theorists critically reflect on the legacy of European conquest, US hegemony, geo-politically uneven development and ethnocentric discourses and writing on cities, gender and development, seeking to recover agency and voice for colonised people.

Case 2.2: Fortress communities in developing countries

Fear of violent crime has led to a proliferation of private security arrangements and gated communities in cities like Sao Paulo and Johannesburg, allowing those who can afford to do so to withdraw from public space (Landman and Shönteich 2002). The privatisation of public space appears to be entirely regressive, with evidence of heightened fear, increased crime, widening gaps in living standards and fractured communities. In Brazil, Landman and Shönteich observe that

> fortified enclaves contribute to higher levels of inequality, fear, suspicion and feelings of vulnerability in those 'outside' the boundaries. Fortified enclaves in Brazil also contribute to the transformation of urban spaces. Some public spaces are privatised (and so prohibit access), while others are neglected, abandoned and relinquished to violence and illegal forms of control.
>
> (Landman and Schönteich 2002)

Borsdorf *et al.* (2007) note that gated developments known as *fraccionamiento* are increasingly widespread in Latin America. They discuss the diverse ways that the Latin American city is being shaped socially, culturally and materially by the construction of privatised highways, large shopping malls with restricted access and enclaves of advanced services. This privatisation reflects not only processes at work within the neoliberal Latin American state but also global influences such as the influx of US and Canadian expatriate retirees who are attracted to life in the sun behind the walls of the *fraccionamiento.*

Historical shifts in thinking on cities

We have selected just four 'paradigms' to represent dominant (typically competing) bodies of theory holding sway at key junctures from the late nineteenth through twentieth centuries to the present day. These four approaches are the 'nostalgia' (or idealisation) of contrast theory; structuralism (notably Marxism); urban ecology (noting the 'other' Chicago school); and post-structuralism. Cross-cutting these (arguably contested) bodies of theory are aesthetic and practitioner-based movements associated with architectural design and technological innovation, as well as changing social norms of behaviour (see Appendix, pp. 297–300). In this sense, these paradigms need to be read as overlays to the previous discussion of urban transformations, alongside the changing epochs which have variously cast cities and societies as colonial, pre-modern, modern and post-modern at various moments in history.

Nostalgia and contrast theory

We observed in Chapter 1 that urban studies texts are littered with historical reference to binary distinctions. This is particularly apparent in the influence of 'contrast theory' on early urban studies. Writing in the nineteenth century the German social theorist Ferdinand Tönnies identified two contrasting forms of association which he conceived as functioning through the environmental differences between pastoral villages and industrial cities. The first, Gemeinschaft (or community), features primary relationships that typify kin groups, clans and village life. The second, Gesselschaft (or society), reflects the loss of intimacy which comes with more complicated and seemingly detached social interactions characteristic of 'the anonymous crowd', or what some refer to as 'traffic relations' based on impersonal commercial transactions.

Mass rural to urban migration stimulated interest in the practical (engineering and planning) means by which new and expanding settlements were to accommodate large numbers of men, women and children, many settling in poorly constructed and unsanitary homes where disease was rife and diet limited. In practical terms, too, academic interest grew in response to concerns to eradicate slum housing conditions, overseen to some extent by the longer established academic traditions of architecture and town planning. In this classic writing, stark contrasts were made between rural 'clean' and industrial 'polluted' societies which tended to depict what went before as 'good', viewed with nostalgia, while the trajectory of rapid industrialisation was viewed more negatively. We define the terms utopia and dystopia on p. 51 and then critically engage with the popular appeal of utopia in Case 2.3.

Elizabeth Wilson notes that critics of urban life contrasted an 'organic rural community' with what they feared was the break-up of traditional order (patriarchy, authority, civilisation): 'the menace for the cities was not only disease and poverty; even more threatening were the spectres of sensuality, democracy and revolution' (Wilson 1995: 60). A clear example of this contrast theory and distinct anti-urban bias can be identified in the classical urban sociology of Georg Simmel, whose 1905 book *The Metropolis and Mental Life* associated increasing population density with depletion and nostalgia for a lost, more harmonious past.

Simmel's writing expresses ambivalence as much as contrast. While he loved the vitality and cosmopolitanism of his native Berlin he also expressed grave misgivings that daily impersonal encounters were eroding his soul and sanity. Simmel's ideas later gained sympathy from others wishing to highlight a sense of ambivalence in detached urban relations in order to promote a 'view from the street' (Keith 2000: 415). Walter Benjamin, for instance, brought to life the mundane desires and fears of ordinary people in European cities such as Berlin and Paris. This shift in emphasis from contrast to ambivalence was combined with renewed interest in 'ordinariness' and 'everyday life', as introduced in Chapter 1. Thus Benjamin's work has been embraced to legitimate the study of hidden, taken for granted norms and discriminatory practices on grounds of gender, class, age, ethnicity and family form. On p. 48 we illustrate how, in the context of contemporary South African cities, Jenny Robinson (2006) draws on Benjamin's work to conceive a 'world of ordinary cities' beyond the hierarchical ordering imposed by notions of modernity and development; a world in which 'difference can be gathered as diversity…without any suggestion that a universal theory of urbanism is possible' (Robinson 2006: 41). Later we expand on this approach in relation to modernity and development and also as an expression of post-structuralism. In turn, Simmel's influence and concern for 'the paradox of the lonely crowd' can be traced down through the years, first to Goffman's notion of social activity as performance (using the spatial metaphors of 'front stage' and 'back stage'), then to Benjamin's urban flâneur and de Certeau's (1984) 'practices' of everyday life (Short 2006: 38–39).

While these social theorists began to shift the focus of interest from the skyline to the street, the view of the street they exposed was an exclusively masculine domain. As John Rennie Short (2006: 39) reminds us, the voyeuristic flâneur was quite specifically a *gentleman* stroller of the city. Elizabeth Wilson observes that the flâneur 'represents men's visual and voyeuristic mastery over women, [their] freedom to wander at will through the city is essentially a masculine freedom' (1995: 65). When feminist scholars evaluate the writings of Simmel, Benjamin and de Certeau on subjects of gender, men, women and the family, they reveal the construction of a masculine understanding of the world in which 'women (are consigned) to the wastelands of [the] philosophical imagination' (Witz 2007: 354; Wolff 2000). Stafford points out, for instance, that nineteenth-century Germany, which provided the context for Tönnies' writing about masculine and feminine social relationships, was 'more patriarchal (some would say antifeminist) than contemporary Britain or America' (1995: 391). In her book *Resident Alien* Janet Wolff (1995) reclaims the 'seductively modern figure' of the 'female stranger' who was so neglected by male social theorists writing in the early twentieth century. She applies feminist cultural criticism to this view of the street, not only to expose the pervasive sexism of this classical writing but also to bring to life the gendered lived experience of intersecting social structures, systems of representation and subjectivities.

Naming the 'problem' of under-development

The influence of contrast theory is similarly apparent in the changing geopolitical context of the early twentieth century. Harry Truman ushered in a new epoch of 'development' in his inaugural speech as president of the United States in January 1949: he was among the first to identify and label the 'problem' of under-development, creating what rapidly became a new field of study and practice (Escobar 1995). The significance of this historical moment has to be understood alongside the concept of modernity. In the context of less developed countries, many of them former colonies, a politically motivated contrast was made between 'modern' cities and 'backward' rural traditions by which urbanisation and modernity were presented as 'good' and rapid change equated with progress – in direct opposition to the distinction drawn by nineteenth-century urban social theorists between the natural rural ideal and the corrupted city. Box 2.1 introduces the concept of modernity, recasting pre-modern cities and society as 'backward' and 'primitive' rather than 'pure' and 'uncorrupted'.

Box 2.1: Modernity

The notion of a modern era and an age of modernity hinges on a rupture with the past. Modernity is variously interpreted as 'progressive', especially with respect to the urban built environment, with the advent of motorised private transport, high-rise engineering capabilities and new media and information and communications (ICT) technologies. The concept of modernity extends beyond questions of technology, however, to distinguish a rupture with 'tradition' and the proliferation of new lifestyle possibilities.

The concept of modernity springs from a narrow model of Western economic development and consequently results in the tendency to categorise cities relative to this model as more/less 'modern', 'developed', 'creative', 'global'. Thus while this concept of modernity best describes affluent industrial cities, in the context of the South African city Jenny Robinson recognises 'the cultural valorisation and celebration of innovation and novelty' (2006: 4) as a function of international 'development' efforts that we go on to outline on pp. 49–50. She claims that 'contemporary Western urban theory' has borrowed heavily from 'the literature of the first decades of the twentieth century, especially concerning understandings of urban modernity. Colonial prejudices from this era have been sedimented into contemporary theory' (Robinson 2006: 4).

The assumptions of colonialism, modernity and a distinct urban bias heavily underscored the emerging discipline of development studies and the wider geopolitical project of 'development'. International interventions and overseas aid closely followed the theory of modernisation and later the strict economic rubric of 'structural adjustment' (see Chapter 3). In turn, widespread faith in a Western definition of 'progress' shaped the course of urban development in the global south, contributing to the precarious conditions of daily life in the 'mega-cities' illustrated on pp. 37–8.

Modernisation theory and stages of growth

Modernisation theory informed initial efforts in the 1950s and 1960s at resolving under-development, just as today an idealised perception of modernity continues to drive the development of urban and rural alike in the global south. It is characterised by the centrality of economic growth and the aim to model less developed nations, economically and socially, on the industrialised, liberal-democratic societies of the West, in order to achieve growth and thus development (Parpart *et al.* 2000; Kabeer 1994). The legacies of colonial thinking and notions of dependency are clearly evident in the framing of indigenous practices as 'backward' relative to the trappings of modern Western 'civilisation'. The traditional cultural practices of non-Western contexts were regarded as barriers to 'growth' (urbanisation, industrialisation, consumption) and particular emphasis was placed on modernising these in order to support development.

Modernisation theory is informed by Parsons' (1951) functionalist explanation of social change and by Rostow's (1960) and Lewis' (1954) neoclassical economic understandings of economic change. Its foundational idea is that development is a natural, evolutionary and linear process of change from traditional social and economic organisational forms to the 'modern', moving through pre-determined stages of growth, with one building from the last on an upward trajectory of progress (Kabeer 1994; Parpart *et al.* 2000). Box 2.2 outlines the assumptions underlying modernisation theory and its indicators of successful development. These highlight the influences of functionalism and neo-classical economics, through the emphasis on individualism and specialisation, and the belief that markets are free, where individuals are judged by achievements versus ascribed identity (gender, race, class) and access is open to all and not influenced by patronage or kinship links. Arguably, by emphasising the modern over the traditional, modernisation theory actively promoted a shift from rural to urban development: indicators of progress were rapid urbanisation and a shift in economic dependence from agriculture to industry.

Box 2.2: Modernisation theory

The assumptions underlying modernisation theory include:

- economic growth will 'trickle down' and benefit all members of society;
- access to markets will improve people's living conditions;

- macroeconomic policies are gender neutral; and
- traditional technology is a constraint to growth and development.

Successful development is indicated by:

- levels of gross national product;
- education levels;
- industrial structure (movement from agriculture to industry);
- urbanisation;
- advancement determined by achievements versus ascribed characteristics;
- reduction in importance of kinship and family ties;
- increasing specialisation and non-personal ties;
- increased individualism; and
- rational self-interest guiding decision making versus tradition or superstition.

Source: Adapted from Kabeer (1994) and Parpart *et al.* (2000).

As the assumptions in Box 2.2 imply, early proponents of modernisation theory paid little attention to women or gender. From its very inception the theoretical discourses underpinning development had a very male, very white face. The assumption was that both men and women would benefit from the economic and social changes brought about through the move to modernity, though in different ways due to men's and women's different roles in society. A division of labour where women focused on reproduction and men on production was 'naturalised' by the assumptions of functional sex-role specialisation and the belief that modernisation would benefit both men and women, in their complementary but separate spheres. It was also assumed that the move from traditional to modern forms of social organisation would benefit women because modern societies were presumed to be more egalitarian than traditional ones (Kabeer 1994). We return to this thread of discussion in Chapter 3 when we dismantle the myth of a steady or complete movement toward gender democracy in any contemporary city.

Feminist critique of contrast theory

Contrast theory has been heavily criticised by contemporary feminist scholars (as have the assumptions of modernisation) not only for the assumptions of a masculinist ontology but also for the sense in which it disregarded the tenacity and powerful ties of working-class industrial communities throughout the nineteenth and twentieth centuries (McDowell 1999: 100). Yet it would be wrong to dismiss this 'classical' writing on alienation and ambivalence as entirely backward looking, individualistic or apolitical. Instead, these and other utopian visions of a better world, where ‘the ills of the present day are banished to another space and time’ (Pinder 2002: 233), deploy nostalgia as a means to provoke transgression and inspire social movements aimed at creating alternative (more collective) modes of human habitation. Thus we find aspects of contrast theory in the otherwise enlightened Garden Cities movement and in the philosophy of its founder Ebenezer Howard who spoke out against the pollution, congestion and social dislocation of the modern industrial city. We will expand on the ideals of inclusive design in Chapter 5. For now it is sufficient to consider the definitions of utopia and dystopia.

> *Utopia* is a term first coined by Thomas More in the sixteenth century as ‘a play on the two Greek words that supply the “u” sound, eu (good) and ou (not). When taken together with topos (place) – the root of the second part of the word – utopia could be construed as a “good place” or a “somewhere that does not exist”’. The question whether utopia (much as inclusivity or gender democracy) is the search for the ideal or a fruitless quest has provided fertile ground for debate in urban studies.
>
> (Gold 2008: 69)

> *Dystopia* is the antithesis of utopia. The dystopian views the social and aesthetic processes of urban transformation in wholly negative terms of alienation, overcrowding, brutality, ill health and as a loss of civilisation.
>
> (Gold 2008: 70)

Case 2.3: Contested utopia: Vanport

Writing from the standpoint of a Christian Feminist, Mary Steward Van Leeuwen (1993) reflects on the sometimes heated discussions that she and her colleagues have shared on the topic of utopian experiments in urban design. She refers in particular to the case of Vanport City, Oregon, which has been extensively written about by Linda McDowell (1983) and Dolores Hayden (1984), among others. Van Leeuwen begins by noting

> the significance of Vanport as the United States’ most visionary and successful attempt to loosen the boundaries between domestic and public life. Its combination of government subsidies, public amenities, and careful city planning made household life much more manageable in terms of schedules, childcare, and home maintenance.
>
> (Van Leeuwen 1993: 449)

Much in the tradition of nineteenth-century philanthropists such as Robert Owen,

> Vanport City was built for Wilhelm Kaiser to house the women who worked in his shipyard during the war, including the iconic figure Rosie the Riveter. Many of these women were married with children and their husbands were in the armed forces. Thus, for the duration, these working women were single mothers and Kaiser instructed his architect to build a town which would make their 'dual roles' as mothers and waged workers as easy as possible. Each individual house was to be near a childcare centre. These centres were open twenty-four hours a day and provided hot food not only for the children but for the mothers to carry away at the end of their shifts, so they did not have to cook when they were tired.
>
> (McDowell 1999: 118)

Van Leeuwen goes on to note that

> To promote the integration of children's school and home lives, the daycare centres were constructed with picture windows overlooking the shipyards, through which the children could watch the launching of the ships that their fathers – or more likely their mothers – had helped build. By the end of World War II, Vanport City housed forty thousand people of white, African American, Asian and Hispanic origins, all living together in largely peaceful neighbourhoods. After the War, however, much of the city was dismantled and the rest was destroyed in a flood. Today, what was once the fifth largest city in the northwestern US is now the site of a park.
>
> (Van Leeuwen 1993: 449)

McDowell observed that 'the men who returned [from the war] were re-established in their "proper" place, both in "their" homes and in the workplace' (1999: 119).

Finally, Van Leeuwen observes that when she and her colleagues 'discussed the Vanport city experiment, some of us thought it came close to being a feminist utopia, while others likened it more to a collective, bureaucratic rabbit-warren' (1993: 449). She makes the point that the whole notion of utopia is not only shaped by standpoint (including gender, class and race) but also contested with respect to 'ideals' of social transformation.

Marx and structural theory

Arguably the most influential writing to emerge out of the nineteenth century confronted conditions of poverty, overcrowding, disease and exploitation with radical socialism, focusing initially on the plight of the English industrial working classes in Manchester and London. In 1844, Friedrich Engels published his survey of Manchester, *The Condition of the Working Class in England*, providing one of the earliest analyses of housing segregation. Karl Marx published Volume 1 of *Capital* in 1867. Marxian ideas undoubtedly went on to inspire the formation of contemporary urban studies and especially in the hugely influential Chicago school in the 1920s. John Rennie Short (2006: 21) notes that 'the city had a special place in Marxist thought…not [as] something separate from the wider society [but] intimately connected to the new capitalist mode of production' (Short 2006: 21).

Revolutionary political possibilities were distilled from the appalling living conditions and collapsing infrastructure observed in rich detail by Engels and Marx. Setting the tone for the macro-economic theorising of the Chicago school to follow, Engels critically dismissed the neighbourhood scale and pragmatic approach of female philanthropists such as Octavia Hill who were also publishing on poor housing conditions at this time (see Robinson 1998; Brion 1995). For Marx and for those who revived Marxist thinking in neo-Marxian urban social theory in the 1970s, notably the geographer David Harvey, class conflict and ownership of the means of production (property, capital) assumed primacy in explanations of urban inequality and the project of transformation. The spotlight was turned on spatial segregation and housing class divisions whereby a homogenous neighbourhood appeared to 'reinforce the tendency for relatively permanent social groupings to emerge within a relatively permanent structure of residential differentiation' (Harvey 1975: 364). Within Marxian analysis, the structural interpretation of class conflict evident in the architecture, civic monuments and systems of housing provision and administration was such that it overlooked structures of patriarchal power. This 'meta-narrative' is evident in most definitions of structuralism.

> *Structuralism*: this approach relates to the logic of systems and relations of production. The object of urban analysis is social structures and not individuals, derived from theory rather than close observation. Structuralism seeks in a formal manner to identify the 'deep' structures (typically historical) that determine 'surface' behaviour.
>
> (Murdoch 2006: 10)

The influence of Marxian thinking was not limited either to the discipline of urban studies or to the growing divisions and inequalities observed across Euro-American cities in the twentieth century. Southern thinkers, particularly in Latin

America and the Caribbean, were also inspired by what Marx had written on inequalities embedded within social relations. They were similarly motivated by Lenin's thinking on how colonial expansion allowed capitalism to overcome problems of overpopulation by using colonies as sources of surplus production and capital. This 'view from the south' led writers such as Samir Amin and André Gunder Frank to turn modernisation theory on its head by casting the urban bias of modern capital development as the source of the problem, not the solution (Parpart *et al.* 2000). Shifting the standpoint from north to south altered the understanding of what drove under-development away from 'low technological advancement and traditional values', exposing instead the exploitative and unequal economic relations that existed between the north and south (what they called the centre and periphery) as driving the process (Parpart *et al.* 2000: 109).

This Marx-inspired structural approach (known as dependency theory) emerged in the late 1960s and dominated leftist thinking on development in the global south through the 1970s. It responded to modernisation theory's expectation that underdeveloped countries would follow a linear path to development similar to that of the more developed countries by pointing to the uniquely differing historical and latent conditions for development of the cities and regions concerned. Also fundamental to this theory was an understanding that unequal core–periphery relations were exploited by colonial and class processes working together to create and maintain the conditions for under-development within and between populations of the south and the north. Urban elite enclaves monopolising the centre of southern cities were seen to work with capitalist interests in the north to manage the exchange relations structuring the north's continuing dominance. The ability of the global north to extract surplus from (exploit) the global south and of elite class interests within southern cities to collude with this process is central to this theory and its explanation of unequal development. Hence dependency theorists call for the overthrow of the elite collaborationist class, a severing of links with the north and autonomous development.

Not surprisingly, liberal and neoclassical economists rejected the foundations of dependency theory as it was so clearly a negative response to modernisation. However, some Marxists also disagreed with it, in part due to the lack of nuance in its arguments against capitalism – it just reversed modernisation theory without thinking further about how capitalism may be used in the global south to support southern development (Warren 1980). It set up an 'us-and-them' binary which perpetuated the same kind of economic reductionism that it found so flawed in modernisation theory. Perhaps most importantly, dependency theory did not take into account countries (such as China, Hong Kong, Malaysia, Singapore, Taiwan and increasingly India, Pakistan, Thailand and the Philippines) that, even at the time Frank wrote, had begun to achieve considerable success. Today, although dependency theory no longer dominates the political economy of development, its threads can be found in debates around globalisation and the new global economic order (Parpart *et al.* 2000).

In short, classical Marxism inspired a wide-ranging neo-Marxian structuralist approach intended to explore the underlying mechanisms of class conflict (notably owners of capital and those who sold their labour power). A useful account of Marxism and the city can be found in the book *Metromarxism* by Andrew Merrifield (2002). This traces the enduring legacy of Marxian-inspired thinking across urban studies and cultural theory and begins to tackle the criticism, notably from within feminist scholarship, that Marxian structural analysis fails to do more than 'add on' race, gender and social difference. The author himself admits that 'Metromarxism delves into an ongoing tale of Marxist ideas that have essentially been propounded by men…looking at how the city fulfils a functional role not only for capitalism, but for Marxism as well' (Merrifield 2002: 6).

Feminist critique of structuralism

Recent decades have witnessed growing confidence among feminist scholars at the margins of urban studies and within development studies debates to challenge the androcentric and ethnocentric assumptions of structural neo-Marxian approaches to urban social and spatial inequality. For instance, David Harvey's failure to recognise the function of gender inequality alongside (or, more specifically, intersecting with) class conflict in symbolic expressions of power and experiences of oppression has provoked heavy criticism among feminist scholars (see, for instance, Katz 1996). It did not engage in any meaningful discussion of women's subordinate conditions with respect to either urban inequalities in the north or colonial exploitation in the south. This neglect derived from the prioritisation of commoditised economic structures of exchange relations which necessarily marginalised the fundamental contributions of women, as most women were engaged in unwaged social reproduction work. Marx himself relegated the reproduction of labour to 'the labourer's instinct of self preservation and of propagation' (1906, vol. 16, cited in Lowe 1995: 91). Yet as Cindy Katz, Donald Lowe and others point out, 'there can be no reproduction of capital without the reproduction of labour' (Lowe 1995: 7).

In short, feminist scholars, including Marxist-feminists, criticise structuralist accounts of urban growth for failing to recognise the 'messy fleshy components of material life' and the reality that 'most of the social reproduction throughout recorded history – that is, the work considered "outside" of production, either unwaged or paid so poorly it cannot serve to reproduce the labourer himself or herself – has been conducted by slaves and their descendants, colonial and post-colonial subjects, children and women' (Katz 2001: 711; Mitchell *et al.* 2004: 11). This critique focuses not only on the neglect of social reproduction in Marxist urban analysis but also the failure of structuralist approaches to closely register the lived experience of urban dwellers – for instance, through urban ethnography and concern for the mundane realities of daily life.

Urban ecology and the 'other' Chicago school

David Sibley (1995) observes that most histories of urban studies view '*the* Chicago School' of Robert Park and Ernest Burgess as the 'fountainhead from which all else flows' (Peach 1975: 1–2, cited in Sibley 1995: 40). Most histories of urban studies consequently date the birth of the discipline from 1920 when the collaboration of sociologists Robert Park and Ernest Burgess at the University of Chicago resulted in a distinctive theory of 'urban ecology', attributing the spatial form of cities (concentric rings and ethnic and class enclaves) through processes of Darwinian evolution and competition between population groups. Unlike the distrust or ambivalence with which cities were viewed by contrast theorists, urban ecologists embraced modernity. Interestingly, Robert Park commented on the emancipatory potential of the urban mosaic of population groups long before the 1980s elicited writing on gentrification and popular associations were drawn between this and gendered and sexualised space. Nevertheless, while the following quote suggests a non-conformist spatial politics at work in the notion of the 'moral region' the assumptions bound up with this theory remain unquestioningly androcentric:

> Because of the opportunity it offers, particularly to the exceptional and abnormal types of man, a great city tends to spread out and lay bare to the public view in a massive manner all the human characters or traits which are ordinarily obscured and suppressed in smaller communities... We must then accept these 'moral regions' and the more or less eccentric and exceptional people who inhabit them, in a sense, at least, as part of the natural, if not normal, life of a city.
>
> (Park 1925: 45–46, cited in Rubin 2000: 62)

It was at this moment in history, in 1920, that a fundamental division opened up between 'abstract' and 'practical' urban social studies. Corresponding with the meteoric rise of the masculine Chicago school was the mass transfer of all the women sociologists at the University of Chicago (active contributors to the Department of Sociology since its inception in 1892) out of Sociology and into a newly created School of Social Services Administration. This restructuring of academic departments set the newly emerging discipline of urban studies on a distinct path which denigrated as 'women's work' the various projects of 'applied research' underway at the time (Deegan 1986: 191; Suttles 1990). Sex segregation in this once radical, progressive, co-educational institution was viewed by these men as an essential means by which to preserve 'abstract thinking' in established academic urban sociology, by distinguishing it from 'practical' dealings with the dislocation of minorities; the poor, the aged, the young, immigrants and women (Deegan 1986: 228–230). This move can be viewed in particular as an act of aggression toward the previously celebrated work of applied sociologist Jane Addams. The women scholars who were evicted from Sociology relocated to a pre-existing philanthropic settlement, Hull House, which went on to actively promote the influence of feminism on mainstream urban studies. Figure 2.5 shows Hull House, which is today a forum for feminist, working-class and minority studies.

Hull House undertook radical and philanthropic research in some of the most deprived Chicago neighbourhoods, while 'the' Chicago school of Park and Burgess was deeply conservative. Park and Burgess viewed the chronic problems of large cities from a top-down macro-scale perspective. In the problems of housing segregation, inadequate sanitation, congestion and sprawl, they saw the

Figure 2.5 ***Hull House, Chicago, 2008***
Source: Chris Hagar

results of unequal competition for scarce resources (premium land and neighbours with shared values) in which class and race defined the axes of power and opportunity: they dismissed the 'mundane' practices of exploitative landlords, racism and sexism which preoccupied Hull House researchers.

Running alongside this hugely significant split in Chicago between 'masculine' urban science and 'feminine' social work was a similar schism in architectural design movements, between pre-industrial 'crafts' and scientific 'modernism'. It was modernism that held the ascendancy for many years, resulting in a long love affair with machines and technology (especially the automobile) influenced in particular by Le Corbusier's 'cities in the sky'. This coincided with the epochal transition to modernity outlined above.

Post-structural theory

From the 1970s onwards writers on the city were observing some remarkably rapid and fundamental shifts in transport, construction, production, consumption and family formation (such as with the impact of contraceptive technology) (see Appendix, pp. 298–300). We consider the wholesale impact of this 'restructuring' of industry, the city and society in more detail in Chapter 3. Visible changes on the ground encouraged talk of a profound rupture with the past and the birth of a new 'post-modern' era. Around the same time the social science disciplines, including urban studies, began to be strongly influenced by post-structuralism or post-structural theory. Post-structuralism (as with structuralism) is difficult to define because, as Jonathan Murdoch (2006: 2) observes, there appear to be 'many post-structuralisms', each accompanied by its own particular set of theoretical and empirical concerns. The definition we settle on here is therefore one of many and best understood as a counterpoint in emphasis and consequence to that conveyed by the definition of structuralism on p. 53. To be clear, post-structuralism gains meaning by reference to theory (coming 'after' structuralism) (Murdoch 2006: 2), while post-modernism, typically conceived alongside post-structuralism, as here, but not to be conflated with this term, gains meaning in relation to epoch as the apparent rupture (cultural, social or technological) with modernism and modernity.

> *Post-structuralism*: as a body of theory post-structuralism influences both the subject of study (increasing the variety of valid topics and removing the 'taboo' of researching previously neglected or controversial subjects) and the method of studying it (raising the profile of qualitative research practices and allowing new, more personal or playful writing styles). A common feature of interest is with spatially situated, culturally embedded people–place relations and a sophisticated reading of 'space', not as a 'container', or site, but rather as a dynamic network of flows. Although still

> recognising deep systems and structures, a plurality of meanings are emphasised which in turn open up new spaces for feminist scholarship – beyond significations of 'woman', 'work' and 'politics', for instance.
>
> (Gibson-Graham 2000: 96, cited in Murdoch 2006: 10)

In urban studies, theoretical interest in post-structuralist thinking took the form of renewed interest in the rich diversity of different groups in society and a shift in scale from 'top-down' grand theory to 'bottom-up' (or grassroots) concern with popular representation. This shift in thinking, together with new modes of ethnographic analysis, promoted greater awareness of cultural influences on the city and ultimately ushered in the 'cultural turn', since which time urban studies has increasingly acknowledged the powerful sense in which cultural analysis is crucial to understanding the words, symbols and practices which constitute the city (Knox and Pinch 2006: 3, 42). A similar shift toward recognising difference and diversity is also apparent in post-structural thinking on issues of gender and development (GAD) in the context of cities in the global south. We will return to consider the nature and extent of GAD contributions to development in Chapter 3.

It is along the lines of post-modern theory and method that traditional urban studies can be said to have been most influenced by feminist scholarship, as we shall see in Chapter 3. Post-modern (post-structural) feminist writing on the city achieved, by an effective sleight of hand, both increased awareness of gender in the absolute sense and at the same time recognition that this identity functions through 'multiplicity and differences between (categories of) women' rather than between men and women (McDowell 1999: 22). Here we can see parallels between aesthetic eclecticism in architecture, and plural thinking through differences in urban lived experience, with reference to such notions as 'hybridity', 'paradox', 'becoming' and 'otherness'. For this purpose, an approach favoured by post-modern urban theory is to 'deconstruct' the many contradictory and paradoxical elements of the city to expose the hard edge of a capitalist, racist, patriarchal landscape (Soja 1989, cited in LeGates and Stout 2007: 167).

Although post-modern thinking originally sought to 'recover' excluded and marginalised urban subjects (space, culture, women, ethnic minorities), it did not seek to bring about constructive dialogue between urban studies and gender studies. In particular, according to Liz Bondi (1990: 156), post-modern thinking ignored the transformative potential of feminism, a criticism which continues to be levelled against architects, urban managers and urban scholars who pursue this course today. This is where we leave this history of urban studies – 'dangling', as it were, at the dawn of the twenty-first century. This apparent vacuum evokes a sense in which an integrated and inclusive urban studies 'hangs in the balance', as an unfinished project of thinking through cities and gender.

Everyday realities of gender, cities and development

In this chapter we have described some of the key transitions from rural to urban patterns of settlement. We have noted alongside this a number of epochal transitions in the way cities and urban social life have been viewed intellectually and politically over recent history. We think it is crucially important for anyone wishing to deploy an explicitly gendered urban theory in contemporary research to first gain a comprehensive awareness of these transitions. At the same time, the primary purpose of the historic developments traced in this and the following chapter is to apply the lessons and skills of critical reflection (witnessing the effects of sedimented androcentrism) in order to explore gendered subjects (and subjectivities, feelings and experiences) in the everyday lived realities witnessed in different cities around the world today. To make the final transition back to the scale, detail, diversity and intersectionality of the everyday, we conclude this chapter with a vignette from a book articulating the 'ordinariness of diversity' which pitches localised resistance to traditional gender roles against persistent sexism in the context of rapid urban growth in Mexico.

Case 2.4: Streets, bedrooms and patios

What all these people have in common (the urban poor and ethnic migrants – transvestite and female prostitutes, *discapacitados* (the physically challenged), gays and lesbians, bohemians and artists and intellectuals) is their attempt to compose their own 'ondas' – that is, their own ordinary, everyday lifestyles in the context of rapid urban growth in the city of Oaxaca, Mexico. We refer to this as the ordariness of diversity within a rapidly changing space. The social actors that we are dealing with represent not groups of exotic people but real people living in the actual material conditions of an Oaxaca that is being integrated into the postmodern world of consumer capitalism. This process over the last thirty years has transformed Oaxaco from a sleepy provincial town with a population of eighty thousand into a sprawling urban area of half a million.

Though the economy is still anchored in the tourist industry, Oaxaca has all the dynamics of rapid urban growth: housing shortages, congested traffic, pollution, unemployment and underemployment, contesting political movements, and the dollarization of the local economy through the twin forces of free trade and migration. (Higgins and Coen 2000: 5–6)

Source: Higgins and Coen (2000).

Comment

The authors go on to consider how, in this context of rapid urban growth, these marginalised groups of residents are challenging traditional barriers of race, class, gender and sexuality by defining new roles and practices and expanding their access to popular parts of the city. Interestingly (in parallels with gentrification), these marginalised groups currently occupy culturally and politically strategic sections of the city and as a consequence they are in direct conflict with 'development' and 'modernity' while themselves negotiating an alternative, emancipatory urban culture. The authors employ ethnographic techniques including oral biographies to explore localised and dynamic experience of racism, sexism, sexuality and class conflict. The authors reveal how such issues play out in the people's daily lives and in grassroots political activism. By doing so, they translate the abstract concepts of social action and identity formation into the actual lived experiences of real people.

This ethnographic snapshot clearly highlights the co-constitution of urban and gender transformation. The way mundane details of daily life are articulated through diversity clearly embraces post-structural theoretical concerns to highlight the fluidity of gender roles and ideals. It neatly captures the tensions we have highlighted in this chapter with respect to the layering of new onto old in patchwork fashion. It also points to the need within gender urban theory to deploy urban ethnography and everyday life perspectives to counter the deadening effect of 'top-down' urban approaches that fail to recognise gender.

Summary

After reflecting on the contents of this chapter, readers should recognise a whole continuum of non-rural settlement types; be familiar with historical trends of urbanisation; recognise competing schools of thought within the disciplines of urban studies and development studies; and be able to make connections between this historical legacy and contemporary urban social issues. The purpose of this historical account has been to promote skills of critical engagement with urban social theory and greater understanding of the geopolitics of 'development' and 'modernity' and the consequences of this multi-scalar inequality for gendered lived experience. This chapter has shown that there are a number of different ways that cities and societies are understood. The persistence of competing theories explaining such phenomena as mega-cities, ethnoburbs and gentrification highlights the importance of standpoint (and the absence of a scientific 'truth' in the social sciences more generally). A common theme to emerge out of these competing paradigms is the neglect of gendered subjects and subjectivities. Recent trends in thinking on cities, notably developments in post-modern theory and method, and the growing influence of feminist critique, have prompted closer consideration of everyday gendered realities and the co-constitution of cities and gender.

Suggested learning activity: mapping social transformations in space

Identify a town or city for which you can get hold of official Census of Population or other secondary data at a fine geographic scale (a 500 property ward, zip-code or super-output area). Use the available data to 'map' (whether graphically or by descriptive statistics) the socio-economic profile of individual neighbourhoods. Try to explore this in terms of family and household structure rather than (or in addition to) aggregated individual population counts by gender, age, ethnicity and class (however defined in the data available for you to use). Where possible, compare a contemporary 'snapshot' with comparable data from an earlier time period (ten or twenty years previously). Note the 'headline' changes with respect to population profile (a rise in single mothers relative to cohabiting or married couples with children, for instance) and spatial segregation (are single mothers concentrated in particular areas?). After completing this desk-based exercise select a single neighbourhood from your mapped area and visit this in person with a notebook and a camera. Note down your observations concerning the architecture (vintage, quality of repair, evidence of personal modification) and the infrastructure (transport, street-life, types of shops, parks and amenities) alongside your knowledge of the population profile (including data on housing tenure and employment status, etc.). What do you learn from this exercise? How useful are secondary data as a means of identifying social inequality and uneven development? What do you learn 'on the ground' that is missed out of secondary data? To what extent is your 'reading' of this neighbourhood shaped by your own standpoint (gender, class, race, ethnicity)?

Notes

1 Interestingly, although new urbanism appears to build on many of her ideas, Jane Jacobs was extremely critical of the new urbanism (Steigerwald 2001: 1) prior to her death in 2007.

2 An example of this in the UK would be the mixed Sikh/Muslim/Hindu Gujurati and Punjabi communities in Evington Valley and Oadby in Leicester.

Further reading

Crush, J. (ed.) (1995) *Power of Development*. London and New York: Routledge.

Dennis, R. (2008) *Cities in Modernity: Representations and Productions of Metropolitan Space, 1840–1930*. Cambridge: Cambridge University Press.

Gottdiener, M. and Budd, L. (2005) *Key Concepts in Urban Studies*. London: Sage.

Hall, T., Hubbard, P. and Short, J. R. (eds) (2008) *The Sage Companion to the City*. London and New York: Sage.

LeGates, R. T. and Stout, F. (eds) (2007) *The City Reader* (fourth edition). London and New York: Routledge.

Short, J. R. (2006) *Urban Theory: A Critical Assessment*. Basingstoke: Palgrave.

Webb, B. (1971) *My Apprenticeship*. London: Penguin Classic (first published in 1926). This book may be hard to get hold of but it will intrigue (and probably enrage) anyone who is interested in Victorian cities and the philanthropic movements (such as the Fabian Society) which emerged out of the work of Charles Booth, Sydney and Beatrice Webb and others.

3 Trends in urban restructuring, gender and feminist theory

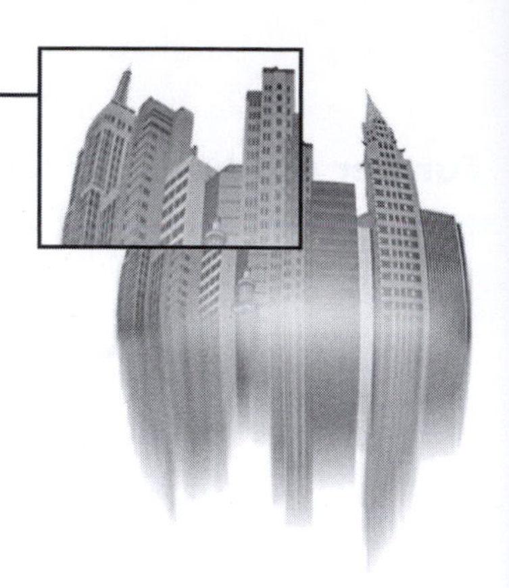

Learning objectives

- to appreciate the significance of global economic restructuring to the co-constitution of cities and gender
- to gain familiarity with historical developments in gender studies and feminist theory
- to move beyond categories of 'women' and 'gender', emphasising the everyday
- to consider historical transformations in gender and development in a range of urban contexts

Introduction

This chapter repeats the format adopted previously, this time exploring the significance of gender in recent trends of urban transformation. In the first part we look at recent trends in economic restructuring and structural adjustment alongside a widespread (but contradictory) 'feminisation' of economic activity, noting the implications of these multiple trends for gender roles and relations in a variety of urban social contexts. In the second part we identify some of the key historical transitions in the development of gender studies and feminist scholarship. The principal learning aims are to gain familiarity with the scope of cities and gender from a feminist perspective, to provoke critical reflection, and to critically evaluate the limitations of narrow disciplinary perspectives and course curricula (refer back to the learning activity for Chapter 1, pp. 29–30).

Just as we have already learned of epochal changes in urbanisation, modernity and development, the past one hundred years have witnessed profound transformations in the way 'sex' and 'gender' are represented and explained. The term 'gender', for instance, is much more recent than the long existence of a feminist movement suggests: 'gender studies' is far younger (and typically marginalised) relative to urban studies, with respect to institutional academic recognition. We note a close association between the relatively young discipline of gender studies and a wealth of research on the social relations and economic implications of gender segregation at home and in the workplace. We have therefore chosen to introduce the key historical developments in feminist thinking alongside key changes and trends of gender at work in urban restructuring.

In the first part of this chapter we introduce and illustrate two discrete trends in restructuring operating on the local and international scales of social and economic processes: changing household divisions of labour and new international divisions of labour. It is important to recognise that while local changes often reflect global flows of capital investment and trade, the local realities of people's daily lives vary fundamentally from place to place. We therefore focus on the micro-scale of restructuring and compare the gendered realities of 'post-industrial' urban labour market transformations in Euro-American cities with the structural adjustment of household livelihoods in cities in the developing world. In Chapter 4 we return to consider in more detail the macro-economic scale of restructuring associated with globalisation. Here our aim is to introduce the key terms, concepts and recent trends in restructuring in order to highlight the pivotal function of gender in these processes.

The second part of the chapter traces the origins of feminism and the rise of gender studies. We also consider parallel concerns within development studies to move beyond categories of 'women' and dependency to 'gender' and empowerment. This conceptual material provides both the context for substantive themes pursued later in this book and a constructive platform from which to cultivate an explicitly gendered urban social theory. This chapter indicates the complex intersection of urban social, cultural and economic restructuring. Because there is limited scope to expand on individual topics in great depth we provide a list of further reading at the end of the chapter.

Urban social and economic restructuring

Recent decades have witnessed fundamental transformations in the global economy, including the increased international mobility of jobs and money and a closer interconnection of people and places around the world. The term globalisation is widely used to describe the growing interconnectedness of and interdependencies between countries and cities on a global scale (Perrons 2004: 1). Economic restructuring is the term more usually applied to the transition to a different phase in the 'what, how and where' of economic production, such as from 'making things' to 'selling services' (Knox and Pinch 2006: 9). We expand on the way economic restructuring is typically defined and understood within urban studies in Box 3.1. It is interesting to compare this understanding with definitions of structural adjustment in Box 3.2. We suggest that readers familiarise themselves with these definitions (and refer back to critical reflections on the 'partial economy' of waged work versus a holistic account of 'multiple economies' introduced in Chapter 1) before moving on to consider the gender implications of restructuring and structural adjustment.

Box 3.1: Economic restructuring

The business term 'restructuring' refers to big changes that are made to production to improve efficiency and to create new markets or adapt to new market conditions. Changes can be made in the scale of production, either 'upscaling' to larger and more efficient, often multinational corporations (MNCs), or 'downsizing' to flexible and specialised production for 'niche' or high-value markets. These changes can result in the relocation of businesses whereby the headquarters of a large MNC might maintain a high-profile address (in London or New York, for instance) while 'back office' or factory functions may be decentralised or outsourced to regions where the production costs (land, raw materials, energy and labour) are lower. By contrast the location of the niche market business may depend on proximity to particular people or ideas such as universities or artistic communities. Overall the trajectory of restructuring in Europe and North America since the 1970s has resulted in a shift from reliance on agriculture and manufacturing sectors to greater emphasis on a high-value 'knowledge economy' and consumer-based service sector industries. This has changed not only the economic make-up of these countries but also the social and gender divisions of labour.

Source: Adapted from Knox and Pinch (2006: 9–10).

Most scholars date the restructuring of the economies of Europe and North America from the first OPEC oil crisis in 1973, when a quadrupling of petroleum prices precipitated global economic downturn. Parallels can be drawn between the 1970s 'oil crisis' and the early twenty-first-century 'credit crunch' but a fundamental distinction is that the 1970s also witnessed a corresponding technological revolution with respect to transport (including international standards for intermodal containerisation) and interpersonal communication. An overview of this intense period of innovation is included in the Appendix (pp. 298–300). The global impact of new technologies such as contraception, the personal computer, the internet, cheaper high-speed transport and satellite and mobile telephones should not be underestimated. The scale of the impact was not simply a function of individual innovations but the cumulative effects of socio-cultural transformation. For instance, the technologies of contraception, women's participation in service sector employment, rising standards of living in housing and domestic appliances – all intersect with a trend of households shrinking in size and the normalisation of dual-income earning couples (see Case 3.3, p. 78).

As a crude simplification, the restructuring of European and North American urban economies has resulted in four distinct employment trends: a new international division of labour; emphasis on 'flexibility'; a 'feminisation' of job descriptions; and a 'casualisation' of the terms of employment, such as with freelance working and zero-hours contracts (McDowell 1991; see also Piore and Sabel 1984). Discrete changes in the nature and extent of the jobs available to (and taken up by) working age men and women have contributed to similarly profound changes in the division of labour within individual households, for partnering, for parenting and the organisation of routine domestic affairs. Manuel Castells points to the co-constitution of social, cultural and economic manifestations of restructuring over this period:

> The structural process of transformation of women's condition, in dialectical interaction with the rise of the feminist movement, has completely changed the social fabric of cities... At the same time, the transformation of households and the domestic division of labour is fundamentally changing the demands on collective consumption, and thus urban policy. For instance, childcare is becoming as important an issue as housing in today's cities. Transportation networks have to accommodate the demands of two workers in the family, instead of relying on the free driving service provided by the suburban housewife in the not so distant past.
>
> (Castells 1993: 483)

While this quotation refers more specifically to the cities of the global north, the interdependence of social and spatial transformations is characteristic of cities of the north and south alike, as we will see with respect to the impact of structural adjustment.

Linda McDowell (1999: 124) observes the impact this deep economic restructuring has had on the social construction of masculinity and femininity in the workplace in affluent 'post-industrial' societies since the early 1970s. She contrasts the loss of jobs in coal mining, shipbuilding, steel or chemicals production, textiles and manufacturing, as well as 'the shared risks and hardship at work and community solidarity' that marked them, with new jobs which have opened up in call centres (now increasingly outsourced to LDCs) associated with the female-typed attribute of 'smiling down the phone' (McDowell 1999: 124; Belt *et al.* 1999). New jobs created in retail, public health and hospitality are associated with an erosion of real wages where the notion of a 'minimum wage' has replaced that of a 'family wage'. Previously, skilled blue-collar jobs which paid a 'family wage' reinforced the idealised family of the male breadwinner and the female home-maker. The terms of employment have also changed, with 'jobs for life' replaced with short-term, insecure or entirely casual employment arrangements. Case 3.1 illustrates the intersections of gender, generation and economic restructuring along with a sense of the neglected costs of market competition for individual men and women and their families and the local communities – in the case of one, now notorious, North of England boot factory. These human costs are frequently overlooked in reports on factory closure over this period.

Although the 'feminisation' of new employment opportunities appears to favour women over men, more women than men are concentrated in poorly paid, temporary and insecure jobs (Somerville 2000: 231). Female dependence on the service sector has increased rapidly: across the European Union, for instance, the service sector accounts for just over 80 per cent of women workers (European Women's Lobby 2000). Outside the home women frequently work alongside other women in jobs which look remarkably like those of caring, domestic and emotion-work long performed unpaid at home. This describes a horizontal segregation of jobs and occupations constructed either as 'women's work' or 'mother friendly'. Vertically too, women are concentrated in jobs outside the boardroom. In a survey of investment banks in the City of London, Linda McDowell (1997a) found that, while women held 36 per cent of the jobs in this sector, men outnumbered women three to one in executive and professional grades. Current estimates suggest (according to the 2006 US Census) that the average full-time woman worker in the USA earns 77 cents to every dollar earned by her male colleague, a differential also distinguished by ethnicity – for instance, African-American women earn 66 cents to the (white) male dollar and Hispanic women 54 cents (BPW USA 2008). This kind of gender gap derives from many causal factors, but is at least partially explained by differences in education, career salience and social expectations of marriage, motherhood and home-making (England 1992).

We see little evidence of women breaking through the 'glass ceiling' into the male-dominated boardrooms of the business world, as witnessed by uniformly conservative all-male portraits in company annual reports from around the world. The tiny minority of women who gain entry to the inner sanctum of the corporate world of 'the city' are held under the media spotlight as deviant. In 1990s Britain, Nicola Horlick was labelled a 'superwoman' as the 'boss' of a male-dominated

Case 3.1: From hobnails to kinky boots

In 2004 British television (BBC2) screened a documentary tracing the misfortunes of the boot and shoe industry in the UK, focusing on the 100-year-old Northamptonshire family firm of W. J. Brooks. The documentary opened with stereotypical images of a 'Victorian' industrial working-class community: narrow cobbled streets of brick factories and back-to-back terraced houses; white, male, working-class machinists; the culture of shift-work; hobnail boots; and the local working-men's club. Footage of life on the factory floor and at management meetings revealed all the signs of gender, class and race divisions, and the factory workers who were interviewed told of parents, siblings and grandparents who had also worked in the Northamptonshire boot industry. The documentary focused on Steve Pateman, the fourth generation of his family to make traditional footwear, who is desperate to do whatever it takes to save this failing family firm. Steve expressed the paternalistic concerns not only of a breadwinner, anxious to support his wife and young son, but also for his 55 employees and ultimately the traditions of village life which relied on the factory for jobs, customers and local identity.

The documentary reported that the firm was failing for a number of reasons associated with global economic restructuring; customers were no longer buying traditional working men's boots because the trades and manufacturing industries defining 'working men' were dwindling relative to white-collar and female-type jobs requiring a different style of footwear. It told a familiar story of global market restructuring whereby labour intensive craft industries, such as boot and shoe making, were being lost to Portugal, Romania and the Far East, where wages and production costs were lower. In this case a common story of job loss and factory closure was made unique because radical measures were being taken to confront difficult times. Looking for new markets to survive, Steve had struck on the idea of shifting production away from traditional boots for industry to the design and fabrication of fetish footwear; to start making thigh-length boots and high-heeled shoes strong enough for 15-stone men as part of a new 'Divine' line (see http://www.bbc.co.uk/programmes/b0077q7l, accessed 2007).

This documentary had unexpected popular appeal; it went on to inspire the movie *Kinky Boots*, released in 2006 by the same production company that made *Calendar Girls* (also set in the stereotypically 'grim' North of England).

Unfortunately, making a documentary and inspiring a blockbuster film failed to save this factory or the jobs and the community at the heart of this story. Today what exists in the place of a craft industry is a virtual, specialist, on-line catalogue business which functions alongside the sex industry, as a marketing enterprise far removed from the original manufacture of boots and shoes.

Readers are encouraged to follow the links below, view the film trailer or watch the film *Kinky Boots*, and to critically reflect on the multiple threads of restructuring (and evidence of persistent stereotyping – of gender, sexuality, modernity and urban space), in this case:

- http://footwearglobal.files.wordpress.com/2008/06/the-real-story-behind-those-kinky-boots.pdf
- http://www.shop.edirectory.co.uk/divine/926/index/
- http: //www.movietome.com/movie/357047/kinky-boots/index.html

asset management firm. Every time the media fixated on this case of a successful woman in the City the 'news' that Mrs Horlick was a mother of five always made it into the opening headlines. The tenor of these articles swung between judgement (mother abandons children for selfish career ambitions) and alarm (City loses hard-headed business sense in dash to kid's carol concert). Missing from the story were the two women working invisibly in the wings (the nanny and the secretary) who made it possible for a woman as well educated and affluent as Nicola Horlick to climb to the top of her profession – against the disadvantages of her sex. Once again this highlights the intersections of class, race and gender and the way *intra*-household inequalities (between husband and wife, for instance) exacerbate *inter*-modes of exploitation between households. Those employed to help middle-class families co-ordinate daily life have fewer resources to reconcile their own pressures of work and the desire to improve their children's prospects (Jarvis 2005b: 141).

At the same time that many educated women have made significant gains in a labour market now valuing stereotypically female attributes of emotional intelligence, many physically strong but poorly educated and socially awkward young men have been consigned to long-term unemployment. In a recent study, Linda McDowell (2003) shifts her focus from women breaking through the glass ceiling to men pushed out of the workforce altogether. She explores the contradictory manifestations of this 'redundant masculinity'. Considerable attention has been paid to the possibility that economic restructuring undermines traditional conceptions of masculinity to the extent that it creates a situation of role reversal or 'husbands at home' (Willott and Griffin 1996; Morris 1995; Wheelock 1990). Perrons (2004: 156) highlights the way films such as *The Full Monty* and *Billy Elliot*, which focus on male unemployment in communities associated with profound restructuring (not unlike those of Northamptonshire in *Kinky Boots* above), promote a sense of nostalgia for a time when 'men could be men'.

Of course we are reminded by the strictures of intersectionality not to reduce the profound transformations of economic restructuring to a feminisation of the labour market for *all* women or to the unequivocal advancement of women's employment relative to men. Who the winners and losers are and where they are positioned within these interdependent, competing processes can only be explained through fine-grained analysis, not only of the work and the workplace, but also of the intersecting identities of gender, sexuality, race, ethnicity, class and cultural capital of the workers themselves. We return to this thread in relation to the cultures of 'fitting in' to workplaces and practices of flexible working in the new economy in Chapter 7. For now it is important to bear in mind, as Diane Perrons observes, that restructuring 'is not necessarily a zero-sum game in which men are inevitably the losers' (2004: 87). Viewed from an intersectional rather than binary perspective, the picture is complex:

> While transitional generations of men may have a sense of loss and women a sense of gain, these changes often become normalized and traditional gender relations can be reborn inside new employment forms and indeed men can reinvent their sense of self-worth and masculinity even within the domestic sphere. Moreover, in some instances women collude in sustaining the more traditional masculine self-image so that unequal gender relations in some ways endure quite significant changes in material conditions. This is not to say that important changes have not taken place. Many women have been empowered by receiving direct incomes but the implications of new employment forms associated with globalization suggest that so far it would be premature to suggest the end of patriarchy was on the immediate horizon.
>
> (Perrons 2004: 87)

Structural adjustment, economic reform and uneven development

The economic crisis of the 1970s was truly global in scale. Following the OPEC oil price rise in 1973 (and again in 1979), large sums of money were made available to less developed countries in the form of low interest loans from international banks who were looking for ways to invest the influx of 'oil dollars'. Oil dependent countries such as Britain and the USA experienced deep recession and a restructuring of their economies (see p. 66), which had a devastating impact on jobless families but over time established 'leaner' and more productive businesses and ultimately paved the way for a consumer boom in the 1980s. As competition for international funds increased, interest rates rose. This contributed to a debt crisis in which heavily indebted poor nations either defaulted or had to borrow more to finance the rising cost of their loan repayments (Perrons 2004: 57). As this crisis unfolded, and at various times since, the supra-national institutions of the International Monetary Fund (IMF) and the World Bank agreed to provide additional funds, but with strict conditions attached. The conditions which less developed nations had to sign up to in order to qualify for loans and assistance were called Structural Adjustment Programmes (SAPs) (more recently couched in terms of economic reforms) (see Box 3.2).

Box 3.2: Structural Adjustment Programmes (SAPs)

Structural adjustment is a term used to describe the requirement of a policy change or the attachment of conditions (conditionalities) in the process of securing new loans from the IMF or World Bank or for refinancing (securing lower interest rates) on existing loans. Conditions are applied to persuade less developed debtor nations to 'develop' according to IMF and World Bank goals to promote economic growth.

According to a summary provided by Diane Perrons (2004: 57), SAPs typically involve: devaluing the currency; opening the economy to free trade and expanding exports; restricting government expenditure; ending price protection and subsidies. Cuts in public expenditure mean reduced health and education provision; it is often assumed that women can fill the gaps but increasingly they are working in the export sectors so their workload intensifies.

Critiques of SAPs: male bias

The three male biases Elson illustrated in her early critique of SAPs were:

1 assumptions under SAPs that the unpaid work done by women to care for the family, including raising children, gathering fuel, nursing the sick, preparing and processing food, will continue irrespective of the economic structure and allocation of resources within it;
2 ignoring the barriers to shifting labour resources between the non-tradables and tradables sector, i.e. treating labour like any other resource and not as an embodied resource (i.e. with a sex, age, ability, etc.) (Jackson and Palmer Jones 1999);
3 assumptions regarding the household and equal distributions of resources within it.

The first male bias rests on an assumption that the unpaid work of women, and more particularly the time they spend on it, can expand infinitely to accommodate new demands related to declining state service provision – this is the assumption of the elasticity of women's time. This assumption results from a complete lack of attention to unpaid work within SAPs, which focus exclusively on productive work in the market, and hence, as Elson argues, these programmes may only be 'successful' due to the expanding unpaid work of women (Elson 1991; Moser 1993).

For some women this increasing burden of unpaid work, taken on to ensure basic living standards for the family, may be coupled with increasing demands for women's participation in 'productive' work (Moser 1993; Elson 1991). Shifts to tradable goods brought about through economic incentives to increase labour intensive manufacturing and cash crop production have changed gendered work patterns. These changes in some locations meant less work for men in urban areas, as export oriented manufacturing work went to women, with their 'nimble fingers' (Elson and Pearson 1981).

Underpinning the IMF and World Bank imposition of SAPs were enduring assumptions that 'development and modernization are enhanced by open markets and free trade – the essence of neo-liberalism; where free markets are understood to allow the factors of production, labour and capital to flow to where they are most efficient, eventually leading to an equalization of factor returns', or 'trickle down' (Perrons 2004: 56). In Chapter 2 we traced the neo-colonial and ethnocentric legacy of modernisation theory. Here we note the consequences of SAP conditions for everyday lived realities on the ground. Later in the discussion we move on to consider how this contributed to considerations of gender in development.

The assumptions and implications of structural adjustment have been strongly criticised by feminist scholars and non-governmental human rights campaigners. They point to the lack of a 'human face' to the partial (quantitative) measures of economic growth and they highlight the unacceptable social cost of the programmes, particularly for women and children (Cornia *et al.* 1987; Sparr 1994; UNDP 1997: 17). Elson's (1991) critique of SAPs led to her contention that there is a clear male bias in development (Box 3.2), something that has been acknowledged by other feminist economists as existing not only in development but more broadly in the discipline of economics (Ferber and Nelson 1993).

The rise of the market and the focus on productive work and efficiency within SAPs led to women being viewed as instruments of development, to be integrated more efficiently for the betterment of development outcomes in general, with little concern for women's own welfare. This was associated with arguments promoting women's work in export factories due to their nimble fingers, more compliant attitude and better work ethic, leading to more output and less trouble for producers, but not always better lives for women.[1]

> *Nimble fingers*: Diane Elson and Ruth Pearson (1981) coined this term in their research on the feminisation of the workforce in the new international division of labour. They noted that women were being recruited for export oriented jobs in assemblyline production, where manual dexterity (nimble fingers) and 'docility' were seen by multinational employers as 'natural' advantages in a female workforce. The research goes on to reveal that wage differentials are often more indicative of 'male bias' (see p. 72) than attributes which they argue are socialised skills.
>
> (adapted from McDowell and Sharp 1999: 185)

A sense of the hardships inflicted by structural adjustment can be understood from the wages puzzle facing African urban families and outlined in Case 3.2.

Arguably, the lived realities of economic restructuring are very different for the unemployed Northamptonshire boot-maker and the urban African family reduced by poverty to one meal a day. At the same time we can draw from these two different cases a common theme of coping with crisis which is rooted in gender and place-based cultural possibilities. For instance, it is culturally taboo in many parts of North Africa for women to handle money in the sale or purchase of goods in public street markets (see Figure 3.1). Rarely are women to be seen associated with trade and commerce. More problematically still, during the Taliban rule in Afghanistan it was deemed unacceptable for women to be visible in public, meaning that many war widows from the different phases of conflict experienced in that country struggled to find a way for their families to survive. Even recently some women 'choose' to appear in public wearing the all-encompassing burkha (see Figure 3.2). Thus, cultural settings influence the coping strategies from which households can choose to create their livelihoods and it is through the development of feminist scholarship that these diversities of experiences across gender, race, age and place became apparent, altering theory and practice across a range of disciplines.

Partnering and parenting

This restructuring of the economic basis on which individuals and families construct their livelihood has been accompanied to a large extent by new and more varied patterns of marriage, cohabitation and divorce, longer life expectancy and falling fertility rates. Together these combine to produce a 'wider palette of family and domestic situations' (Buzar *et al.* 2005: 414). This is

Case 3.2: The wages puzzle facing African urban families

The evidence for the terrible hardships inflicted on African urban families as incomes from wages slipped completely out of line with the minimum income required to keep a family (or in many cases even an individual) fed, let alone sheltered and clothed, abounds in the literature. Wide-ranging analyses of the devastating impact of structural adjustment programmes in the third world was provided early on in Cornia *et al.* (1987), and the vulnerability of the African urban poor was often highlighted. In Uganda, for example, the minimum monthly wage by the early 1980s could only buy about one week's supply of food (Jamal and Weeks 1994, cited in Potts 1995: 249).

The huge gap between wages and minimum necessary expenditure has been termed the 'wages puzzle' by Jamal and Weeks (1994). The puzzle is answered by the reality that people are surviving by the widespread adoption of two well-documented coping strategies. The first involves a great increase in informal sector activity, with previously non-earning household members (notably women and children) entering the petty commodity sector, as well as wage-earners taking on supplementary cash-earning activity. The second involves the development of food growing by urban households on any available arable land within and around the urban area. Neither of these strategies can be undertaken without significant costs for the household. The burden of extra work is usually unevenly distributed among household members, with women shouldering a very significant additional burden. This has important implications for childcare and the women's own welfare, and in turn women's and children's health. Both urban agriculture and informal sector work can also bring the participants into increased conflict with urban authorities (Potts 1995: 250).

Source: Potts (1995).

increasingly true for the poor cities of the south as well the post-industrial cities of the north. A common trend which is evident to some extent across the urban world is a tendency for households to shrink in size. The implications and manifestations of this trend are indicated in Case 3.3. In the global south, harsh economic conditions have contributed to a growth in female-headed agglomerate families and a globalised dispersal of the household on a national and international basis (see Chapter 6). These trends are also associated in the north with a reduced working age population counterbalanced by the flows of migrants from the south, and a situation in which the rate of growth in the number of households

Figure 3.1 ***North African street market***
Source: Helen Jarvis

through fragmentation and migration (and pressures on land for housing and associated infrastructure) exceeds that for the total population. This is because more households accommodate only one person – each typically using a larger family dwelling.

Greater variety of generally smaller families and households reflects a number of interdependent socio-cultural and demographic trends, including growth of women's employment, delayed marriage and non-marriage, delayed child-bearing and remaining childless, lone-parent and blended family households. Case 3.4 illustrates the dimensions of transnationalism among families, a trend rising in response to pressures from economic restructuring and global economic forces.

For some scholars these transformations in partnering and parenting have led to the conceptual extension of notions of the household. While, historically, most definitions rest with kinship and co-residence, there is evidence that the co-determinacy of 'family' and 'household' and 'home' and 'household' is being eroded. This is evident both in the proliferation of 'blended' or step-families,

Figure 3.2 ***Women out in the streets of Kabul in full burkha***
Source: Paula Kantor

where two families are 'blended' through marriage, resulting in a fragile web of domestic and parenting arrangements (Smart and Neale 1998) and in the proliferation of transnational households (which we explore along with 'living together apart' arrangements in Chapter 6). It is similarly evident in the term 'fictive kin', a term used to indicate the emotional significance of friendships in situations of dislocation and isolation, where new migrants replace absent kin with close-knit social networks and cultivate within them the kind of mutual trust and support conventionally associated with bonds of blood and marriage (Stack 1983; Jarvis 2006).

Another trend which is common to countries across the post-industrial north is a shrinking working age population relative to elderly and other non-working populations (Evans 1992: 132). Nancy Folbre (2001: 102) points out that while in 1930 the proportion of the US population over age 65 represented 5.4 per cent, in 1990 this had risen to 12.5 per cent and is projected to constitute 20 per cent by the year 2050. Meanwhile, in Europe the magazine *Management Issues* suggested in 2004 that if the increased number of pensioners in the UK were all to continue receiving the basic state pension (then at about £80 per week) the UK would need to attract 10 million new migrants by 2025. This demographic trajectory combines with the

Case 3.3: The shrinking household

Across the world we find a trend of shrinking household size. In 1990, average household size in developed and developing countries was 2.7 and 4.8 persons, respectively. By 2050, it is projected that the range may be as little as 2.6–3.4 (O'Neill et al. 2000, cited in UNFPA 2001: 80). Multi-family households are rare in the UK and USA but this household type is common in the developing world and an important source of welfare in old age. In India, the proportion of single-person households currently stands at just 4 per cent, while in urban areas 35.6 per cent of households comprise 3 or 4 members (28.9 per cent in rural areas) and 20.8 per cent comprise 7 or more members (26.4 per cent in rural areas) (Census of India, 2001). There is nevertheless evidence of a shift towards nuclear families and more non-family households in rapidly developing countries: extended family arrangements are being undermined by pressures of rural to urban and international migration, the emergence of a new middle-class urban elite and a cultural shift towards individual achievement over collective welfare.

In 2001 the British household averaged 2.4 people in size (compared with an average of 2.6 in the USA). The dominant trends of change in household composition over recent decades include a huge rise in the proportion of one-person households relative to single-family households, as well as significantly increased numbers of single-parent (lone-mother) households. Because most adults in couple households are from the same occupational class, the population is increasingly divided between affluent 'work-rich' dual-career households, 'multiple-job' low income households and 'employment-deprived' no-earner households (McRae 1986; Jarvis 1997). Children in lone-mother households in the USA are five times more likely to grow up in poverty than children in households with two adults of working age (USCB 2004: Table C8), and because of the concentration of this household type within spatially segregated black and minority ethnic neighbourhoods, poor households are further disadvantaged by limited scope for support from the state or from employers (Jarvis 2006: 353).

neo-liberal state in the assumption of a 'universal worker' model of citizenship (Lewis 2002). In order to maintain national productivity, governments depend increasingly on the continuing (uninterrupted) participation of 'prime-age' women and migrants in paid employment and thus rising numbers of dual-income and multi-job households. To make sense of these multiple trends we now turn to the growing body of feminist scholarship to explore competing explanations for the different and dynamic gender roles and relations bound up in processes of urban economic restructuring around the world.

Case 3.4: Transnational families

Honduras is one of the poorest nations in Central America, hit by both structural adjustment and a massive hurricane in 1998 which devastated the economy. Unemployment is high, at an estimated 28 per cent in 2004 (CIA 2008), pushing men and women alike to seek opportunities to the north, in the US, to provide for their families. This is not a strategy of choice or accumulation, but one of necessity and survival. It is important to scrutinise what this mobility means for family forms and the way families function socially and geographically.

Schmalzbauer (2004), in a two-year, two-nation study of transnational Honduran families and their survival strategies, interviewed transmigrants in the Boston area and members of families left 'back home'. She documents the strain of these survival strategies on all parties involved, emotionally and financially. While family back home are clearly doing better financially than those without a member away, this comes at the cost of separation, conflict at times of visits (infrequent due to costs) and emotional stress due to young children not understanding why a parent has left and asking during weekly communications about a return which is not economically feasible. Those away face their own survival pressures, to get by on low wages, sending as much money home as possible. They are often bombarded by requests for financial help, given media portrayals of life in the US and the riches available. Family back home do not understand the life that also exists to support this life of ease – the low wage service workers who struggle day to day.

Comment

The growing trend of female migration noted above has given rise to multiple studies of 'care chains' and the movement of women into domestic service work in wealthier countries for middle- and upper-class households, leaving their own children and families in a care deficit, relying on 'other-mothers' (grandmothers, elder daughters, aunts, neighbours) to raise their own children (Ehrenreich and Hochschild 2003; Parreñas 2001; Schmalzbauer 2004). Pyle (2006) also notes the care deficit of the women migrants as well, adrift in a new place, working very hard, with few opportunities to form networks of support. Of note here is the way that when women are the movers the mothering and care-giving roles are passed on to other women, giving further testimony to the rigidity of gender roles in care work.

Tracing the origins of feminism

A body of ideas and political movement known as feminism pre-dates the formal academic study of gender inequalities by nearly two hundred years. This is an important indication of the difference between the study of cities and the study of gender, whereby the latter has been party to a very long struggle for recognition as a subject worthy of scholarship and public concern. Paola Monzini notes that it was the English feminist movement which raised awareness of the 'white slave trade' by which 'girls would be lured to England by newspaper advertisements offering them employment as domestic servants, or else approached at railway stations as soon as they arrived in a continental city' (2005: 4). In a move anticipating the philanthropic 'action research' of Hull House in the 1920s, educated and affluent English feminists in 1880 'hired a London lawyer, T.W. Snagge, to carry out a thorough investigation (of sex trafficking)', providing the first robust empirical evidence that women and girls were being pressed into prostitution against their will as a direct consequence of their lack of autonomy, including the right to vote or to have control over their own existence (Barry 1979, cited in Monzini 2005: 4). This case attests to the enduring history of feminist direct action.

It is popular to describe the existence of three 'waves' of feminist thinking and politics. Held and Grew suggest that this 'wave analogy is useful in so far as it alludes to the successive diffusion and churning over' (2007: 5) of academic and activist discourses and political agenda. Crucially, each wave makes a discrete contribution, rooted in dominant issues of the moment, without fully replacing that which went before. The maritime metaphor is less welcome to Ann Oakley, who attributes its use to 'many people's worst fear – of the absolutely engulfing tidal wave (of women's liberation)' (2002: 53). It is nevertheless useful to outline these waves of feminism as a 'movement', as a frame within which to relate competing aspects of debate in feminist theory, as well as challenges to urban studies, dependency theory and modernisation and the origins of gender studies as a discipline.

Three waves of feminism

A first wave of feminism focused on the issue of women's formal exclusion from a sex-segregated elite male 'franchise'. Historically, women were denied the right to vote in local and national elections, barred from owning and inheriting property and typically barred from taking up or maintaining employment after marriage. In historical terms this wave is fundamentally associated with late nineteenth-century campaigns of suffrage (by suffragettes) across the English-speaking developed world. Action included marches, hunger strikes, both peaceful and militant protests, as well as individual essays and countless subtle efforts at domestic insubordination by educated wives and daughters at this time. Arguably the major achievements were political, rather than scholarly. The goal of suffrage focused on immediate issues without challenging the monolithic category 'women' as other to 'men'. Among the most prominent first-wave feminists were Mary Wollstonecraft, Jane Addams, Simone de Beauvoir, Marion Phillips, Marie Stopes and Virginia Woolf.

A second wave of feminism can be traced to the upsurge of campus political activism from the late 1960s onwards in the USA and Britain (Rowbotham 1997). It mobilised around struggles over women's rights over their own bodies, including issues of contraception, abortion and campaigns against pornography. It was in this period that feminists gained the hackneyed 'bra-burning' stereotype. Classics of this period include: Kate Millett (1971) *Sexual Politics*; Ann Oakley (1974) *The Sociology of Housework* and Shulamith Firestone (1972) *The Dialectic of Sex*. The archetypal slogan 'the personal is political' embodied a powerful campaign (of early mainstreaming) to signal that politics did not stop at the kitchen sink or the bedroom door. Women's experiences – in relation to men, work, family, public and private life – informed an expanded view of what counted as political entitlement. Second-wave feminism typically explained women's subordination by emphasising structures of constraint in the law, religious institutions, social norms and patriarchal practices, generally with reference to paradigms of structuralism, theories of Marxism, patriarchy or dual systems. We expand upon each of these aspects of feminist theory later in this chapter.

Yet by the late 1970s anything like a formally organised 'movement' began to fragment along lines of race, class and sexuality, as exemplified by bell hooks (a black African-American feminist) and her seminal publication *Aint I a Woman?* (hooks 1982). Divisions occurred between radical and moderate feminists, with socialist feminists heavily engaged in single-issue campaigns such as the Greenham Common women's campaign against US nuclear bases in Britain as well as radical protest in support of the 1980s miners' strikes. The popularity of a 'radical' second-wave feminism in the 1970s is reflected in the editorials and journalism of two magazines: *Ms* magazine, founded in 1972 by Gloria Steinem in the USA; and *Spare Rib*, formed by Rosie Boycott in the UK in the same year (see Chambers *et al*. 2004 for more on feminist journalism).

By the time a third wave or 'new feminism' emerged in the early 1990s the self-identified label feminist was being widely rejected, either viewed as dowdy and outdated by 'post-feminists', or considered colonial and bourgeois by black and lesbian scholars, who shifted their attention to post-colonial and queer theory. The thorny question 'What is feminism?' is explored in a text of this name by Chris Beasley (1999) (see Box 3.3). Associated with the new feminism was a third '*post-structural*' challenge to malestream urban studies which has been ongoing since the 1990s. We expand on this body of theory on pp. 88–9.

The 'new feminism' that emerged centred on individualism and a politics of difference – rather than a common cause shared on grounds of sex. According to Nagar (2004: 31) the 'painful splits' created by racism and homophobia in the 1980s women's movement (especially in the USA) forced the white feminist mainstream to step back from assertions of similarity and homogeneity to examine instead questions of plural identity, difference and diversity. This comprised 'an uneven, divisive, and slow process of shifts' that remains incomplete in both theoretical and political terms today (Kaplan 1994: 140). Prominent scholars underpinning the third-wave split include Judith Butler, Donna Haraway, Inga Muscio and bell hooks.

Box 3.3: What is feminism?

The term feminism is difficult to pin down to an exact or fixed definition. Rather than contributing to inclusivity this ambiguity tends to assume the narrow characteristics of contemporary Western feminism. The term is by popular use, rather than by definition, ethnocentric. Notwithstanding this narrow world-view, feminism entails a variety of widely differing approaches, including differences in understanding 'equality' (as different but equal, as sameness, as rooted in capital, or interpreted as respect).

The history of feminism offers a series of responses to the 'critique of misogyny' and different ways of 'challenging women's subordination'. Charting the boundaries of feminism typically rests with claims of distinction from the mainstream/malestream, as innovative, inventive and rebellious. In particular, feminists 'see their work as attending to the significance of sexual perspectives in modes of thought and offering a challenge to masculine bias' (Beasley 1999: 3). Resistance is variously against misogyny (literally hatred of women) as well as androcentrism (assumptions of male superiority and centrality) whereby non-feminist social and political theory is 'written by men, for men and about men' (Theile 1986: 33).

Source: Beasley (1999).

Comment

Feminism recognises that gender forms one of the major axes of social difference, but debate continues *within* feminism concerning the *nature* of gender (as something born into, as a matter of being, or something socially constructed, by 'nurture', or as a matter of doing) and the extent to which gender can be attributed to one or more 'systems' of constraint. Fundamentally, gender is distinguished from sex where the latter is narrowly understood through biological characteristics. Explaining the differences between male and female lives by recourse to sexual (genital) characteristics is widely dismissed as a form of *'essentialism'* whereby male and female characteristics (dressing up in blue or pink, fixing things or talking about them, exhibiting strength or nurturing) are deemed to be inherent or 'in essence' 'already there' from birth.

While sex at birth is not always clear cut, as witnessed in the trauma of various intersexed conditions and ambiguous genitalia, it is the *social construction* of gender which post-structural feminists claim provides huge scope for diversity through multiple positioning of masculinities and femininities. A significant recent shift in emphasis from 'being' (born male or female and as such 'trapped' into conforming to a 'given' body-type) toward 'doing' (constructing and reinventing gender identity) focuses attention on the performance, language, meaning, manifestation and popular representation of different forms of masculinity (such as through 'macho', 'yob', 'new man', 'camp', 'nerd' labelling and performance). Similarly, 'doing' femininity is increasingly understood to take multiple, complex forms through the interweaving of roles (daughter, wife, mother), norms (of 'respectability'), practices (of earning and partnering, for instance) and forms of labelling ('girlie', 'girl-power', 'butch', 'bitch', 'slapper').

The emergence of gender studies

The history of the feminist movement introduced above charts a subtle but incredibly important transformation in the way the 'problem' of men's and women's different lived experience has come to be defined. Initially the goal was to raise awareness of 'women's issues'. Experimental classes in women's studies were first taught at US universities in the 1960s, when university administrations and the public were persuaded that feminist issues and politics were serious subjects to which funding should be made available for professorial appointments and departmental space. Initially women's studies developed along the lines of ethnic studies, Latin-American studies, peace studies and the like. Similarly, development of a movement for gender equality emerged alongside a parallel movement for race equality. This can be observed in the 1966 formation of the National Organisation for Women (NOW) in the USA to scrutinise US law and regulation on similar lines to that of the National Association for the Advancement of Colored People (NAACP) for black Americans.

From the 1970s debates arose *within* feminism, *about* feminism, which splintered the movement. A powerful critique suggested that focusing on 'women's issues' perpetuated dependency and notions of 'the weaker sex'. It was not sufficient to add women's issues to an already established intellectual agenda. Instead, all that went before (malestream urban studies, sociological theory and philosophy) had to be rejected. The new starting point was the 'gendered body' (Beasley 1999: 5). It was at this stage that the separatism implied by the title women's studies, which tended to exclude male authors and development studies scholars, became untenable for feminists who wished to acknowledge multiple and diverse inequalities associated with masculinity as well as femininity. Dissidents of the second wave started to mobilise around the newly defined discipline of gender studies.

Also associated with this shift in thinking was the redefinition of the topic 'women and transport' (or women in cities) to that of gender and daily mobility (Law 1999: 567). Similarly, as we explore on pp. 87–9, paradigmatic developments in feminist thinking influenced the shift in development studies from discourses of women in development (WID) to women and development (WAD) and finally to gender and development (GAD). In the summer of 2008, women's studies vanished as an undergraduate option at British universities. At the same time, prospective students witnessed a proliferation of gender-related programmes of study on offer. On many curricula today students can read not only gender studies but also post-colonial studies (see Chapter 2) and subaltern studies.

> *Subaltern*; *subaltern studies*: originally the colonial military term for someone holding a low or subordinate rank, this term has been mobilised within post-colonial discourse specifically to refer to the perspective of people from regions and groups outside the hegemonic (Western) power structure. Subaltern studies began in the early 1980s as an offshoot of South Asian studies. It is closely associated with post-colonial feminist theory in development studies.

This period also saw the emergence of a critical men's studies and publication of influential texts highlighting contemporary challenges to masculinity (such as with de-industrialisation and male unemployment) and a diversity of masculine identities, not least those reflecting cultural pressures to repress homosexual orientations or to 'come out' against dominant heterosexual norms (or heteronormativity) (Connell 1995, 2000; Morgan 1992; Mac an Ghaill 1994; van Hoven and Hörshelmann 2005: 8–9).

From structure to agency and beyond

From the complex theoretical transitions which punctuate the progressive waves of feminist theory identified above, it is possible to identify distinctive paradigms broadly differentiated as 'structural' and 'post-structural'. This paves the way for developing an explicitly gendered urban theory by drawing parallels between the 'top-down' structuralism of urban ecology and that of a dual systems theory of persistent gender inequality, as compared with the 'bottom-up' post-modern interest in popular urban resistance and the emphasis on difference and diversity in post-structural intersectionality.

Structuralism: Marxist feminism, patriarchy and dual systems theory

Structural approaches to gender studies have emerged periodically in response to evidence of systematic discrimination and constraint: misogyny, patriarchy, exclusion from property ownership and denial of human rights. A defining characteristic is the emphasis on a shared oppression among women which is connected with notions of a 'sisterhood'. While those who pursue a structural approach today typically deny accusations of gender essentialism, it is on this basis, and by neglecting differences between women on grounds of race, place, class and identity, that significant critique has developed since the 1980s.

This body of theory initially drew on Marxist analysis alongside feminism. A straightforward reading of Marxist feminism points to the aim of rendering women and the sex/gender system 'visible' in debates on capital and class conflict where 'women do not appear where they should' in mainstream Marxism (McDowell and Sharp 1997: 23). With reference to Simone de Beauvoir's claim that 'one is not born a woman, one becomes one', differentiation between gender and sex emphasises the way that structural theories of women's oppression distance themselves from an essentialist interpretation of sex as biology and oppression as being defined by characteristics of birth. The aim was to expose and explain the reproduction of dominant interests (capital, misogyny) through the functions of systematic structures of subordination and oppression. This analysis emphasised the interweaving of gender and class, largely to the exclusion of race; representing sexual oppression and the subordination of women, relative to men, as a dimension of class power (Beasley 1999: 60). A particular focus of oppression for Marxist feminists was the institution of marriage and the presentation of this as 'legalised prostitution'. Because 'obligatory heterosexuality' (exchanging sexual services for material survival) was central to the oppression of women, withdrawal of women from the marriage was a potent symbol in the politics of withdrawal from men and androcentrism. To be a subject in the Western sense meant reconstituting women outside the relations of objectification (as gift, commodity, object of desire) and appropriation (of babies, sex, services) (Rubin 1975; McDowell and Sharp 1997: 31).

In an article at the centre of much Euro-American Marxist and socialist-feminist debate, Hartmann insisted that patriarchy (literally 'rule of the father') was not simply an ideology, but a material system that could be defined 'as a set of social relations between men, which have a material base, and which, though hierarchical, establish or create interdependence and solidarity among men that enable them to dominate women' (Hartmann 1981: 14). Patriarchy is most often used to explore the sexual division of labour across public and private spheres, emphasising control over women's bodies through motherhood and housewifery (Walby 1990). In the Egyptian context Sarah Loza endorses a benevolent patriarchy by arguing that 'in a society that accepts separate spheres, equality has no meaning' (Tinker 1990: 12). Different systems of production and kinship represent distinct kinds of 'patriarchal bargains' which not only shape 'women's culturally specific sense of sexed subjectivity' but also act 'as a powerful determinant of women's potential for adaptation or resistance in the face of change' (McDowell and Sharp 1997: 32). Within this frame, Hartmann attempted to explain the partnership (or *dual systems*) of patriarchy and capital.

The neglect of 'race' by Marxist feminists as well as critics of patriarchy and the dual systems of capitalism and patriarchy has been a major source of division within the feminist movement since the 1980s. This critique, alongside post-socialist political realignment (symbolised by the dismantling of the Berlin Wall in 1989) has led to a situation in which 'few feminist theoreticians continue to describe themselves as Marxist feminists. Nevertheless, a rich seam of contemporary feminist theory and political activism, especially that relating to public policy, welfare, work/life reconciliation and material technologies, arguably continues to highlight fundamental structures of constraint insofar as they intersect with plural identifications rooted in postmodernism' (Beasley 1999: 59–60).

The universalising theories of the sex/gender system and the analytical split between public and private spheres were deeply influential for many Euro-American feminists in the 1970s and 1980s. Yet this attracted harsh criticism, especially from women of colour, as part of the ethnocentric and imperialising tendencies of European and Euro-American feminisms. The category of gender was said to obscure or subordinate all the other 'others'. Efforts to use Western or 'white' concepts of gender to characterise a 'Third World Woman' often resulted in reproducing orientalist, racist and colonialist discourse (Mohanty *et al.* 1991: 36). Carby (1987) points to the way black women in the United States were not constituted as 'woman' in the same way that white women were. Instead they were constituted simultaneously racially and sexually: 'If kinship vested men with rights in women that they did not have in themselves, slavery abolished kinship for one group in a legal discourse that produced whole groups of people as alienable property' (Spillers 1987, quoted in McDowell and Sharp 1997: 37). Ultimately Haraway defended the position of feminist Marxian analysis of sex/gender systems. She concluded that 'gender' was developed as a category to explore what counts as a 'woman' so as to problematise the previously taken for granted (McDowell and Sharp 1997: 39). It did not preclude recognition of multiple axes of exploitation.

From WID to WAD

In the context of development studies, the economic crisis of the 1970s cast the spotlight on women's roles in development. This period marked the first international women's conference in Mexico City in 1975 and the first UN Decade for Women, 1975–1984. It also witnessed the publication of Esther Boserup's (1970) groundbreaking scholarly publication *Women's Role in Economic Development*. Although this publication persisted with dominant neo-classical models of economic analysis, it thoroughly unsettled development thinking at the time by making the case for women's productive role in development. This moved beyond modernisation theory's paternalistic notion that women's key contributions to development were through motherhood and care-giving.

Boserup's book inspired two approaches to bringing women into the development process: equality and anti-poverty. Both of these approaches reflected the tenets of WID. These marked the first focused efforts to consider how women were included in development. They nevertheless remained limited for just this reason: they did not question orthodox (androcentric, ethnocentric) models of development, modernisation and growth. The focus on women's economic inclusion was a direct result of Boserup's work and her conclusions that women had greater social position and freedom of movement in contexts where they were more active in the productive sphere; hence equal opportunities to participate in the economy were identified as key to women's greater benefit from development. This emphasis on equal opportunity for women merged liberal feminism with modernisation approaches from an arguably Western-centric standpoint.

Not surprisingly, the WID approach was not successful in creating sustained changes in women's social positions. Kabeer (1994) notes two major failings. First, this approach relied on the market as the means of reducing gender inequalities, without recognising the role of institutions, including the market, in perpetuating such inequalities. Second, it was based in liberal feminist ideas; valuing rationality, individualism and downplaying and devaluing reproductive work. It tended to essentialise women as a category, ignoring the real differences in material welfare and power between women (Kabeer 1994). This led to unrealistic notions of how women could be integrated into development as it tended to ignore a major component of the work women do, as well as ignoring the political struggle over reallocating household work burdens in a more egalitarian way.

Radical feminism influenced debates around women's roles in development and led to the WAD approach. The WAD approach highlighted women's differentness and distinctiveness, in terms of knowledge, experiences, work and responsibilities, and calls for these to be recognised and valued, particularly in the development process (Parpart *et al.* 2000). Thus, women's distinct roles in development were to be celebrated, not overlooked and taken for granted as they had been previously. WAD generally focused on local community-based action, recognising women's different needs in national structures. The debate between integration and separation continually arose, with the risks attendant on integration within male-dominated mainstream institutions weighed against the marginalisation and scale limitations of maintaining separate women's organisations, often due to lack of funds to scale up initiatives. The latter has been a key weakness of WAD through a tendency to marginalise these separate women's groups and organisations because of their continued small scale and low impact. However, its arguments served as an important counterpoint to the WID approach's unquestioning contention that the liberal state could be used to redress gender inequalities, omitting consideration that the state itself may create and maintain such inequalities (Parpart *et al.* 2000).

Post-structuralism and the 'cultural turn'

In the context of feminist theory, the 1980s saw a significant shift in the emphasis from discussion of sex differences to the reproduction of gender relations. Whereas sex depicts biological differences (male and female), gender describes socially constructed characteristics (masculinity and femininity). It is in this respect that post-structuralism, while mounting a critique of Marxist feminism, patriarchy and dual systems, continues to owe a debt to structuralism (Beasley 1999: 89). This debt relates to the sex/gender distinction. Linda Nicholson (1990) uses the analogy of the coat rack to explain this relationship, whereby sex or biological difference is the basic frame onto which different societies, in different historical periods, have hung various coats – the socially defined arrangements of gender characteristics. While pursuing with the sex/gender distinction, post-structuralism 'describes a point of departure from structuralism' by emphasising the 'contextual fluidity and ongoing production of meaning' (of gender and forms of power relations) (Beasley 1999: 91). By recognising locally constituted cultural practices this approach enables gender relations (and inequalities more generally) to be spatially variable across a range of different scales (from the body to the nation-state, across rich and poor countries) (see, for instance, Butler 1993).

This shift in emphasis hinged on post-structural critiques of universalism (essentialism), ultimately unsettling the monolithic categories 'woman' and 'man'.

> *Essentialism*: most commonly understood as a belief in the essence of things, as a belief in the existence of 'invariable and fixed properties which define a given entity'. In second-wave feminism such arguments appear in accounts which appeal to notions of a pure femininity and in terms of universal female oppression.
>
> (WGSG 1997: 21; Fuss 1989)

Criticising the essentialist position and drawing heavily on historical and cultural studies, post-structural feminist theory took account of both masculinity and femininity to highlight the constructed and contested nature of gender differences and identities (Panelli 2004: 69; WGSG 1997: 21).

The inclusive language of difference and diversity associated with third-wave feminism can also be identified as the bedrock of more recent developments in queer theory, post-colonial theory and transnationalism. Queer theory seeks not only to question essentialist frameworks of male/female but also dominant *'heteronormativity'*. In this context, 'queer' signals links with radical 'anti-assimilationist' elements within gay politics in the West, seeking to celebrate eclectic sexed identities and desires, in complete opposition to the mainstream (Beasley 1999: 98).

From WID to GAD

The influence of a post-structuralist feminist critique of essentialism is similarly evident in a shift from WID to GAD approaches to development in the 1990s. Frustration with the 'add Mary Wollstonecraft and stir' approach (Grimshaw 1986: 10–11, cited in Beasley 1999: 5) of simply adding or including women in development led to efforts to provide space for consideration of structural bases of inequality alongside recognition of women's diverse potential to resist or act against these structures.

The GAD approach had two main influences; first was publication of the book *Development Crises and Alternative Visions* (Sen and Grown 1987) by Development Alternatives with Women for a New Era (DAWN), a consortium of women and women's groups established prior to the 1985 UN conference on women; second was the work of the Subordination of Women Group at the Institute of Development Studies in England, who sought a middle way between liberal and Marxist feminism. Because they are rooted in recognition of global and gender inequities GAD and related empowerment approaches are fundamentally political in orientation. GAD argues that women's material conditions and their positions in global, regional and national economies define their status, and that these are influenced by the structures of patriarchal power within society and the resulting norms and values embodied in a society, which vary across time and place, meaning women's conditions, positions and status also vary (Sen and Grown 1987).

The shift from WID to GAD shifted the frame of analysis from the isolation of women's issues from those of men, to specific recognition of the social construction of gender. This recognised not only gender as the construction of masculinities and femininities but also the culturally specific lived realities of urban women around the world. This approach recognised that what it meant to be a woman in the United States in 1998 was very different from what it meant to be a woman in Afghanistan in 1998 under Taliban rule – women are not the same due to shared biology but have different social roles, constraints and opportunities depending on time and place. Thus GAD and the subordination of women group mobilised the term gender relations, and moved development studies' concern with women in development to consideration of how men and women relate to each other in a given society and what this means for women's condition, position and power in society.

The political orientation of GAD means it has largely been popular with southern non-governmental organisations (NGOs) and less so with mainstream international development agencies. GAD seeks to facilitate a process of women's empowerment, an alteration of the structures framing inequity, through action – hence it views women as central agents in the development process and in creating the social and political conditions to end gender inequity (Moser 1993; Parpart *et al.* 2000). In short, GAD combined interest in practical gender needs with strategic political interests, as illustrated in Box 3.4.

Box 3.4: Practical and strategic gender interests

Practical gender interests (PGIs): interests women or men identify from within their prescribed social roles. These correspond to women's and men's condition in society and do not challenge the social structures defining gendered social positions. They tend to arise out of an immediate need, often related to deficiencies in living conditions, i.e. lack of access to water, housing, sanitation, employment.

Strategic gender interests (SGIs): interests women identify based on their subordinate position in society. Addressing strategic gender interests involves challenging the structures creating and maintaining inequity and hence is a political act. SGIs vary by social context in relation to the power structures at play and resulting gender inequities that prevail. Examples include ending domestic violence, achieving equal pay and changing the gender division of labour. PGIs can be achieved in ways that address SGIs, for example, by increasing access to child care in locations convenient to men's workplaces.

These interests, or prioritised concerns, are translated into needs through development planning and practice, with needs being the means through which the concerns are met. For example, a strategic gender interest may be for increased equity which can be met through the strategic gender need of ending the gender division of labour.

Note that 'gender interests' is used instead of 'women's interests' to recognise the heterogeneity of possible interests developed by women or men, influenced by the range of social factors defining gender in a social context.

Source: adapted from Molyneux (1985); Moser (1993).

In keeping with other post-structural feminist approaches, GAD opens up space for women's different voices and experiences. In her seminal work *Under Western Eyes*, Chandra T. Mohanty (1988) criticises the 'othering' of southern women in development discourse. She claims that static and often ahistorical accounts of women's experience of everyday life and its struggles cast southern women as victims of oppressive traditions who lack agency and expression (Mohanty 1988; Mohanty *et al.* 1991). As previously noted in Chapter 2, such criticisms found resonance and formed a part of post-colonial writing on development, including interrogations of the development discipline itself and the power relations embedded within its labelling of countries and regions as 'underdeveloped' against a Western standard of 'development' presumed desirable by all (Escobar 1995; Sachs 1992). It is in this context that closer collaboration has come about between gender-related sub-disciplines, such as with the strengthening of subaltern studies and post-colonial theory. Seminal texts in this regard include Edward Said's *Orientalism* (1978) and Spivak's *Can the Subaltern Speak?* (1985). It is at this point that the intersecting threads of cities, gender and development converge in the growing recognition and valuation of diversity and difference.

Post-structural intersectionality

Post-structural emphasis on multiple and fluid identities (and identifications) highlights a crucial theme to have emerged in recent feminist theorising, known as *intersectionality* (Rose 1993). According to McCall, 'intersectionality is the most important (feminist) theoretical contribution so far' (2005: 1771). Notions of difference and diversity are suggested in the term itself (Fraser 1997), not least in terms of a culturally diverse understanding of what equality means (hooks 1984; Smith 1990). This way of thinking about and viewing gender recognises 'that gender identity operates in complex relationships to other social identities such as race, ethnicity, class, national, First World–Third World, religion, sexuality, age, health' and so on (Martin 2004: 22). It deconstructs the way 'women' and 'gender' as categories of normative connotation intersect with other identities. The key difference between this approach and previous efforts to recognise contextualisation and interconnectedness is a conscious attempt to 'map multiple discriminations and geometries of oppression' (Valentine 2007: 10). From this perspective, axes of difference are construed as ways of 'doing' contingent and discontinuous identities through 'clash, conflict, contention, controversy, debate, disagreement, discord, and dispute' (Fernandes 2003, cited in Valentine 2007: 14). In some sense then we can see a return to the direct action and transformative potential of the late nineteenth-century feminist political movement.

The significance of gender

We recognise that alongside this shift toward the intersectional interweaving of identities (of gender and sexuality) with other social differences there is a growing reluctance in some academic quarters to use gender as an analytic category (Bordo 1990: 135). Yet in this book we explicitly foreground gender for just this reason: the perception that gender pervades the mainstream in practice once again renders it invisible (and taken for granted). We argue that gender remains a crucial everyday political realm within which to measure social justice and human rights.

As a theoretical framework, intersectionality suggests the ideal canvas for the 'intertwined' approach of an explicitly gendered urban theory intended for this book. A recurring problem for feminist debate has been a tendency to present gender discrimination and unequal gender relations as a category above and beyond other defining characteristics of the human condition. In particular, black feminists have long maintained that the very terms 'women' and 'gender' are routinely used in ethnocentric ways to privilege the specific and limited concerns of white middle-class women as a purportedly shared matter of 'sisterhood' (Spelman 1988; hooks 1984). It is a recurring problem too that popular use of the term gender does not routinely encompass masculinities as well as femininities. The majority of research on fatherhood, for instance, tends to be conducted in sociology, family studies or critical male studies rather than as an explicit strand of gender studies.

Intersectionality has the potential to overcome this separatism by focusing in-depth on the nuanced complexity of individual and household biographies and detailed, site-specific case studies. Likewise, critics point to insufficient recognition of poverty and class in contemporary gender debates. Widespread class condescension is increasingly evident in practices of 'labelling', bullying and humiliation in Euro-American cities, both in confrontational reality TV shows and in the management, surveillance and policing of public spaces and social order, differentiating between class-specific notions of 'rough' and 'respectable' moral worth, for instance (McRobbie 2004; McDowell 2006). Recent initiatives include banning young people from wearing 'hoodies' and anyone taking photographs from shopping malls. Similarly, a 'zero tolerance' approach to urban regeneration assumes a very narrow socio-cultural definition of moral order. Again, intersectionality represents a promising lens through which to critically reflect on the puzzles and paradoxes of gendered urban identity.

Summary

After reflecting on the contents of this chapter, readers should be able to appreciate the significance of global economic restructuring to the co-constitution of cities and gender; recognise the critical distinction between sex and gender in early feminist work; be familiar with historical developments in gender studies, development studies and feminist theory; and critically engage with different and competing paradigms, making connections back to previous discussions of urbanisation, modernity and development.

This chapter set out to make sense of multiple trends, historically and geographically. Evidence of the gendered processes and consequences of economic restructuring and structural adjustment brought to the foreground the cultural, time-and-place-specific construction of gender roles and relations, including locally constituted language and meaning alongside institutions and material systems of regulation and representation. As suggested by the case studies presented, the significance of locally constituted language and meaning can be witnessed in a variety of local practices relating to gender. In the North East of England, for example, numerous colloquial terms are used to greet women (especially when unaccompanied), such as pet, hinny, love, darling and sweetheart. Use of such language tends to be fiercely protected as 'part of the local culture' but there is no escaping the fact that these terms are sexist in their aim to infantilise women. An equivalent vocabulary of diminutives does not exist for men, whether young or old. In other parts of the world, such as Afghanistan, women's movements outside the home remain severely restricted by cultural taboo, where language intending 'respect' for women equates less to 'regard' than it does to confinement and concealment (where extreme seclusion is purdah). These examples help explain why orthodox accounts of urban economic restructuring are arguably gender blind: to fully understand how cities and urban societies are constituted, we must acknowledge and account for the locally embedded cultural construction of gendered power relations.

This chapter has contributed a largely chronological history of ideas on gender and feminism as a frame of reference for the contemporary case studies introduced in subsequent chapters. Although an understanding of difference and diversity within and between women's and men's urban lived experience is now widespread across the social sciences, we believe that the lessons of history remain crucial to understanding cities and gender as contested and dynamic subjects. There is no room for complacency! Of course, as Chris Beasley (1999: 20) points out, contemporary emphasis on diversity (as opposed to unity in the face of adversity) does complicate the question 'What is feminism?' If the category gender is lost to multiple categories of the self, what can a gendered urban theory contribute that is unique to the study of urban social inequality? Our response, as stated above, is to stress that gender remains a crucial everyday political realm within which to challenge the social, cultural, institutional and economic barriers to inclusive 'cities for all' (Beall 1997). It is the purpose of subsequent chapters to strengthen this argument by reflecting on discrete aspects of urban daily life, drawing on case studies from around the world.

Suggested learning activity: three waves of feminism in TV/film

Research, read about and/or watch extracts from the following three US television shows. Discuss the ways that these media might be said to reflect three discrete waves of feminism. What does this analysis say about 'progress' towards gender equality in the social organisation and cultural expectations of girls', boys', men's and women's lives? How well do popular visual media provide a window on contemporary attitudes and experiences?

- *A Tree Grows in Brooklyn*, US film released in 1945, based on a book of this title, set in Brooklyn, New York, in the 1900s.
- *Peyton Place*, first US soap opera, opened in 1964, contrasting the lives of two women from Peyton. Alison pursues a lifestyle of family domesticity, while her friend Betty ventures to New York as a single city girl, tasting newfound 'liberation' and transgressing sexual taboos.
- *Sex and the City*, US sitcom, 1998–2004, following the 'lifestyle choices' of four ambitious, attractive career women. The show highlights the contradictions of

being a (white, middle-class) woman in the 1990s and the impossibility of 'having it all'.

Further reading for this activity: Gurley Brown, H. (1962) *Sex and the Single Girl*. New York: Harper.

Classroom exercise

Read the following quotation from Beasley and then go on to describe your own 'typical' feminist:

> When you consider that images (said to represent feminism and feminist) may refer to styles of dress, haircut, and ways of behaving, attitudes and so on, you can probably conjure up a number of graphic pictures yourself. It is interesting that these easily evoked images are more often associated with pejorative views of feminism. However, the images also support an impulse to tie feminism down to something and to ignore considerable differences over the characteristics of feminism.
>
> (Beasley 1999: xiii)

Note

1 Note that women may be financially better off through such factory work, compared to other opportunities available – so relatively better off than before this work was available. But, in some cases, if compared against an absolute standard including work conditions, rights to organise, time burdens, etc. the welfare outcomes can be questioned.

Further reading

Blunt, A. and Rose, G. (1995) *Writing Women and Space: Colonial and Postcolonial Geographies*. New York and London: Guilford Press.

Chouinard, V. and Grant, A. (1995) 'On not being anywhere near "the project": ways of putting ourselves in the picture'. *Antipode*, 27: 137–166.

Connell, R. W. (1995) *Masculinities*. Berkeley: University of California Press.

Escobar, A. and Harcourt, W. (2005) *Women and the Politics of Place*. Bloomfield, CT: Kumerian Press.

McCall, L. (2005) 'The complexity of intersectionality'. *Signs Journal of Women in Culture and Society*, 30.3: 1771–1802.

McDowell, L. (1992) 'Doing gender: feminisms, feminists and research methods in human geography'. *Transactions of the Institute of British Geographers*, 17: 399–416.

Ruddick, S. (1996) 'Constructing differences in public space: race, class and gender as interlocking system'. *Urban Geography*, 17: 132–151.

Web resource

The Women and Social Movements in the United States, 1600–2000, Document Project http://alexanderstreet6.com/wasm/wasmrestricted/DP48/intro.htm.

4 Scale, power and interdependence

Learning objectives

- to be familiar with different ways of understanding globalisation
- to explore intersections of global and local processes of gendered power
- to think about the concept of power as being more than material
- to offer insights into the expression of gendered power and practices across multiple scales within and between cities around the world

Introduction

This chapter concludes the first part of the book, on approaching the city. In it, we provide further bridging concepts that together serve as a crystallising 'lens' through which to explore the co-constitution of cities and gender in subsequent chapters and case studies. First we seek to define cities and urban daily life with respect to scale – indeed, multiple scales. Starting with the 'macro' scale we consider different ways of understanding globalisation, including discourses which privilege 'world' cities. Then we turn our attention to the 'micro' scale, notably the 'embodiment', of gendered power relations. In the course of moving through these scales we expose not only the hazards of defining 'a scale' (where the local is in the global and the global is inscribed in cultures of embodied performance, for instance) but also the interdependence of power within scale. From this we critically examine concepts of power and empowerment in different urban social contexts, considering how relationships of power change across scales as they reside within the city. Building on these theoretical conceptualisations of interdependence we conclude by highlighting global integration within local specificity in the gendered experience of urban daily life.

As suggested by the title of this book, we wish to raise awareness of particular geographical scales and contexts, urban spaces and the people – men, women, children, young and old – who live and make their living in this realm on a day-to-day basis. This fusion of 'people in place' signals a particular set of interactions and possible scales of analysis – from homes, streets and neighbourhoods to cities and regions. Arguably, cities represent an ideal site within which to read the gendered tendencies and consequences of globalisation –in local situations of uneven development which reside in the nexus of individual state law, regulation and welfare regimes and the circulating flows of people, products, capital, ideas and ideologies, as well as state laws, regulation and policy.

Considerations of scale and power also remind us of the importance of history, geography and standpoint. Where people live in the world (and the agency they are able to exercise in their own nest of networks) has a significant bearing on their health, life expectancy and the choices and opportunities available. Equally important is the question of how an individual is positioned, in terms of income, property, power and status relative to others competing for scarce resources (housing, jobs, education, marriage partners) within a defined area or population. It is not sufficient to observe uneven development between rich and poor countries or cities, for instance, when we find evidence of rich and powerful individuals thriving in poor countries and poor people excluded (or impoverished by their exclusion) from the widespread opportunities of more prosperous countries. Arguably, questions of uneven development and gender inequality call for analysis of the intersections of placement, position and power – across multiple scales. Neither is it sufficient to observe highly spatially segregated residential locations. With the advent of increasingly global integration of people and places we have to consider the way distant events and international systems of interconnectedness impact on local everyday experiences. In the sections to follow we explore these puzzles and paradoxes through a gender-relational lens.

Scale and urban definition

The city is a concentration of socio-economic ebbs and flows in which ideas of scale are constantly challenged, remoulded and revisited by the actions and agency of the individual, family, group and network. Thrift's declaration that 'there is no such thing as a scale' (1995: 35) refers to scale as a social construction whereby, as with other social constructions (such as gender and the family), it changes over time in a multi-scalar fashion. It is nevertheless useful to define scale in terms of size, level and relation (Howitt 1998), while at the same time recognising within this framework a 'sense' of scale in place (or scaled places) as 'the embodiment of social relations of empowerment and disempowerment and the arena through and in which they operate' (Swyngedouw 1997: 169).

Definitions aside, because this book actively promotes new ways to approach the city through everyday life and urban ethnography, we wish to emphasise the effects of scale and multi-scalar processes in the social construction and reproduction of gendered power relations.

We are particularly interested in the pressures exerted by relationships of power, both in constructing and remoulding social networks, and how relationships of power change over space and time. A good example of the pressures that shape gendered power relations, historically and geographically, would be government policies which promote and regulate marriage, fertility and parenting. In both the UK and USA, for instance, the family and household have repeatedly been held to account by neo-liberal government policy as a 'moral scale' by which to measure citizenship against idealised (arguably patriarchal) behaviour and values. This is evident in the 'Back to Basics' campaign of John Major's government in the UK between 1990 and 1997, in which traditionalist 'family values' were alleged to constitute a central plank of government policy, and, in the US, the increasing importance over the last three decades of a traditional view of the family as defined by more conservative Christian beliefs, and the consequent furore over the idea of gay marriage. The fact that substantial numbers of politicians, ministers and leaders of all kinds in both countries have themselves been unable to live up to the demanding standards set for 'normal' citizenship does not seem to have diminished the iconic importance of these ideas.

The idea that moral (relational) scales of gendered power also coincide with 'nested' (or hierarchical) social and spatial scales is reinforced by Doreen Massey's work on power geometries – webs of relations of domination and subordination (Massey 1993). Recent research mobilises this concept in an urban studies context and largely draws attention to direct, physical, arguably masculinised, webs of relations: flows of money, maps of connectivities (IT and labour-market-based) specialist market teleconnections and stock market interrelations (but see Jane Tooke (2000) on the role of institutional power geometries in persistent assumptions about masculinity in 'garbage work'). This suggests a geometry which is rooted in the capital economy. The 'soft' infrastructure networks that we are interested in (such as networks of reciprocity and exchange) (see Chapter 5) can also be local, regional, national, transnational and multinational in scale, mapping uneasily onto these 'hard' networks, and yet frequently both are intimately interconnected and interdependent. As Massey *et al.* (1999) point out, networks differ in terms of the kind of social power they carry as well as the numbers of people participating in them and affected by them. Different cities are foci of different numbers and types of networks, giving them distinct and different kinds and ranges of influence. An example of this would be the popular understanding of a 'digital divide' reflecting the uneven nodes and networks of internet connectivity and the distribution of broadband and reliable energy supply (Norris 2001; Zook 2001). These are not just academic considerations, as we go on to illustrate.

This variation means that the half of the world population that is defined as 'urban' contains a myriad of different lived realities. Only a relatively small proportion, for instance, live in a place that would be strictly defined as a city – by the presence of a cathedral, for instance, or by close proximity to a central business district. Around the world, we find towns and cities fused together in conurbations (a term coined by Patrick Geddes in 1915) and metropolitan regions, and more recently polycentric and megalopolitan land use patterns (Fishman 1990). Some of these represent 'superstar' regions of high-value productivity and wealth (Perrons 2004). Others, notably in the mega-cities of the global south, attract global recognition by their sheer scale rather than productivity or wealth. Before we explore the lived realities of these differences it is important first to critically examine this notion of global scale relations with respect to globalisation.

Globalisation, cities and gender

Recent years have witnessed globalisation held up as a powerful narrative in relation to wide-ranging social, economic and political transformations in the post-modern era (see Box 4.1). The term is popularly (often crudely) applied to a 'cultural homogenisation' thesis which suggests, for instance, that shopping streets tend to look increasingly the same, with the ubiquity of global brand names such as Starbucks and McDonald's. This argument might be made, for instance, with respect to the global brand names which can be seen, removed from any sense of local cultural context, in the Guatemalan shopping centre pictured in Figure 4.1. Similarly, the term is loosely applied to the opening up of international markets (for labour as well as consumer goods and cultural services) and the common language of market competition whereby this 'neo-liberalisation' erodes the distinction between cities and patterns of social and economic development. Of course the manifestation of this erosion (the causes and consequences of which remain contested) is highly uneven, socially and geographically, within as well as between cities around the world. Consider, for instance, the enduring local cultural specificity of the market stall in Figure 4.2. The woman dressed in indigenous costume selling vegetables outside the Mercado Central marketplace in Guatemala City is similarly bound up in processes of globalisation (notably neo-liberalisation) but with markedly different local effects to those represented by the modern shopping centre less than three miles away.

Criticism has been variously levelled against the way globalisation has been used as a catch-all term with limited consensus on causal attributions; as an explanation of

Box 4.1: Determinants of globalisation

Held and McGrew (2007: 4) identify five 'deep drivers' underpinning the nature and pace of contemporary globalisation (domestic and transnational forces). These are:

- rapid and fundamental information and communications technology (ICT) developments globally – highly connected cities in the north act as hub gateways to an unevenly developed global infrastructure;
- the development of global markets in goods and services (again unequal);
- the development of the new global division of labour – where the structures of inequality and hierarchy are determined in large part by the power of large multinational corporations but also domestic policies (migration and labour law);
- the proliferation of a climate of 'neo-liberalism' (the end of the Cold War and the diffusion of democratic and consumer values) as well as marked reactions to this;
- the growth of migration and the movement of peoples, linked to shifts in patterns of economic demand, demography and environmental degradation.

We would add a sixth 'deep driver' which reflects our approach of bringing gender issues into the mainstream:

- the development of movements for and by women on an international scale (and, associated with this, the emergence of an international environmental movement), emerging out of the questioning mood of the 1970s and diversifying and fragmenting according to the rapid blending of 'universal' and 'local' interpretations of feminism.

These feminisms share a universal concern with empowerment through the questioning and challenging of the local articulations of patriarchal norms and an increasing ability to raise awareness of forms of gendered discrimination through the use of artful media campaigns and an ability to act at the global level.

Theories of globalisation can be identified in the form of three overlapping waves:

- *hyper-globalisation*: the nation-state is increasingly irrelevant, rootless multinational corporations travel where they will to maximise profits;
- *sceptical globalisation*: there is nothing particularly new about globalisation, which is predominantly a 'rhetorical effect' (Pemberton 2001: 169);
- *post-sceptical globalisation:* globalisation is not viewed as a process or end-state but instead as a tendency to which there are counter-tendencies (Hay and Marsh 2000: 6, cited in Martell 2007) (Holton 2005: 5).

Again, as with feminism, some scholars suggest we are now entering a new wave or paradigm of 'post-globalism'. This latest 'turn' reflects the widespread belief that the terrorist events of 11 September 2001 marked a turning point in modern world history and a return to 'normal' self-identified territorial and cultural interests (Held and McGrew 2007: 1). Evidence of 'radical de-globalisation' can be witnessed in the reassertion of religious fundamentalism and local ethnic dress. It can also be demonstrated in social symbols of ethnic distinction and the strengthening of spatial and cultural separation (such as with gated communities and ghettos) within cities. If we are to accept this idea of a new paradigm of parochialism there are serious implications for renewed gender segregation and the subordination of and violence against women as well as against those who transgress locally sanctioned codes of masculinity and femininity (in dress and behaviour, for instance).

Figure 4.1 ***Global brand names, Centro Comercial Los Procere, Guatamala City***
Source: Jonathan Cloke

Figure 4.2 ***Woman in indigenous costume selling vegetables outside the central market, Guatemala City***
Source: Jonathan Cloke

social change; as a description of social reality; and as a political project to idealise a particular metric of 'social progress' (Held and McGrew 2007: 2; Brenner and Theodore 2002). All three applications have implications for the representation (and widespread neglect) of gender in urban theory. For instance, Linda McDowell (2004: 3) argues that use of this umbrella term has overemphasised the image of a world becoming one and the same, and of local differences being wiped out or abandoned. Instead she points to evidence of a continued and even intensified sense of locality, with local communities asserting their own identity, customs, practices and language. This suggests an alternative thesis whereby local identities and cultural distinctions are intensifying alongside the uncertainty and discrimination of 'the uneven integration into the global system of different areas of the planet (countries, regions, cities and the flows between these)' (Castells 1993: 481).

Nevertheless, the vivid language of global flows of information, finance and technology has led the scholar Manuel Castells (1996) to argue that the 'logic' of fixed locations, localities and places has weakened over recent decades, so that it makes more sense, he believes, to view cities as being in a state or scale of flux. One suggested consequence is the subjugation of local affairs and localised representation, notably the everyday concerns of women, children, young and elderly people, to powerful global coalitions (Horelli 2002). This argument is illustrated by evidence of a 'global intelligence corps' which is manifest in a multinational construction consortium in Case 4.1. Reinforcing this thesis of unequal representation, Ann Phillips argues that no system can claim to be democratic if it does not recognise the legitimacy of popular control and popular equality. In practice she notes in democratic countries that this does not translate as popular or equal representation by gender, class, race or ethnicity. A quick glance at any elected assembly of politicians illustrates this point (Phillips 1995: 27). While gender equality is now an explicit aspiration of many political systems, women still lack a visible role in political life and 50/50 equal political representation of men and women remains elusive (Paxton and Hughes 2007: 10).

This narrative of the domination of global coalitions and unequal political representation tells only a partial story of course. There is equally compelling evidence that grassroots resistance is far from extinguished (see, for instance, examples of feminist environmental groups and mothers acting to combat gun crime in Chapter 8). We also point to the uncertain influence the organisation of the state has on different possible future outcomes – such as more (or less) gender-neutral or family-friendly urban policy. Recognising the paradox of this 'glocalisation', Held and McGrew (2007: 7) call for greater emphasis on conceptualising multiple globalisations and the dismantling of any singular umbrella term. They point to the way that global trends of economic neo-liberalism and cultural Westernisation are frequently conflated, and to the way that different globalisations collide in the form of vociferous local resistance, often to a perceived global threat, whether to liberal cosmopolitanism, religious fundamentalism, environmental activism or multitudinous migrant 'hordes' taking up local residence (see Glassman 2001 on scales of resistance).

Globalising cities

In Chapters 2 and 3 we noted that urbanisation has occurred alongside globalisation – where multinational economic restructuring and the structural adjustment programmes of supra-national organisations have determined the pattern and process of city growth to a considerable extent. If it is accepted, as many commentators suggest (see, for instance, Badcock 2002: 24), that the formation of a global market economy began with the fall of the Berlin Wall in 1989, then within this reading of globalisation some cities are singled out as 'globalising cities' (many of which are found alongside rapid growth in the south) while others (mainly in the global north) are assigned the status 'world' or 'global cities' (but see Robinson 2006 for a critique of this hierarchical labelling).

Case 4.1: The 'global intelligence corps'

The effects of globalisation can be observed not only in the apparent erosion of a culturally distinct architectural aesthetic but also in the powerful influence of multinational corporations in large-scale construction projects around the globe. Peter Rimmer (1991: 74) points to the influence of an elite segment of the building professions (architects, engineers, information technology specialists), which he calls the 'global intelligence corps' (GIC).

According to Stephen Cairns (2004: 29), this entanglement of architecture with the movements of capital comprises a very small number of elite architectural and planning firms that undertake the most prestigious commissions. These firms tend to be synonymous with high-profile charismatic men for whom architecture and city building are both a professional and a personal pursuit, a pursuit that fully consumes them throughout their lengthy careers (Olds 2001: 142–143). The charismatic male bonding of the corporate business environment enables firms to acquire architectural work, especially in the context of the patriarchal nature of Chinese society (Olds 2001: 144).

The GIC segment of the architectural profession amplifies the patriarchal hegemony of the profession. 'Big build' is a heavily masculine business, as the all-male line-up (of 'signature architects' and their associates) implies. Professionally, this situation has arisen in part because of the gendered natures of architectural education (which we discuss further in Chapter 5) – the selective 'great monuments, great men' approach, one that 'isolates and objectifies the designer and the work' (Ahrentzen and Anthony 1993: 15; see also Greed 1991; Colomina 1992; Grosz 2001).

Comment

Observations in this case, concerning intensive, macho, corporate and design-studio male bonding, resonate with the findings of other 'cultures of work' research – such as the male-dominated world of merchant banking (McDowell 1999) and of creative self-exploitation in new media and computer software design (Ullman 1997; Pratt 2002; Perrons 2003).

Again, according to Kris Olds,

> architectural schools continue to train students to worship the 'masters'; the all male 'geniuses' covered *ad nauseum* in architectural history books. These texts, virtually all written by male architectural critics, construct Architectural History in a manner that renders females as spectators, or else excludes their achievements

> as producers. In the business arena, architecture and property development at the mega-scale is managed by some of the most powerful members of society (regardless of which nation you focus on) and power over property capital and urban development processes is primarily held by men. Indeed, the construction of masculine identity for some of these men is associated with their power to produce urban space that other people take notice of.
>
> (Olds 2001: 143).

The idea that the landmark skyscrapers are representative of the personal desire (or ego) of male property developers for tall (phallic) buildings is widespread (see, for instance, Fainstein 1994: 4; McNeill 2006: 48).

The neo-liberal market-led city is obviously far more open to globalising processes than was the case with previously centralised, state-regulated economic regimes. A classic example of the 'opening up' of a city is the 'Big Bang' in London in 1986, in which (among other things) fundamental changes were made to the way in which stocks and shares were traded that allowed foreign ownership of UK brokers for the first time. The extent of interconnectedness helps to intensify these globalising processes, and planning which is now left to the functioning of an idealised market tends to become fragmented, tuned to economic and political constraints (Knox and McCarthy 2005: 515–516). In this way, the capital market and financial planning itself are gradually divorced from perceptions of public interest. They are instead geared towards the needs of the producers and consumers of the city. In this context planners and policy makers are expected to be increasingly entrepreneurial. Gradually the interstices in planning and government in which the lived realities of the mass of 'unglobalised' city inhabitants take place expand, as municipal and urban governments direct more and more effort towards helping the city to compete for a place in an apparently inevitable and unstoppable process of globalisation. This is reflected in the very narrow discursive framework of 'urban competitiveness' by which cities are ranked according to instrumental 'hard' indicators such as jobs, money and new construction (see, for instance, Begg 2002). Even the superficially attractive idea of the 'creative city' rests with the capacity of a place to attract and retain high-earning new media and entrepreneurial workers (Florida 2002). This leads us to critically examine the privileging of global and world cities.

World city discourse

The term 'World City' originated in 1915 with Patrick Geddes, a leading figure in the evolution of city planning, but it could be argued that the modern development (post-World War II) of concepts such as world city and global city key into far older myths of the iconic city or city-as-ideal. Atlantis, Jerusalem, Camelot and El Dorado have all acted at various times and in different cultures as evocations of the culmination of societal ideals, the perfectly planned polis containing a civil society and social hierarchy in perfect harmony, and more latterly the fount of abundant wealth as well. It is not our intention to suggest that global and world cities theorists necessarily seek to represent an ideal city or utopia (as defined in Chapter 2). Instead, the characteristics identified in Box 4.2 suggest both an eye to the future and a foothold in the past.

Box 4.2: Characterising a world city

Reed (1981) and Friedmann (1986) returned to this theme of the 'world city' in the 1970s and 1980s in an effort to describe the ways in which cities were evolving due to the economic and technological transformation going on globally. Friedmann (1986) in particular noted some key characteristics which he believed delineated the global city:

1. The ways in which a city integrates into the world economy and the role it plays in the New International Division of Labour (NIDL) must be decisive for any structural changes in that city.
2. World cities are key cities used by global capital as basing-points in the spatial organisation of production and markets and occupy key places in a hierarchical network.
3. There are global control functions exercised by these cities which are explicit in one or more production/employment sectors.
4. World cities are major sites for the concentration of international capital.
5. World cities are destination points for large numbers of migrants.
6. World cities tend to show features of the major contradictions of industrial capitalism; for instance, they may demonstrate spatial/class polarisation.
7. World city growth tends to generate social costs exceeding the fiscal capacity of the state to deal with.

Source: Friedmann (1986: 69–71).

Explicit in the idea of world and global cities are relationships of power. In this book we understand gendered structures to be, above all, material, institutional and moral expressions of power relations, and as such we are interested in how that power is exercised in relation to these (indeed all) cities. Friedmann (1986) and other writers have examined the ways in which world cities, for instance, tend to have increased control over the production and transmission of news, information and culture; have marked economic and social polarisation frequently expressed as spatial segregation (particularly in terms of 'ghettoising' large numbers of migrants); and have heightened tensions over control and possession of space and the mechanisms that go with it, such as local government, governing bodies of public spaces and civil society organisations.

As Knox and McCarthy (2005: 158) point out, these tensions, contestations and fragmentations are a direct consequence of the ways in which world and global cities establish their place in the hierarchy of cities. Types of fragmentation include not only spatial divisions according to inequalities in wealth and of access to the means to create wealth, but also the development of enclaves of international banking, finance and business centres (City of London, Wall Street), enclaves of internet and information technology development (Silicon Alley in New York), technopoles or other areas of high-tech development (Silicon Valley, California), foreign direct investment zones with customised infrastructure and legal/tax exemptions (Porto Alegre in Brazil), Data-processing/call centre enclaves for e-commerce and offices (Sunderland in the UK) and 'logistics' enclaves – airports, ports, export processing zones (El Paso, USA).

Set in historical context, the concept of global/world cities provides a vital tool for analysis and research because analysing the individual city has no meaning without an understanding of its historical context and supporting network. Nevertheless, as feminist critics point out, a global or world-scale analysis based on cities as actors in a hyper-politics of capital tends to reify 'hard', quantifiable, formal economic networks, thereby rendering invisible the 'soft' networks, conduits and flows which essentially reproduce the city's ability to create its own socio-political realities. In order to explore the multi-scalar processes of gendered power relations, therefore, we turn to consider alternative scales and tools of analyses. For this purpose we reflect first on a household perspective and then on the insights that can be gained from thinking through the embodiment of cities and gender.

Cities of households

Throughout this book we stress that, alongside the city, one of the most useful scales of analysis is the household, together with the harder to define, but pivotal, 'binding properties' of social networks and community associations, to which we return in Chapter 7. As Bob Connell observes, 'we live most of our daily lives in settings like the household, the workplace and the bus queue, rather than stretched out in relation to society at large or bundled up in one to ones' (1995: 29). In this respect the dynamic institution of the household provides a more realistic setting within which to explore the local effects of gender restructuring, rather than individual biographies for women and men. While the household and locale typically represent a shared system of dispositions, this cultural coalescence does not have a unifying effect: individual agents circulate within and outside the arena and as such introduce new ideas and practices to the group at the same time as the group influencs the receptiveness of that individual to make changes in his or her routines (Jarvis *et al.* 2001: 88–89).

As globalisation has increased across scale and by degree of interconnection (albeit unequally and unevenly), the household in particular has taken on an increasingly transnational character. For example, large numbers of Philippine households depend on female family members working as agency nurses in the UK; Nicaraguan households depend on female members working as domestic employees in Costa Rica and male members working as agricultural labourers in California; Bangladeshi families living in Sylhet construct hotels built on the remittances of male household members working as cooks in Indian restaurants in the UK. We consider these ideas in more detail in relation to transnational migration in Chapter 6. The task for now is to consider the effects of global interconnection associated with a dominant neo-liberal (and again, we would suggest, highly masculinised) view of the political economy – which has acted to substantially shift cultural taste and demographic transformation.

We noted previously that running up against a straightforward thesis of increased global connectedness is contradictory evidence to suggest that most people continue to pursue highly localised daily activities. The concept of the household 'locale' is particularly relevant in this respect: it draws attention to the need for a detailed, urban ethnographic understanding of gendered power relations in everyday life.

> *Locale*: a setting or context for social interaction. The term was proposed by Anthony Giddens in his development of structuration theory, suggesting that the flow of human agency 'binds' time and space. The term 'locale' differs from 'place' in the way it characterises physical settings associated with different types of collectivities (such as family members, friends at leisure, parents chaperoning children and the like).
>
> (Johnston *et al.* 1994)

While the scale-ideal of the locale 'may range from a room in a house, a street corner, the shop floor of a factory' (Giddens 1984: 118), Nigel Thrift (1983) highlights the paradox that 'a locale does not have to be local': it also extends to towns and cities and to the territorially demarcated areas occupied by nation-states. Nevertheless, the kinds of social interactions which serve to define a particular locale (such as parents expressing concern for road safety) usually require co-presence or, in the case of mediated contact (such as telephone and email), a high degree of familiarity and trust. Case 4.2 usefully illustrates the significance of the

Case 4.2: Isabel Dyck: feminist geography, the 'everyday' and local–global relations

It is a commonplace to note that the effects of globalisation play out differently for those at the 'centre' of the economically developed (Western) world than those in the 'developing' countries of the periphery. However, such a dichotomy is misleading. Work on caring amply demonstrates the impact of global processes on spatially disparate locales, which, however, may be connected at the level of the everyday through caregiving activities. Configurations of care are reworked under conditions and contingencies deriving from locally felt effects of economic restructuring and migrations accompanying the unevenness of globalisation (Rankin 2003). These processes affect the composition and 'spatiality' of family households. And while most research has focused on the commodification of care, it remains that most care is provided in family households in both the West and developing countries.

Complex 'chains of care' become visible, as the effects of globalisation and neoliberalism combine. Take, for example, the well-known figures of the live-in childcare worker and cleaning woman who perhaps are from the Philippines and Mexico and work in the homes of their middle-class employers, most often in the urban centre (Pratt 2003). As they support the economic and status achievements of their professional single or dual-career household employers, their own children or elderly parents may be cared for in their home countries by other family members (Pratt 2004; Stasiulis and Bakan 2005). Further, they may be sending remittances back to support family projects, such as building a home or paying for the education of another family member. There is a dual place-making, contributing to the maintaining of households in two countries, with these in turn linked to the economic activity sustaining the strategies of nation states in the wider context of globalisation.

Source: Dyck (2005: 239).

Comment

Isabel Dyck sums up here what has become increasingly fertile ground for feminist critics with respect to the androcentric and ethnocentric assumptions and partial perspective of 'grand narratives' on cities and globalisation. She points to the multiple and contradictory expressions of global interconnectedness. Alongside the transnational flow of people and technological innovation we find the reassertion of national, religious and ethnic cultural separation. Women paradoxically gain economic opportunities through the dismantling of trade barriers, while simultaneously being restricted to some of the most degrading and disposable jobs (Brodie 2003). These processes impose new structures of inequality and hierarchy at the same time as they reproduce long-established and persistent inequalities.

locale in terms of local–global 'chains of care' linking economic activities with unpaid social reproduction through the household–locale nexus.

The paradoxical quality of local–global interdependencies suggests that cities are potentially both emancipating (enabling) and confining (disabling and elitist). All cities are suffused with a social environment which brings pressure to bear on gendered norms; in other words, because gendered constructs and processes are social and therefore vulnerable to change, the vivid and intense nature of life in the city forces a quicker pace of change and development of alternatives. Globalising and world cities, however, experience not only an intensification of social process but an increased exposure to different cultural and social norms, because of their very openness, and therefore have the capacity to induce gender liquefaction – melting, blending and re-solidifying new gendered norms, themselves subject to re-melting and re-formation. A global or world city, therefore, has to have the capacity to create and maintain different gendered forms that can only thrive in this rich, diversified urban environment, else it may become the stage for a conservative backlash. These contradictory social processes are highlighted in Case 4.3 in the murder and mutilation of women in Basra and Guatemala who transgress strictly prescribed gender norms.

These extreme cases offer perhaps the most convincing argument for a shift in the scale of cross-comparative urban cultural analysis from national or regional level to household and locale. It is, after all, the local milieu within which notions of 'the right thing to do' (as well as 'the wrong things') are negotiated or dictated, rather than there being an identifiable 'regime' of ascribed duties for men and women 'out there' in the national psyche.

Bodies and embodiment

Another way of thinking about these multi-scalar power relations is to reflect on the effects on the body of the physical and emotional urban landscape. It is not uncommon, for instance, to feel dwarfed by tall buildings or exposed and vulnerable in empty, wide open spaces. In numerous ways people can be diminished or 'buoyed up' in direct response to their surroundings, by the feelings of being an outsider, being shown very little respect, or when threatened with physical violence or verbal abuse. Approaching the city through the human scale

Case 4.3: Murder and mutilation in Basra and Guatemala

Gendered norms (whether in global/world cities or not), filtered through the politics of identity, can, on occasion, be highly, pathologically resistant to change. The melting process in cities not infrequently induces a violent reaction which can take a variety of forms, some of them deadly. In the city of Basra in Iraq, in the year 2007–2008 the Organisation of Women's Freedom in Iraq (OWFI) reported that at least 133 women (but probably many more) were killed and mutilated. In the main they were 'PhD holders, professionals, activists, and office workers' (Susskind 2008). OWFI suggests that these killings are in the main carried out by self-proclaimed 'Propagation of Virtue and Prevention of Vice' squads working with the Shia militias in the city, as a way of pushing women out of the social and political spaces they now occupy, back into the home.

In Guatemala in the three years up to 2004 approximately 1,500 women were abducted, tortured and murdered, an average of about 41 a month (Grais-Targow 2004) – a direct legacy of the bloody civil war in Guatemala until 1996 which has been visited on women labourers moving into the city to work in the *maquila* zones, but more fundamentally derived from the racism, paternalism, and violence endemic in Guatemalan culture following the war. A similar situation pertains in Ciudad Juarez in Mexico where hundreds of women working in the *maquiladoras* have been murdered, and Mexican authorities suspect that similar activities are taking place in other cities such as Chihuahua. In all of these cities poor, fragmented and dysfunctional societies have been created, frequently in a post-conflict situation (and it is suggested here that the activities of the drug cartels in those Mexican cities constitute a conflict as bloody as anywhere), in which an absence of local and national government control, corrupt and inefficient police forces and the effective non-functioning of judicial and legal systems allow a pathological and extreme performance of the logical consequences of uncontested patriarchal power.

Source: Grais-Targow (2004: 29–31); Susskind (2008).

of embodiment has become a recognised aspect of urban ethnography in recent years. A good example of how to witness the embodiment of 'fleshy possibilities' in urban space (Sennett 1996) is to reflect on the experience of riding through the city in crowded public transit (whether a stuffy subway train, packed with commuters, or an open truck or city bus filled, beyond safe limits, with passengers, livestock and cargo); people react differently to this 'too close' encounter of imposed intimacy; the experience is shaped by gender identity, position and 'stature'; there are those who manage to 'command' more space and those who get pushed about or, worse still, grabbed and groped (see Chapter 6).

Making connections between the body and the city is of course potentially controversial – there are the hazards of essentialism, 'both in term of seeing gender as being centrally about bodily differences and, more specifically, in terms of constructing women as being, in some ways, more embodied than men' (Morgan *et al.* 2005: 2). We wish to avoid this danger by emphasising both the human scale of gendered power and the embodiment of everyday routines and practices, such as the 'limits' (capabilities) of different individual bodies' ability to 'go on', and the interdependence of this embodiment with multiple scalar processes of gendered power relations. Moreover, as Neil Smith observes, 'the politics of the body are not delineated by gender alone. Bodily style and clothing mediate personal construction of identity, within regional, national and global cultures and provides access to the body by the international fashion industry' (1996: 103).

> *Embodiment*: humans are always located somewhere and at some time, and our awareness is profoundly influenced by the fact that we have a body. The process of embodiment reflects the way the individual body is connected into larger networks of meaning at a variety of scales; the production of social and cultural relationships through and by the body simultaneously with the 'making up' of the body by external forces.
>
> (Cresswell 1999: 175–178)

Feminist scholars have expressed growing interest in the ways in which bodies become marked as male and female in particular urban contexts, such as the crowded bus, as well as how stylised images of femininity and masculinity are mapped onto the urban landscape (Nash 1996; Longhurst 1995; Cresswell 1999). Rachel Pain (1999) recognises the body as a site of individual expression, of 'becoming', beyond the containment and limitations of history and location. By this token there is potential both to control our own bodies (particularly in the West) but also to have them controlled or regulated by others, as in the contested position of abortion and of sexual preference (Pain 1999, cited in McDowell and Sharp 1999: 19).

The urban landscape reproduces masculine and feminine identities in numerous ways, both visible and more subtle. The signs and texts of popular culture (from music, film, advertising and designer labels to graffiti) play a significant role in this process. In the West, for instance, idealised masculine and feminine identities are heavily influenced by the proliferation of a media-driven celebrity culture. Some sense of the iconic status the celebrity body can achieve and the influence of this on gender role identities (especially among the young) is depicted in Figure 4.3. This image shows a wall of pink heart-shaped messages addressed to the singer Kylie

Figure 4.3 ***Kylie-mania at the Victoria and Albert Museum, London***
Source: Helen Jarvis

Minogue – whose petite body was publicly celebrated in 2007, in the unlikely venue of London's Victoria and Albert museum, through an exhibition of the singer's performance costumes. The exhibition drew explicit attention to her evolving image as a 'show-girl', symbolic of 'girl power' (Laurie *et al.* 1999).

Another manifestation of the embodiment of gender in the city can be observed in the way the 'protected spaces' (Sennett 1996) of cafés and pubs cultivate powerful local identities (such as 'biker', 'single', 'gay', 'lesbian', 'yummy mummy' and the like). Sites of recreation and entertainment also frequently reproduce stylised gender identities, either by attracting permissive or exaggerated 'festival bodies' (Sennett 1996), or by idealising hyper-masculinity and the sexually available female body. For instance, Figure 4.4 shows the flamboyant display of male body-builders and bikini-clad girls splashed all over the fairground equipment at the annual Hoppings funfair, Newcastle upon Tyne[1]. This is an event which attracts thousands of young people each year to its white-knuckle rides and motorcycle stunt shows, including the 'Wall of Death'. A travelling fair was first invited to the city in the nineteenth century to promote family entertainment (as a temperance measure); sideshows included

Figure 4.4 ***Representing the fun of the fair with body-builders and bikinis***
Source: Helen Jarvis

'the world's ugliest woman', 'the reluctant lady' and 'Anita, the world's smallest woman'. Arguably, while fashions and technologies have changed, the message inscribed in this urban pleasure environment that 'sex sells' remains the same.

> *Identity:* a complex, contested term which is central to the recognition and articulation of differences. Notions of identity typically distinguish a sense of self as distinct from variously culturally constructed others. Feminist theories of identity argue (against essentialism) that all social identities are culturally constructed, changing from time to time and from place to place. Recent theories have asserted the problems and possibilities of a more 'hybrid' notion of identity, typically associated with 'lifestyle' choice and a range of social movements (including around gender and generation, ethnicity and the environment). While identities may be plural and dynamic, they remain subject to social regulation through cultural norms and expectations.
>
> (McDowell and Sharp 1999: 133)

On power

Of all the core themes tackled in this chapter, power is perhaps the hardest to conceptualise and pin down. Extreme cases of murder, rape, genital mutilation and sex slavery aside, gendered power relations most routinely function in subtle ways that rarely equate to literal interpretations of female subordination to patriarchal control (but see Pateman 1988). Arguably, then, there is a need both to critically engage with the literature on power and to pay close attention to the way power is routinely reproduced in the practices of everyday life.

Power is conventionally conceived as the ability of one actor to impose his or her will on another. This imposition of will is typically presented in a structural framework of relative power between political states, classes or ethnic groups (Nagel 1975; Foucault 1980). Power is also generally reduced to economic or functional determinants (Blau 1964). A common failing in existing conceptions of gender and power is the false identification of material resources (such as money and status) with power itself. From this perspective money (from wages) is taken to represent the ability to make choices and obtain preferences as an expression of 'bargaining power'. Morriss describes this as the 'vehicle fallacy' (Morriss 1987: 19). In an alternative thesis, Morriss suggests that we understand power in the household 'as a dispositional concept, [as] neither a thing (a resource or vehicle) nor an event (the exercise of power), [but as] a capacity' (1987: 19) which may or may not release an event in any given situation. Viewed as a suite of potential practices, power can be 'traced' through individual discourse, social interaction and a web of locally embedded social networks (Jarvis 1999b). In this way, everyday routines of engagement effect the socially constructed stratification of gender, class, education and cohort (Edley and Wetherall 1995) within reflexively reproduced positions of influence and status (Clegg 1989: 142).

Reference to power in the specific context of household gender divisions first appears in structural-functionalist (often biologically determinist) paradigms from the 1950s onwards. This perspective views power as a function of social constructions of 'instrumental' male and 'expressive' female sex-roles (Parsons and Bales 1956, reviewed in Segal 1990). Power is seen as normatively secured (through male-dominated macro-economic practices) in gender-specific spheres of behaviour and typically transmitted via state economic structures and the separation of 'private' and 'public' spheres. A similar degree of determinism is also found in literature explaining divisions of labour in terms of sex-role specialisation based on explanations of 'economic efficiency', 'time availability' and normative socialisation (Parsons and Bales 1956; Becker 1981; Hartsock 1983; these approaches are reviewed in Hiller 1984). Much of the early Marxist and feminist literature can also be criticised for reductionism by ascribing primacy to cultures of patriarchy such that women are 'always' seen to be exploited (Oakley 1974; Hartman 1981). By contrast, from an everyday life perspective, we argue that men and women experience the operations of power in diverse ways in a multiplicity of sites.

To a certain extent these narrow conceptions of power have been dismantled by recent feminist literature which allows for plural masculinities and femininities, with renewed emphasis on gender difference leading to 'power struggles' in the home (Segal 1990). In her book *Gender and Power in the Workplace* (1999), for instance, Harriet Bradley sets out a multidimensional model of different types of power resource which could be used by women and men to secure dominance in different contexts. It is not a spatial or explicitly urban model but it does lend itself to an understanding of the 'multiple positioning' and potential for 'multiple disadvantage' recognised by our previous framework of intersectionality (Bradley 2007: 190). Bradley (1999) highlights the contested concept of gendered power within feminist scholarship. On the one hand she notes that Foucault's understanding of power as a property of historical and geographical context has been widely embraced by post-colonial feminist theory. Yet she advocates instead a resource-based approach, such as those of Bourdieu or Giddens. Foucault (1980) notes that power is not readily incorporated into a single theory because it is not a 'thing' which exists independently of historical and geographical context (Foucault 1978: 82; Dreyfus and Rabinow 1982: 184). Both perspectives argue that power needs to be viewed as enabling as well as prohibiting whereby expressions of power can result in non-events and non-decisions as well as behaviour that is revealed.

These recent theoretical developments have shifted the focus of attention in feminist urban social analysis to the micro-level of bodies and their capacities, considering how power constitutes certain types of bodies which in turn become vehicles for the transmission of power. We noted these effects with respect to media representations of fearful bodies in Chapter 1 (McDowell and Sharp 1999: 218). Also in Chapter 1 we saw how supermum Nicola Horlick's busy, affluent lifestyle exploited the relative resource poverty of other women whereby chains of feminised domestic and caring work (both paid and unpaid) function through uneven development and the intersection of race, class, gender and resource endowments (see also Case 4.2). These commonplace examples highlight significant variation in the power relations that operate among women, between households, alongside those which have become socially, culturally and economically entrenched between men and women.

Case 4.4 illustrates a mundane household context in the USA in which power in decision making can be observed to operate in 'more than material', non-functionalist ways, in stark contrast to neo-classical assumptions that 'money

Case 4.4: Observing gendered power in household practices in the USA

It is widely recognised that women's (and men's) everyday experiences of home–work–family spatial arrangements vary by age, race, ethnicity, class, sexual orientation and situation within particular household structures (Katz and Monk 1993: 12). Less widely recognised are the myriad ways that gendered power may be expressed and practised in ostensibly similar, relatively advantaged, heterosexual, Western family households. Evidence from a study of 100 working-family households highlights a spectrum of gendered power as manifest through repertoires of moral and emotional as well as cultural and economic practices and strategies of coping (Jarvis 2005a).

Consider, for example, the case of Bob and Gina Summer, who live in the North East Seattle neighbourhood of Kenmore. Bob is a merchant mariner, while Gina describes herself as 'a single mother with a part-time husband'. Bob sails four weeks together then returns home for two weeks before returning to his ship. When he is away contact is limited to faxed messages and 40 minutes of talk-time by cellular phone per week. Tightly regulated opportunities for conversation mean that Gina has a pivotal role in household decision making. At the same time, everyday practices function to ensure that despite his absence Bob retains the power to veto decisions which undermine his masculine identity. Gina observes:

> I get all of the mail and I have a big outward basket and the mail that comes for him during the time (while he's at sea), if it's important I open it, if it can wait I put it in the basket and then go through that and rate it and some mail I send to him. We always, for twelve years now, each time he leaves I've sent him a package, you know; hand prints, pictures, a pound of coffee, so I send him unimportant things, you know, that don't have to be paid or what have you, and then I take care of all the bills and then I leave the stubs of the bills that have been paid for him to file when he comes home.

Just this brief glimpse of routine prioritising in one household reveals a great deal about the 'doing' of gender and power. In this case gendered power is not carried, possessed, or worn, but rather is expressed through discursive practice – by doing rather than being. The key observation here is that it is the terms and conditions of Bob's absences as well as Gina's routine coping strategies, not simply the income generation of one spouse as 'breadwinner', which regulate the multiple directions and flows of power in this context.

Source: Jarvis, unpublished case material from 1998–2001 ESRC funded research (grant number R000271085).

is power'. It is instructive to compare this with a parallel situation in India in Case 4.5. While in the latter case the women interviewed have gained income and status since taking up paid employment outside the home, this does not equate to increased 'voice' or empowerment in the context of still powerful cultural norms circumscribing 'deferential' femininity. These social and cultural inequities in power relations have been much debated within development studies, notably within the GAD perspective, with respect to conceptualising and facilitating empowerment – processes which we turn to consider in the next section.

Empowerment: more than rhetoric?

In less developed economies, the concept of empowerment lies at the heart of efforts to reduce inequities of gendered power. While initially conceived of as a political process whereby those without power go through both individual and collective processes of consciousness raising to gain power (in a positive sense), the term empowerment has come to be so loosely applied today that there is an argument that it has been reduced to no more than rhetoric.

In its political manifestation, empowerment is conceived of as a process to be facilitated, through which women and other marginalised groups gain social, economic or political power in society. Empowerment and power are not defined as 'things' to be given to women but as something that must be struggled for through action and internal reflection. Neither is empowerment about women gaining power over others, a zero-sum game where men lose because women win – fears that this is the goal form part of the resistance to empowerment among men and other more powerful groups. Instead, empowerment is conceptualised in terms of four different types of power and the social changes which support women's abilities to gain the first three, in order to challenge and resist the last:

- power to: generative power, not dominating but action oriented;
- power with: collective action;
- power within: a sense of one's own agency and self-worth – motivates action to resist;

- power over: controlling or dominating form of power – able to make others do what you want, against their own wishes.

Empowerment has become a development buzzword. Particular examples include the tendency by international development agencies to equate empowerment with access to resources and participation, and not with control over resources and the content, conditions and outcomes of participation (World Bank 2001a, 2001b). As a result, the power dimension of empowerment has tended to disappear. In development practice, efforts to bring about empowerment can be witnessed in the establishment of microcredit programmes, but these typically focus on women and not gender, and this 'backwardness' clearly indicates the lack of insight into the complexities of power in the empowerment approach. Hence programmes may state that women's empowerment is an aim, but then do little to achieve this apart from providing women access to money and the opportunity to participate in groups.

The Grameen Bank's approach to microcredit and its focus on female clients was strongly criticised in a study by Rahman (1999) as reinforcing gender inequities rather than working against them, due to the instrumental use of women as clients in order to facilitate prompt repayment. In the Bangladesh context, women are generally the prime carriers of family honour, with their behaviour representing that of their family. The implications are that families do not want to risk their reputations through female clients' late payments and corresponding penalties – which is why women make ideal clients from the Grameen Bank's perspective. Similar findings are emerging from research in rural Afghanistan on microcredit and rural livelihoods, where some lending programmes with a stated goal of empowering women make few concerted efforts beyond providing small cash loans (Kantor and Andersen 2007). The example in Case 4.5 similarly illustrates how expectations around women's access to income via participation in paid employment are not on their own sufficient to facilitate a process of empowerment among women living in the urban slums of Lucknow, India. Arguably there is continuing need for a gendered urban theory to critically examine both the aims and the outcomes of such policy, which may be intended to address the negative effects of uneven development and social inequality but which may have unintended consequences for the most marginalised.

Summary

This chapter has tackled themes of scale, power and interdependence, reflecting in particular on the ways that these have come to be articulated in patterns and

Case 4.5: Household 'voice' in Lucknow, India

The conditions under which women's access to paid work leads to their empowerment are a focus of much research, largely because the work types and locations in which women work, particularly in the urban informal economy, tend to be quite marginal in terms of earnings and opportunities to challenge existing gender norms, such as around female seclusion and occupational segregation. Research conducted in 2001–2003 in 12 urban slums in Lucknow, India, sought to examine whether or not women's work in the city's informal economy led to increased voice in the household, one empowerment-related outcome. It did so by testing whether working women in each of six employment status categories (salaried, domestic service, casual wage, own account work, subcontract production and unpaid helper) differ from non-working women in their voice in three key household financial decisions – whether to save, how much to save and large household purchases. The expectation is that because of how women are integrated into work and their often low returns, working women will not necessarily achieve more voice in financial decisions because the characteristics of their work are not sufficient to break down gender norms regarding who should be involved in such decisions.

The study's quantitative results showed that for the most part working does not increase the likelihood that women are involved in household financial decisions. Only salaried workers show some difference relative to non-working women in their likelihood of being involved in both of the savings-related decisions. For large purchase decisions, women's access to work has no effect across all employment status types. Thus, the negative terms under which women access work and their relatively low returns from work are insufficient to effect positive change on women's voice in the household. These results are supported by focus group discussions on women's work in Lucknow, with these statements being representative of others:

> It is not so that women who are earning are respected more. I am earning now and am not respected more.
>
> (Female FGD participant, Haddi Khera)

> It cannot be that after we are working outside then we should stop respecting the elders, mother and mothers-in-law. It is not this way that if we are working and earning we should start dominating. This [respecting elders] is the custom of our society and we cannot override that.
>
> (Female FGD participant, Haddi Khera)

Source: Paula Kantor's unpublished field data (2003).

processes of urbanisation through the profound effects of globalisation and deep economic restructuring. In this respect the processes explored here should be recognised not as abstract ideas but as the dynamic systems binding the urban transformations outlined in Chapters 2 and 3: outward migration to suburbs and the emergence of a variety of urban enclaves and hybrid development types; alongside greater diversity in partnering and parenting and the widespread implications of increased female participation in the labour force. Pieced together, these map the key processes and changing patterns of gendered urban development, highlighting interdependencies and intersections, as a prelude to scrutinising in more detail the fine-grained realities of discrete sites and activities of change underway within this broad picture.

First, we provided an outline of globalisation and the privileging of world cities, and then subjected these forms of urban analysis to feminist critique. This led us to draw into our discussion of the intersections of global and local processes the key concept of identity and the scale of the body. We argued that in order to develop an explicitly gendered urban theory the study of cities had to consider alternative scales and tools of analysis to the globalising or world city discourse. Working as always toward an everyday life perspective, we first suggested a household perspective and then combined this with insights that can be gained from theorising the embodiment of cities and gender. We critically examined the notion of 'locale' with respect to the negotiation and reproduction of everyday embodied performances of masculinity and femininity. At the same time, case study evidence of the sometimes violent reactions to women's empowerment that have erupted in the face of competing global–local identity politics served as a reminder of enduring androcentrism.

Making connections between the body and the city, alongside fluid and dynamic gender identities, performances and representations, enabled us to mobilise the tools of urban ethnography with respect to the many manifestations of gendered power in practice. First, we applied recent feminist theory to critique the way power is conventionally reduced to economic or functional determinants. We went on to assert that power is more than material; it is performed and expressed through multiple economies and resources, not least powerful emotions such as guilt, fear, love and duty.

Finally, the learning activity below cultivates the skills and theories accumulated over the first part of this book. The aim is to become an urban ethnographer who observes and experiences the city as a gendered subject, with acute awareness of the continually negotiated co-constitution of cities and gender.

Suggested learning activity

The aim of this activity is to explore a street or neighbourhood that is familiar to you (where you live, work or study, for instance) within an established town or city. In doing so you are to engage in 'new ways of seeing' through gender awareness and then to communicate your observations and findings clearly to others. You are to write a set of instructions and provide a lively commentary (ideally with images and possibly historical cameo characters or vignettes) as a means to guide a visitor along a route of your devising. For the content of your journey, taken as an urban ethnographer, you can scrutinise the local environment and lived experience in the present day, from a particular social or demographic perspective, or as the accumulated legacy of historical events and personages (agents). Think of this as an 'alternative travel guide'. Your travel guide should clearly demonstrate the influence of an intertwined 'cities and gender' theoretical perspective.

Note

1 The name of the Hoppings derives from the hopping or dancing which occurred at the original temperance fair held on the Town Moor (common land) in 1882. The Hoppings is a major event in the North East; showmen families travel from all over the country to attend. It is one of the largest non-permanent fairs in the world because, unlike many similar fairs held in streets, this fair is located on many acres of common land close to the city centre.

Further reading

Bales, K. (1999) *Disposable People: New Slavery in the Global Economy*. Berkeley and London: University of California Press.

Fincher, R. and Jacobs, J. M. (eds) (1998) *Cities of Difference*. New York and London: Guilford Press.

Howitt, R. (1998) 'Scale as relation: musical metaphors of geographical scale'. *Area*, 30: 49–58.

Paxton, P. and Hughes, M. (2007) *Women, Politics and Power: A Global Perspective*. London: Pine Forge Press.

Staeheli, L. A., Kofman, E. and Peake, L. J. (eds) (2004) *Mapping Women, Making Politics: Feminist Perspectives on Political Geography*. London and New York: Routledge.

Taylor, P. (1995) 'Beyond containers: inter-nationality, inter-stateness, inter-territoriality'. *Progress in Human Geography*, 19: 1–15.

Suggested learning activity

Part II

Gender and the Built Environment

5 Infrastructures of daily life

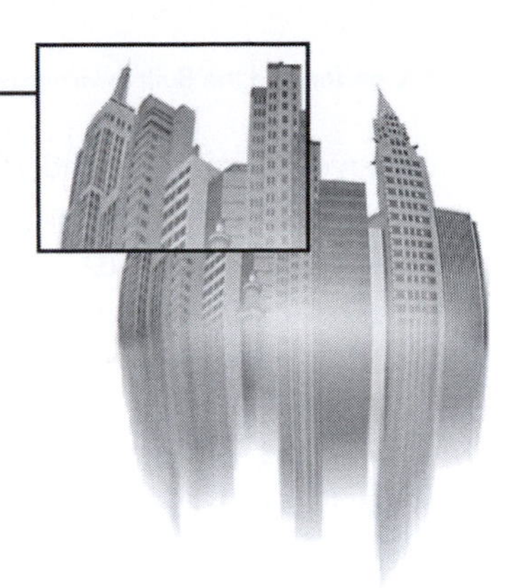

Learning objectives

- to think about the concept of infrastructure as being more than material
- to critically examine the way cities are shaped by gendered assumptions in the design and management of the built environment
- to identify multiple intersecting urban infrastructures and appreciate their gendered distribution and cultural role

Introduction

We begin this chapter (and Part II of this book) by exploring the origins of androcentrism and ethnocentrism in the built environment. We build on the bridging concepts and theory introduced in Part I, adding to this an understanding of the cultural constructions underpinning men's and women's engagement with the built environment. Alongside questions of androcentrism 'by design' we also explore the legacy of neo-liberalism with respect to a politically motivated language of market competition, individual 'choice' and fiscal prudence. This discussion is framed from the outset by an understanding of multiple, intersecting infrastructures of daily life alongside a feminist critique of the narrow definition and prioritisation of 'care-less' competitiveness (Jarvis 2007a). In Chapter 1 we introduced the physical manifestation of highly unequal access to networked infrastructures, pointing to the image Graham and Marvin (2001: 7) present of electric power lines that rip through poor towns and villages of the global south without serving any of the people who live in their shadow. In this chapter we move beyond this understanding of 'engineered' infrastructures to acknowledge a variety of hugely significant but neglected, less visible, highly gendered infrastructures of constraint, including, for example, cultural expectations, guilt, love and obligation. We consider the way persistent inequalities are reproduced by the uneven distribution and gendered cultural associations of multiple infrastructures underpinning everyday routines and practices, historically and geographically. We point to feminist research and methods which expose the gendered cultural barriers influencing the acceptability, respectability and use of urban amenities and resources (such as transport, childcare, sports and recreation). This discussion indicates that proximity and availability alone are not sufficient to determine whether individuals have equal access to publicly funded amenities. We illustrate this point through a number of case studies on domestic architecture, cycling and mental maps of fear. We conclude the chapter with accounts of a growing 'care deficit', suggesting that measuring progress in terms of GDP (as discussed in Chapter 1) systematically exploits and diminishes those unpaid activities and caring values most commonly ascribed to women, or to men and marginalised groups who adopt or find themselves cast in 'feminine' caring roles.

Gendered infrastructure networks

Travel to any city in India and you will find two common images: women lining up with pots of various shapes and sizes waiting for water; and men and children defecating in the open (women have to do this under cover of darkness). The basic services of clean water and sanitation have still to reach millions of people in India even as it boasts of an accelerating rate of economic growth.

The crisis is well illustrated by a visit to one of the many slums that dominate the landscape in India's commercial capital, Mumbai. Almost half of the city's 14.5 million people live in slums or dilapidated buildings. They are located on open land, along railway tracks, on pavements, next to the airport, under bridges and along the city's coastline. Although there are variations, generally this half of the population gets little water and has even fewer lavatories. The water, when available, is often unclean. And the lavatories that exist are usually filthy, broken down and generally unusable: it is safer to defecate in the open than to use some of them. So few are functional that open defecation is anyway the only alternative for millions of people.

Mangal, who used to live along the railway tracks, described her daily experience to two researchers working with the urban poor in these words:

> For lavatories, we had to use the railway tracks. There were public lavatories, but they were some distance away – about half an hour walk. They used to be so dirty that we did not feel like using them. And there were such long queues! Instead of using those filthy lavatories, we used to go on the tracks after ten at night or early in the morning at four or five o'clock.
>
> (in Sharma 2007)[1]

The basics of survival such as water, toilets and housing are taken for granted by many in the global north. Yet, for many poor households living in the mega, medium-sized and small cities of the developing world a large part of the daily routine is committed to basic survival. Women feel these deprivations directly as well as indirectly, as those responsible for meeting household needs (see Figure 5.1). Degrees of risk and uncertainty which characterise the lives of poor urban residents are highly gendered: men, women and children often experience these risks and hazards in different ways with respect to personal insecurity, ill health and exploitative working conditions. In such conditions, planning for the future is often a luxury and getting through each day is the paramount struggle. Many poor women living in cities awaken only to ask: 'How safe will my daughters and I be on our morning "toilet" trip – will men be watching, will we be harassed?' 'My child is sick – the city tanker water was contaminated again. If my husband doesn't find work in the chowk today, how will I feed the family, let alone buy medicine?' 'What can I do to feed and care for my family today?'

Most of us in the West are acutely aware of this physical manifestation of highly unequal access to networked 'grid' infrastructures of water, sewerage, gas and electricity, even though we are personally detached from the experience by distance. At the same time, there is less widespread recognition that the nature and extent of uneven infrastructure networks go beyond pipes and cables to

Figure 5.1 ***The daily work of fetching water in Kabul***
Source: Paula Kantor

encompass institutional, moral and emotional conduits. By extending the concept of 'infrastructure' we do not wish to diminish the importance of the basics of survival such as access to water and power. Indeed, we qualitatively extend not only the definition of what constitutes the infrastructure of daily life but also the way questions of 'access' are understood, recognising that alongside geographic proximity, distribution and affordability it is also necessary to appreciate cultures of meaning, social mores and network utilisation.

Urban planners and geographers have long been concerned with the distribution of fixed assets such as homes, schools, hospitals and shops, as well as bus stops and railway stations. These are fixed in the sense of their location as well as their concrete quality and by virtue of the fact that they are generally costly to build and replace. The costs involved in building roads and laying track for trains are such that these projects are typically paid for publicly, out of the collection of taxes. Although these infrastructures are taxed as a 'public good', the distribution of the amenities they deliver remains highly uneven. This is because powerful coalitions operate at multiple scales (constructing homes, managing cities, and controlling regional and multinational markets) to extend pre-existing influence. We explored this process in Chapter 4 with respect to the idea that city–regional networks conform to and in turn reproduce 'geometries of power' (Massey 1993; Graham and Marvin 2001: 11). Here we extend this understanding of scale and power to consider the influence of city leaders and the processes driving competition and growth.

A parallel development in the debate on urban amenities points to renewed criticism of 'the masculine metaphor of cutthroat competition' (Logan and Swanstrom 1990: 21; Jarvis 2007a). This is associated with feminist scholarship highlighting a holistic (or pluralist) approach to urban flows and networks – recognising formal and informal (cash in hand) work, remittance economies, domestic food production and self-provisioning, reciprocity and the 'economy of regard', of unpaid care-giving and emotion-work (Gibson-Graham 1996; Williams and Windebank 2000; Pavlovskaya 2004; Bakker and Silvey 2008). Essential to appreciating this complexity is the concept of 'multiple economies', which we introduced in Chapter 1. Although we are deeply committed to this project ourselves, we would argue that this call for a 'more feminine image of nurturing the strength of local context' (Logan and Swanstrom 1990: 21) is currently far from actually realising a more socially just, gender-sensitive, women-and-child-friendly urban model for the foreseeable future. However, recent debate does offer some hope among campaigning groups, some of which we introduce here, and again in Chapters 7 and 8. Taking a particularly optimistic stance, Ellin (2006) suggests, for instance, that a 'revolution' is taking place in architecture and urban planning whereby:

> we are witnessing a gradual reorientation toward valuing slowness, simplicity, sincerity, spirituality, and sustainability in an attempt to restore connections that have been severed over the last century between body and soul, people and nature, and among people.
>
> (Ellin 2006: 1)

On the whole we are rather less sanguine about the immediate prospect for genuinely progressive architecture and planning, largely because the culture of these professions remains dominated by androcentric and ethnocentric assumptions, as we demonstrate on pp. 135–6. Nevertheless, faced with the major challenges of climate change, ageing populations, rising levels of household debt, surveys which expose the paradox of unhappiness amidst affluence and excess, obesity and alcoholism, together with a widening gap between rich and poor (within as well as between global regions), debates on the future of cities are potentially more receptive than they once were to radical and progressive interventions. We suggest that feminist understanding of diverse economies and multiple intersecting infrastructures (material, institutional, moral, emotional) provide a constructive route by which to bring these key issues into the foreground, where the private costs of these are typically borne through highly gendered divisions of labour and sacrifice. We make a similar argument for bringing 'feeling' back into classroom teaching in Chapter 10.

A useful tool to begin this integration is provided by the 'infrastructure of everyday life' framework: this encompasses all that it takes in a practical sense for individuals and households to 'go on' from one day to the next (see Box 5.1). Within this framework, structural constraints determine what is actually available (by the distribution and affordability of amenities) as a possible course of action in a given situation. Capabilities include the temporal and spatial 'grain' of provision, including when and where the buses run, when the shops are open, how safe the streets appear, parking restrictions, traffic congestion and so on. In this respect it is useful to think of Bourdieu's 'logic of the situation' as an expression of what is possible given the closely bound nature of spatial arrangement and temporal ordering (Jarvis 2005a: 116).

It is important to remember that capabilities are not only situational but also embodied – bound by an individual's physical, cultural and moral limitations of mind and body. Institutional infrastructures encompass all manner of regulation from those functioning within the household to that of the state and the extent to which it regulates behaviour or subsidises private markets. Moral infrastructures reflect local or regional norms of behaviour such as 'good enough' parenting (Winnicott 1960) and cultures of work and consumption (with regard to the corroding effects of such normative cultures, see Reeves 2001; Bunting 2004).

Chapter 2 introduced the foundational literature on everyday life, highlighting within this the functions of time, space and an ethic of care. In this chapter we take forward time-geographic frameworks of institutional and moral infrastructures in particular.

The distribution and use of these infrastructural networks arguably vary from person to person and in direct relation to the activity concerned. For instance, the moral and practical imperatives which confront a government official journeying from home to work each day are very different from those confronting a child-minder; even more so if resource endowments and capabilities are compared. The well-paid executive who can purchase a plane ticket remotely by hand-held computer has considerably greater geographic reach, effectively 'shrinking' distance, than the single parent living on benefits whose movements are confined to short distances travelled on foot. The opportunities of the latter may then be further diminished by the strategic withdrawal of local bus, banking and post-office services where these market-led services are no longer deemed viable in low-income neighbourhoods. Moreover, if the executive is single, male or unencumbered by the immediate needs of dependent kin then his or her movements do not require close coordination with others.

Attending to what is possible in the spacing and timing of daily life highlights the contingency of these routines and strategies of coping. It exposes the potential for sometimes fragile daily routines to be disrupted by external events, constrained by the need to coordinate movements with others and made uncertain by gaps or delays in infrastructure networks. Such 'gaps' may include unreliable transport, limited childcare opening hours, or a sick child unable to attend a school or nursery, traffic congestion and general disruption to the timing and spacing of carefully planned and choreographed daily events. This bounded nature also reflects points of friction in the built environment caused by micro-architectural barriers (steep steps, narrow doors, poor lighting), forms of 'spatial apartheid' that include the lived experiences of fear, identification and belonging. There are parallels here of course with concerns for inclusive design to cater for the needs of people with a whole range of potential mobility impairment/disabilities as well as an ageing population (see, for instance, Imrie and Hall 2001).

Box 5.1: Infrastructures of everyday life

The inspiration for an understanding of integrated infrastructures resides in the Nordic feminist housing and building project 'New Everyday Life' (Forskargruppen) (Gullestad 1991), where the shared vision was of a more harmonious, creative and just society in which children's and women's needs and the social reproduction of all people and natures are valued as central motives for action. Recent examples are also found in Urban Ecology Australia, whose Child Friendly Cities claim: 'That which makes cities good places for children to live in, makes them good for everyone' (Urban Ecology 2008).

Other advocates of progressive planning include EuroFEM, a loose affiliation of European women scholars committed to promoting and implementing gender mainstreaming in planning policy, job creation and local initiatives, models of participatory engagement, and the reorganising of everyday life around housing, drawing on historic ideals of collective living (Booth and Gilroy 1999; Darke *et al.* 1994; see also Matrix 1984). The EuroFEM Toolkit is a collection of participatory and culturally sensitive methods and stories taken from women's projects across Europe, from southern as well as northern contexts, which can be used to build capacity for collaborative planning (Horrelli *et al.* 1998, 2000). These groups share in common a critical distancing from the criteria of profitability, competitiveness and engineering efficiency traditionally applied to urban development and planning.

Androcentrism by design

Contemporary cities and urban lived experience are to a large extent man-made, shaped by androcentric planning and design cultures and by centuries of gender division and conflict. The early industrial European and North American cities were constructed to a large extent through clear architectural distinctions between residential areas and sites of industry, commerce and government. Residential areas were spatially separated, designed for (not by) women as the domain in which 'respectable women' were expected to display feminine skills of home-making – subject to the authority of the husband. We saw this in relation to suburbanisation in Chapter 2, where we also stressed the growing variety of settlement types associated with the privatisation of space as well as larger, smaller and hybrid settlement types. Mass, home-based consumption was constructed to underpin this starkly gendered division of urban work and home space (Graham and Marvin 2001: 68) as the modern ideal of cleanliness, comfort, convenience and economy of effort (Shove 2003). Appliance producers directly targeted female 'housewives' in their advertising. In turn this drew on social and cultural norms of competitive cleanliness to drive a treadmill of consumer expectation and norms of ideal womanhood (Rose 1995: 200, cited in Graham and Marvin 2001: 70).

Here again we can observe the intersections of gender, class and resource endowment. Bev Skeggs (1997) claims, for instance, that attitudes toward housework and cleanliness are highly gendered and classed, so that housework is less about hygiene than about creating an impression of respectability. The promotion of domestic appliances and the Western ideal of the decorative home both reflect and reproduce these notions of respectability, especially among working-class women for whom alternative expressions of status are limited by a lack of material resources and cultural capital (Skeggs 1997).

This raises the question of to what extent androcentrism has been produced 'by design', through the assumptions and bias of male architects, planners, designers and advertising executives. On the one hand, Elizabeth Wilson argues that town planning originally organised itself as a profession to 'exclude women and children, along with other disruptive elements – the working class, the poor, and minorities' (1991: 6). Architects and planners were complicit in this marginalisation both as 'products and carriers of the flux of ideas and power relationships inherent to particular stages of urbanization' (Knox and Pinch 2006: 138). Women were 'kept in their place' through comprehensive plans and zoning ordinances that were sometimes hostile, often merely insensitive to women's needs (Knox and Pinch 2006: 140; Ritzdorf 1989; Bondi 1998).

Nigel Whitely claims that goods and services have been badly designed in relation to female-users because 'the vast majority of products are designed either by men (as producers) for women (consumers), or by men for men' (1993: 141). We can see the same systematic bias in the built environment with respect to other marginalised groups. Buildings for disabled users are generally designed by able-bodied architects; homes for an assortment of family types are designed from a 'pattern book' of data collected and managed by people whose executive lifestyles are typically unlike those of their home-oriented customers. Understood this way, marginalised groups do not function as a statistical concept but instead are those 'who are treated unequally because of their physical or cultural characteristics' (Oakley 2002: 11).

At the same time, Cynthia Cockburn questions whether this bias is conscious or cultural. She notes in her work on technological processes in the print industry that machines and systems which exclude women and marginalised groups 'need not be conspiracy' but instead the 'outcome of a pre-existing pattern of power…Women vary in bodily strength and size; they also vary in orientation, some having learned more confidence and more capability than others. Many processes could be carried out with machines designed to suit smaller or less muscular operators or reorganised so as to come within the reach of the "average woman"' (Cockburn 1985, cited in Whiteley 1993: 142). This suggests that prevailing patriarchal power structures militate against adjustments being made to enfranchise women users.

History reminds us of the paradox that Enlightenment ideas of early gender equality (such as movements in Euro-American cities in the 1930s to reform men's and women's dress to facilitate gardening and bicycle riding) often go hand in hand with potentially backward-looking nostalgia for pre-modern village life. Indeed, similar tensions between elitism and democracy, nostalgia and innovation, endure today in popular expressions of the new urbanism. Yet, rather than suggest a monolithic agenda, the history of architecture and design suggests that difference, dissent and competing schools of thought have always been present in the profession. Knox and Pinch, for instance, describe the origins of the modern British town planning movement as an essentially reactionary movement which 'grew from a coalition of sanitary reformers, garden city idealists and would-be conservers of the countryside and architectural heritage' (2006: 139). At the same time, a drive to maintain an established (patriarchal) social and moral order appears to have stifled periodic efforts toward progressive alternative visions of city and society. We saw this process at work in Chapter 4 with respect to the influence of a 'global intelligence corps'.

Marion Roberts (1991: 6) maintains that cities are indeed 'man-made' because of a historically entrenched mix of state policy and cultural expectations regarding family life, gender relations, public health and motherhood. Moreover, it can be argued that even in the twenty-first century Euro-American cities remain quite literally 'man-made' because architects and related professions are dominated by middle-class white men. Membership data for the American Institute of Architects (AIA) shows that the profile of its members remains anything but representative of a general, college-educated, working-age population. In 2006, 71 per cent of AIA members identified themselves as white Caucasian (1.4 per cent African-American, 4.3 per cent Asian, 3.1 per cent Hispanic and 19 per cent undeclared). At this time men constituted 83 per cent of all members (compared with 90 per cent in 2000) (AIA 2007).

The picture is very similar in the UK, where research by the Centre for Education in the Built Environment highlights a problem not only with recruitment but also with retention. While 38 per cent of students accepted onto architecture courses are female, women make up only 15 per cent of architects practising the profession. Similarly, 12 per cent of UK architecture students come from minority ethnic backgrounds but this group constitutes only 2 per cent of registered architects (CEBE 2006: 9). The result is that white (male) students from 'traditional (middle-class) student backgrounds' are four times more likely to achieve first class honours than their minority ethnic, female and non-traditional background student counterparts (CEBE 2006: 9).

In 2008 Ruth Reed became the first female to be elected president of the Royal Institute of British Architects (RIBA), 174 years after the institute was granted its Royal Charter. While it is too soon to suggest a shift in culture, this appointment was made in the context of growing unease that women, ethnic minorities and those from low-income backgrounds are underrepresented and less likely to remain in the profession.

In large part, the problems of retaining underrepresented students can be attributed to an uncompromising studio culture which promotes personal sacrifice and commitment to the project through late nights, long working hours and the promotion of conflict through individual competitiveness. The CEBE *Guide to Supporting Student Diversity in UK Schools of Architecture* highlights the process by which this culture is reproduced and intensified in professional practice:

> those who succeed [at university] carry unsustainable studio practices into their professional lives, and so contribute to working practices that further erode underrepresented groups in the profession. As a result, a profession that aspires to be more inclusive is deprived of members who might make a valuable contribution to the developing practices of architecture.
>
> (CEBE 2006: 8)

This resonates with what we go on to discuss in Chapter 7 as the gendered implications of 'workaholic' cultures and pressures to be always present in a place of employment. We point to the effects of 'fugue-like' creative industrial innovation, and particular workplaces, such as 'sand-pit' buzz-groups, which serve to perpetuate 'macho' cultures of work (see, for instance, Jarvis and Pratt 2006; on design training, see Clegg and Mayfield 1999).

These trends of androcentrism and ethnocentrism have not gone without comment. The architect of the Garden City movement, Ebenezer Howard, was himself a radical social reformer with some truly revolutionary ideas that were never fully realised. Howard was fascinated by the socialist feminist ideals of inclusive urban design popularised by Charlotte Perkins Gilman. Inspired in part by Gilman's ideas, Howard proposed a form of 'cooperative quadrangle' to release women from household drudgery in the private home, arranging garden apartments around a collective kitchen, dining room and open space. Several of these remarkably innovative projects were built between 1911 and 1930, designed specifically for single female professionals, elderly and two-earner couples, although they never became standard provision in the garden cities (Hayden 1984: 90).

This emphasis on the relationship of gender roles to domestic architecture reflects a long-established feminist critique of male bias in the design of housing, in terms of residential location and domestic technologies. This is illustrated in Case 5.1 in the context of Victorian feminism, in which 'the house, like the vote, was a highly charged political issue' (Addams 1996: 129). Over the years, gender assumptions in modern housing have been scrutinised in-depth by Dolores Hayden (1981, 1984), the London feminist design collective Matrix (1984), Marion Roberts (1991) and others. At the same time, this tendency of feminist critique of the built environment in the past to focus quite narrowly on housing provision and the home indicates a fresh need to incorporate gender awareness at all levels of urban planning, design and management, to build cities fit for everyday purpose.

Technological transformations

Arguably, both capital and domestic modes of production are unrecognisable today from their pre-industrial (and industrial) forms. Turning once again to the chronology of urban, social and technological developments included in the Appendix A (pp. 298–300), it is evident that over the period 1900–1960 households in the global north gained access to a whole range of modern infrastructure networks (water, sanitation, general transport, gas, electricity, car, telephone, televisions) (Graham and Marvin 2001: 68). Not only did this transform the activities of women as home-makers, by liberating them from labour-intensive drudgery, but it increased household income requirements to keep up with a 'treadmill' of consumer expectations: larger homes; more frequent and exotic package holidays; gadgets and technology aimed at making it possible to do more in less time (hence intensifying rather than simplifying daily life through convenience). Again this illustrates the interdependence and intersection of global trade, neo-liberalisation and deep economic restructuring with profound changes in gender relations and local moral cultures, not least with respect to rising numbers of dual-income families and a parallel commoditisation of care.

At the same time, it is crucial to remember that a majority of the world population living in the global south do not take convenience, comfort and competitiveness for granted in the same way that the Euro-American consumer has, for instance, come to expect and indeed assume the right to certain levels of central heating and air conditioning, for example (Shove 2003; Guy and Shove 2000). Similarly, rapid technological advancement in recent decades in fields of transport (increased speed, reduced cost, larger and smaller sizes), information and communications (mobile phones, portable computers, internet, broadband, Bluetooth) and human reproduction (contraceptive pill, IVF, donor conception)

Case 5.1: Perspectives on domestic architecture: from Victorian feminism to big-box consumption

Annmarie Addams highlights four aspects of androcentrism in Victorian housing. Her historical analysis is unusual in the way that she recognises domestic architecture to be a 'more than material' political concern. Each of these four historical moments conveys both the persistence of structures of patriarchy alongside fluid and changing social conventions and institutions. This dynamic helps to explain how the transition was made from Victorian feminism to the concern 'big-box' retailers now have to appease mothers and children – while the basic design of contemporary 'brick-box' housing continues to fail families in all manner of respects.

1. *The separation of mothers and their children in the typical middle-class home:* descriptions of London houses in the women's press were filled with despair: these houses contained on an average from seven to ten flights of stairs. Mothers complained that they could neither see nor hear their children in other rooms of the house (in contrast with the horizontal layout of continental apartments such as in Paris at the time).
2. *A 'feminine' understanding of the house boosted by the rise of interior decoration as an appropriate occupation for middle-class women*: the introduction of 'feminine point of view' perspective drawings marked a contrast to the 'scientific' (anatomical) sectional drawings. Exterior views of architecture rarely appeared in the women's press. A proliferation of decorators, department stores and exhibitions (such as the International Health Exhibition) at the end of the nineteenth century cultivated a form of 'spatial feminism' by encouraging women to view their houses critically and to rearrange them to suit their needs.
3. *The construction of purpose-built housing for women in London*: the journal *Work and Leisure* sponsored an architectural competition for 'Erection, Arrangement, and Management of a block of Associated Dwellings adapted to the needs of single women' which resulted in several designs for the construction of housing for women. One of the resulting blocks on Chenies Street (opened in 1889) contained distinct sets of rooms, each with its own front door, together with shared kitchen and dining facilities to promote 'cooperative housekeeping'. The construction of these buildings marked an enormous victory in the nineteenth-century women's movement and marked a turning point in the recognition that large cities attracted a new breed of 'unmarried, independent women'.
4. *The acceptance of women as professional architects*: in 1898 Ethel Charles became the first woman member of RIBA. The eventual acceptance of women

architects in England was based on their proven abilities in two key debates of the time: sanitary reform and social reform.

Source: Addams (1996: 129–162).

Comment

Writing in the mid-1980s the pioneering women's design collective Matrix argued that modern housing design revealed a lack of respect for the importance of mothering. They noted that halls were too narrow for a normal pram or pushchair (which then had to be unloaded and folded in a confined space). Today we might contrast this with growing awareness in the design of public spaces that it makes financial sense in the commercial interests of restaurants and retail to cater for the needs of children and their parents. The European retailer Ikea, for instance, provides a well-resourced free crèche service, not out of sympathy or benevolence, but to give parents two hours free from distraction in order that they focus their energies on maximising consumer spending.

may have transformed the frontiers of possible human behaviour, but the combined effects of unequal purchasing power, state regulation and public moral discourses function to restrict availability and acceptability in highly uneven ways. In affluent societies, for instance, there is frequent reference in the media to a 'postcode lottery' in fertility treatment or elected surgery for gender reassignment and the like. As Graham and Marvin (2001) stress, the near-ubiquitous diffusion of a particular technology does not ensure widespread access. Scientifically, it is possible through the technologies available to virtually eliminate infant mortality and extend through drug therapy the life expectancy of anyone suffering from HIV/AIDS across the globe, but this knowledge has very little impact on the reality of daily life for people in the global south who lack entitlement (resource purchasing power) to this rationed quality of life.

A practical example of the uneven utilisation of a near-ubiquitous technology is provided in the case of the simple bicycle, which is widely available and certainly more affordable and environmentally preferable to private motorised transport. Case 5.2 introduces the (fictional) case of Yasmin's daily travel dilemma to illustrate how the cultural construction of approved behaviour for men and women in different contexts shapes the use that is made of everyday transport infrastructures.

Case 5.2: Gender, ethnicity and bicycle travel: Yasmin's story

Yasmin is a young Bengali Muslim woman attending university in East London. She lives on one side of a large park, across which spans a beautifully landscaped cycle path leading directly to the university campus. Yasmin has chosen to travel each day on an overcrowded and infrequent bus service. Theoretically, she could replace what is a 30 minute crowded, stop-and-start bus ride with a 10 minute exhilarating bike ride. But for Yasmin the use of this cycle path is not simply a matter of proximity nor even of safety and fitness. This is because her mode of dress (full-length loose cloak and hijab scarf) and behaviour (not drawing attention to the shape of her body through vigorous movement) are governed by powerful cultural codes and standards of decorum. Consider, for instance, the following quotation from an Islamic website concerning the proper dress of a Muslim woman:

> The outer cloak is of a dull colour and extremely loose so that the Muslim lady is fully covered from head to toe and does not draw any attention to her femininity when she is outdoors. However, she can dress as she pleases for her husband indoors.

This website goes on to state:

> The *kuffaar* [Western non-Muslims] say that Islam has enslaved its women by enforcing them to stay in their homes and cover themselves when they come out. How wrong they are! What have the *kuffaar* given their women; prizes for displaying their bodies and exposing them to the risk of rape and molestation; insecurity of losing their husbands to beautiful mistresses; constant battle with nature to look young and pretty enough. On the contrary, Islam elevates its women to the place of respect and saves a woman from disgrace and humiliation, giving her a chance to be treated like an honourable human being and not a mere sex object.
>
> (http://www.inter-islam.org/actions/Hijbdu.html#How)

Overcoming restrictions on women's and men's use of available infrastructure (such as cycle paths, parks and public spaces) requires an understanding of diverse cultural sensibilities. Again, our student Yasmin might be encouraged through the social milieu of the university (by which conduits of information serve a discrete 'social infrastructure' of daily life which we discuss further on pp. 149–50) to visit the nearby Jagonari centre in Whitechapel, East London, to find out how she might ride a bike in full Muslim dress. Yasmin logs on to the website from the university library to find out more.

East End 'Wheelies' – Jagonari Cycles, London

Jagonari is a women's centre which provides a range of learning opportunities (IT, yoga) including bicycle training for local Bengali women. Launched in March 2005, the bicycle training project responds to the specific needs and interests of the local community, such as by providing female cycle trainers rather than men. Funding was applied for from the local London Cycling Campaign group Tower Hamlet Wheelers. A pool of bright orange Dutch-style bikes were provided by the 'good thing' campaign to offer free cycle training to women in the local community.

The aim of these initiatives is to give women in ethically diverse communities real choices in the form of an easy mode of transport and an enjoyable way to keep fit. The website includes the following testimonials from a woman who attended the 'beginners' bicycle training course.

> I was really inspired by an older friend who's a teacher. She wears full Muslim dress but still manages to cycle everywhere. By learning to ride bikes we're bringing down social stereotypes and encouraging other women to take part.

Extracts from http://www.everyactioncounts.org.uk/en/fe/page.asp?n1=6&n2=84, accessed on 17 May 2007.

Consumer citizenship

Throughout this chapter we stress that the life-choices people have open to them rest to a significant extent with actually available resources; education and healthcare are good examples of this. In many poor countries, for instance, children's education and access to healthcare and hospitals are rationed by ability to pay because the costs of building a school (or clinic or hospital) are frequently recovered by imposing charges for use. This can be seen to co-opt aspects of the household economy into the formalised, profit-/tax-based economy in ways which seek to extract revenues in monopolistic settings with little or no concomitant improvements in quality of life. Bound up with this rationing are illicit bureaucracies built on taking advantage of gendered activities at a time when women in particular are under increasing pressure to increase breadwinning activities as well as to deal with the majority of household reproductive work.

A good example of the complex interactions between the healthcare and education systems and the impacts of cost and affordability is provided in Case 5.3 drawing on evidence from Nicaragua's participation in the UN's programme for the Promotion of Educational Reform in Latin America and the Caribbean (PREAL).

Case 5.3: Nicaraguan bureaucracy – a poverty tax?

In 2004 the Promotion of Educational Reform in Latin America and the Caribbean programme (PREAL) exposed the poor quality of Nicaraguan education. It highlighted problems of insufficient numbers of schools, high rates of dropout and a number of other indicators of decline and increasing inequity across the education system – illiteracy among 10-year-olds was 31 per cent among the non-poor, 60.2 per cent among the poor and 83.9 per cent among the very poor (Nitlápan-Envío 2004). Levels of public investment in education in Nicaragua at that time were on a par with Africa, while Nicaragua had the lowest percentage of children in school of any country in Latin America. One reason for this is suggested by a 1995 article (Faune 1995) in which it was reported that at that time between 10 per cent and 28 per cent of family income for families below the poverty line was being earned by children, and for poorer Nicaraguan households children were contributing as much as one-fifth of household income, a situation that the succeeding 10 years of economic hardship changed little, except for the worse.

Behind these indicators lay the educational 'reforms' brought in as a result of a series of structural adjustment programmes begun by the first accord signed with the IMF in 1990. The public sector experienced massive cuts in all areas, including education and health, and over the next 10–15 years a series of cost-cutting measures designed to devolve educational costs and administrative responsibility onto families ensured that inequity increased and that poor families, already burdened by the need to use their children as income-earners, were in addition unable to meet the increasing costs of sending their children to school. Initiatives such as the proposal by the international financial institutions for 'school autonomy' were in fact little more than a means of devolving costs and responsibility for running schools onto already overburdened parents (and substantially onto women). The democracy envisaged by the 'autonomy' programme came, for instance, with 'voluntary' fees of about $1 per pupil per month, little enough in itself except when taken in tandem with increased costs for mandatory school uniforms, payments for books and other necessities. These costs were the reason 40 per cent of parents cited for their children dropping out of school (Nitlápan-Envío 2004).

These patterns of system- and bureaucracy-induced inequity are both self-replicating and have extensive impacts outside the immediacy of the educational system. For instance, it was reported that for women in Nicaragua who complete secondary education the average number of children was two, whereas for women with little or no education the average was six (Equipo Nizkor 1999). Therefore, in a country in which the healthcare system is suffering from a similar lack of investment and the same devolution of costs as the educational system, the impacts of maternal mortality, post-natal mortality and all of the other costs associated with frequent pregnancy are borne predominantly by those women excluded from the educational system, who are increasingly likely to take their own children out of school as a result of worsening poverty, thereby perpetuating and exacerbating the problem on a national scale.

Other aspects of making education and health systems more 'economically efficient' frequently include a substantial increase in corruption in such sectors. Across South Asia, Transparency International (2002) reports a bewildering range of 'fees', 'commissions' and outright bribes necessary to secure the most basic services in hospitals and clinics. These range from admission payments directly to doctors or persons of influence, hospital staff or political contacts just to get into hospital, to payments for receiving prescribed medicine, gaining access to a bed, getting an X-ray taken, getting a pathology test done, getting a blood test done or for a blood transfusion; being forced to buy medicines from designated pharmacies, being forced to get tests done in designated centres, being forced to pay for proper food, being forced to pay for prescribed treatment. Such corruption falls across all of the poorest but, because of the vulnerability of women before, during and after pregnancy in poorer countries, hospitals and health clinics represent a growth industry in what is maternity taxation – everyone from the gatekeeper who charges the family admission to visit to the doctors who prescribe brand-name drugs instead of cheaper/free generic versions because they have been bribed by a drug company representative to do so is part of a malignant bureaucracy thriving on the reproductive role of women.

Add in the difficulties for women in applying for or receiving land titles, gendered aspects of access to officialdom (government/local government jobs are scarce and privileged and the domain of men far more than of women), the necessity for a woman to have a 'husband's permission' to conduct basic functions such as applying for licences, even travelling in some countries, and globally we are then confronted with a wide range of socially constructed gendered infrastructures which greatly increase the financial/labour hour costs of vital functions and processes for an already marginalised group. These are ex-officio socio-cultural gendered infrastructures that arise from and operate through the gendered structures of the built environment, encouraged by a particular take on modernity and 'modern' systems whose failure to comprehend the gendered socio-economy guarantees the growth of opportunistic and parasitic shadow bureaucracies.

The co-ordination of everyday life

Feminist scholars have widely adopted an everyday life perspective because it 'makes room' for the experiences, meanings and practices of women as well as men (Vaiou and Lykogianni 2006: 733). The concept is used to convey taken for granted 'habitual' connections of time, space, place and codes of conduct as well as a means of recognising a situated embodiment in the social and organisational dimensions of urban life. This is in direct contrast to the way these 'soft' networks have been previously sacrificed to the privileging of engineering and environmental determinants (see also Jarvis *et al.* 2001: 3). Feminist accounts of everyday infrastructures renew the foundational ideas of everyday life associated with the work of Lefebvre (1971, 1991), de Certeau (1984), Hayden (1981) and Gortz (1980). Lefebvre's work on rhythm analysis, for instance, draws attention to the rhythms of 'city time' and how this privileges certain types of knowledge, contrasting the formal professional expertise of city planners with the practical, often improvisational, knowledge of household coping.

This body of theory reminds us that while some aspects of everyday life may be familiar to us, this does not mean that we understand the everyday realities of everyone everywhere. This subtlety is brought to life in the following two case examples. Case 5.4 reports on a survey of how older people gain access to amenities in and around the city of Newcastle upon Tyne. It highlights differences between the way a particular 'user group' (in this case older men and women) 'know' their city compared with how officials (such as transport operators and town planners) 'know' the technical systems of the city. Questions of access and inclusivity highlight the importance of temporal rhythm (time and timing) in the design and management of urban infrastructures: it points to the enduring appeal of time-geographic analysis as a tool for scrutinising the urban lived experience of different user groups. This is suggested in Figure 5.2 by the way that access to parts of the city is restricted to certain hours of the day. Similarly, Figure 5.3 shows how places may get a reputation for being hostile toward particularly marginalised groups – such as by the explicit message 'no gays' spray-painted on street corners through one neighbourhood of Guatemala City. Case 5.5 shows how the language of time-geography can be used to make sense of generic 'dilemmas' associated with the distribution and 'spacing' of jobs, housing, schools, transport and childcare services within a particular area and the options actually available to utilise these within given time and personal capability constraints. This case also shows how restructuring and global interconnectedness (introduced in Chapter 4), reflected here in the decentralisation and gentrification of retail premises, can significantly alter the distribution and availability of urban amenities in ways that increase

Figure 5.2 ***Park closed at night, Chicago***
Source: Helen Jarvis

Figure 5.3 ***Signs of exclusion, 'no gays', Guatamala City***
Source: Jonathan Cloke

Case 5.4: An older-person-friendly city centre

The Elders Council is a group of older people which was set up to give older people the chance to speak up about issues that are important to them. A Working Group (eleven women and four men) carried out a survey of Newcastle city centre between 2002 and 2006. The first stage focused on basic facilities in the city: public transport into the city centre; pedestrian access into and around the city centre; streets and street furniture; open spaces. The second stage focused on buildings in the city centre which are important in the commercial and cultural life of the city: shops and shopping malls; banks, post offices; recreation and leisure; museums, art galleries, libraries, theatres, cinemas and concert halls.

The method used throughout the study was for older people themselves to assess the streets, the buildings and the amenities of the city through direct observation, using a checklist of the points of interest. Wherever possible visits were unannounced but followed up by letter if necessary.

The survey identified problems, for instance, with the length of time people have to cross the road at pedestrian crossings. In addition, they raised awareness of gaps in the provision of public transport, lack of seating in the street and in shops, inadequate provision of public toilets, and limited access to public buildings and cultural life. In the case of inadequate provision of public toilets it was found that, in the absence of any legal requirement to provide public lavatories, local authorities tend to encourage the private sector to open up their provision to non-customers – but older people (especially women) are reluctant to go into a busy pub to ask to use a toilet.

With respect to seats in shops, valued by older people needing to rest their feet and their parcels for a short time, the survey revealed that one major department store had reduced from 39 to 11 the number of seats provided. Recent developments by retailers also indicated that an increase in selling space over circulation space was having a negative impact on older shoppers – for whom walking distances, lifts and escalators, and provision of seating made a real difference to the enjoyment or otherwise of a routine shopping trip.

Source: Elders Council (2006).

the exclusion of already-marginalised user groups, such as mothers chaperoning small children, non-car owners, the mobility impaired and the frail elderly (see Jarvis *et al*. 2001: 52).

Case 5.5: Anne Karpf's watch battery dilemma: introducing time-geography

Life is precarious in Kashmir, the Middle East seems insoluble, and street violence is on the increase, but I'm preoccupied with my watch. It needs a new battery, but there's no place to get one on the daily journeys I make. Nothing in the shops round the corner, nothing near my children's schools (where small repairers have increasingly been edged out by aromatherapy parlours or sofabed showrooms). The nearest place that could replace it would require a special journey – a round trip of nearly an hour-and-a-half if you include waiting for the bus or trying to find somewhere to park – whereas I could get a new watch for £15 within spitting distance.

Ditto for having boots reheeled, road tax renewed, and all those other things that come under the rubric neither of work, leisure or childcare. With the lack of time (most of us have a daily orbit that's almost impossible to stray from) and local facilities more than a mile away (the local post office/library/shop ceases to be local), the reproduction of daily life has become increasingly difficult. They talk of time–space compression, but the internet can't change my watch battery for me. It may be a global village but it doesn't seem to have any watch-repairers in it. Caring for children or parents, indeed the ageing process itself, all have a major impact on the time–space axis. About five years ago the council tried to close down my local library, insisting that it wouldn't take more than 20 minutes to reach the nearest surviving one. Twenty minutes for an able-bodied person who goes to the gym, perhaps, but for a less mobile older person, waiting for an unpredictable bus, carrying a shopping bag, in the rain, you could easily double that.

Source: Karpf (2005)

Comment: on time-geography

The dilemma described above provides a useful, if seemingly trivial, example of time-geography at work. The origins of this powerful theoretical understanding and method of tracing daily routines and movements reside in the pioneering work of Torsten Hägerstrand (1916–2004) and the 1970s 'Lund school'. It remains influential in urban geography today, as a means of mapping and measuring location-accessibility problems, but it has lost favour among feminist scholars who criticise the use of 'space' and 'time' as empty abstractions of distance and duration.

Hägerstrand's efforts to capture the 'choreography of existence' builds upon the way Georg Simmel understood urban dwellers to circulate within restricted webs of social interaction. The terminology of time-geography views the most mundane 'projects' of daily life, such as chaperoning a young child to school on the way to work, as a function of multiple 'pockets of local order', the path or trajectory of which is governed by three simple but fundamental contextual limitations. The first of these, the capability constraint, concerns physical limits to movement, including the inability to be in two places at once. Second, a coupling constraint describes situations which compel people to come together at certain times and locations such as for face-to-face service delivery, family celebrations, medical appointments and the like. Third, authority constraints exist in the form of legal sanctions and regulations such as those restricting entry to places selling alcohol. The most arresting image associated with this choreography is a three-dimensional prism, mapping an individual's path in space (annotated movements across the *x*- and *y*-axes of a morphological map) and relating this to a time budget (the 24-hour day) recorded on the *z*-axis.

The prism renders complex situated behaviour with a simplicity which is appealing but heavily criticised for the tendency to reduce webs of interaction to quantified nodes or intersections. Sweeping structures appear to displace individual agency and erase the significance of cultural practices and local knowledge. Feminist scholars point out that time is not a gender-neutral, quantity-based resource which is equally available to everyone as a measure of the calendar or clock. The passage of time and the pace of life will vary, just as people's experience of space and place is shaped by subjectivity: power, fear and physical capability. (But see Figure 6.2, p. 163.)

Despite this criticism, there are ample examples in contemporary research of the constructive application of time-geography to questions of work–life balance. Moreover, it is now possible to redeploy Hägerstrand's principles to combined GIS-ethnography initiatives, with the aid of sophisticated satellite tracking devices. For example, researchers have exploited advanced tracking technologies to study pedestrian spatial behaviour, whereby it is possible to incorporate a log of perceptual experience (fear, discomfort, intimidation, landscape aesthetic and the like). Future time-geography research calls for a more integrated, materially embedded theory of everyday coordination. Mei-Po Kwan of Ohio State University has used space – time GIS models to show disparities in gender accessibility – even from the same household – that would be rendered invisible in traditional models of residential location based on distance alone (Corbett 2006: see Kwan 1999, 2000; Carlstein *et al.* 1978; Jarvis 2007b).

A growing care deficit?

Feminist research shows that emotion-work, altruism and 'economies of regard' are important components of the infrastructure of everyday life. For example, local cultures of school-year social interaction (parents gathering in the school yard) function as 'lively conduits' of learning and behavioural constraint (Pratt 1998: 28; Jarvis *et al.* 2001: 88–89). Moreover, recent developments in feminist theory (see Chapter 3) have opened a debate concerning the question of whether these networks are being undermined by urban restructuring or whether we are witnessing their growing significance as sites of moral surveillance. On the one hand there is the claim that alongside the competitiveness associated with consumer-citizenship we are witnessing an absolute care deficit (see Folbre 1995). As families shrink in size and move away from extended kin networks, and in a climate in which more adult offspring and parents participate in paid employment, often for longer hours (in dual-career and multi-earner, multi-job households), and indeed as older generations remain longer in employment and take up active retirement (which precludes them from looking after grandchildren), the sense is that there are fewer opportunities for intergenerational collaboration.

An alternative thesis suggests that powerful cultural codes of 'responsible' parenting, fuelled by widespread media coverage in rare cases of child abduction, increase the pressure on parents (notably mothers) to be hyper-vigilant in supervising their children or enlisting commercial providers of structured play and learning (music lessons, creative development, sports and 'hot-housing' academic skills such as languages and maths) to do the same. In a climate of media-induced risk aversion, these circuits of behavioural moderation (or simply peer pressure) serve to propagate fears of 'stranger danger' as well as an unwillingness to allow (or be seen to allow) children to walk, cycle or cross busy roads unsupervised. The result is that women's time availability and action-spaces can be seen to be culturally as well as materially circumscribed (by conscience as well as car availability). In turn, children's geographies and experience of childhood are tempered by these prevailing norms such that the transition Euro-American children make toward independent mobility in cities and their move out of the parental family home as independent adults now occurs many years later than it would have done in previous generations. Again we note the combined effects of social demographic and labour market changes in household structure together with local cultural norms and media representations.

In Britain the moral imperatives of good mothering contribute to very high rates of part-time employment, especially when children are of school age (Duncan and Smith 2001). This cultural confinement to 'school hours' employment explains, in part, the persistent gender gap in wages and pensions entitlements as well as the heavily circumscribed daily movements of mothers committed to chaperoning their children to and from school each day according to the short day and regular breaks of the typical school calendar.

In the United States higher rates of full-time employment among professional women tend to be associated with a proliferation of pre-school, after-school, 'tween' and 'teen' childcare settings, many of them replacing unstructured play with instrumental skills acquisition or risk-averse competitive sports. Jarvis (2005a: 121), for instance, describes how public sector employers in Seattle subscribe to special children's day-care facilities to address the problem of time lost to working parents when a mildly ill or infectious child is excluded from their regular day-care. This perk emphasises the way gender equality and family-friendly initiatives in the USA are predicated on productivity gains rather than an ethic of care or scope for a parent to 'be there' in person for their child when they are unwell. Interestingly, in Japan, where the pressures on working age men (and more recently women) to conform to a corporate culture of long working hours are well known, government initiatives seek to promote economic growth through higher levels of female employment by extending childcare services for overnight, round-the-clock and sick-child cover (Ben-Ari 1997). By contrast, in social democratic regimes more generous leave entitlements encourage fathers as well as mothers to take paid leave for sick dependants. We will return to the issue of gendered state policy and planning in Chapter 8.

Mental maps of fear and fatherhood

Ever since the seminal work on mental maps in the 1960s (Lynch 1960; Gould and White 1992; Orleans 1973), it has been widely understood that people carry around in their head a cognitive image or mental map of the city which is made up of multiple sensory landscapes, of sights, sounds, places and feelings (Ginsberg 2004). All these 'memories' contribute to a subconscious geography of safe and fearful places and sites of interaction. The formal analysis of sketch maps drawn by subjects for the purpose of analysing this subconscious cognition is no longer widely used in the social sciences. Nevertheless, a long tradition of research on the fear of crime takes close account of the development of defensive, prejudicial and altered behaviour, and this can be viewed as a direct legacy of mental map research (Ginsberg 2004). In this sense cognition also represents an infrastructure of everyday life.

A consistent finding is that images of the city differ markedly according to the social class, resource entitlement (such as car ownership), gender, age, race and ethnicity of the map-maker. For example, well-resourced middle-class white residents tend to hold a more comprehensive image of a metropolitan area beyond their immediate neighbourhood, compared with working-class or black residents (Knox and Pinch 2006: 224). Another common feature is that residents typically share a common understanding of perceived danger points in their local area and in this way they reproduce (through street conversations and networks of interaction, as suggested previously) strategies to avoid stigmatised 'no go' areas (near gang hang-outs, abandoned buildings and places where drugs are peddled) (Ley 1974; Knox and Pinch 2006: 228). This process in turn reinforces the labelling of particular areas, housing estates or a social grouping as dangerous, deviant or otherwise stigmatised.

These processes of mental mapping and labelling are illustrated in Case 5.6, drawing on Kevin Roy's (2004) interviews with 40 low-income African-American fathers in Chicago's South Side. Many of the fathers in this research are involved with their children from a position of 'absence' or separation from their children's mother, in some cases following periods of incarceration or involvement with gangs or drugs. Confounding the stereotype of absence, these fathers have developed elaborate strategies to interact with, care for and protect their children and wider kin network (Burton and Snyder 1998; Danziger and Radin 1990). They use their own knowledge (mental maps) of dangerous spaces and sites of potential discrimination or harassment (gang activity, police presence) to confine their children's movements and limit their trust in others in order to protect their children as 'good fathers'. The fathers are observed to construct 'three-block' safe spaces for family interaction, relying on paternal kin, making use of neutral spaces and managing complex negotiations with their children's mother and maternal kin (Roy 2004: 528).

A feminist perspective can effectively be mobilised to shed light on both perceived and real dangers for men and women across the city. It is widely recognised through existing research, for instance, that perceived fear of particular crimes and sites (such as child abduction or sexual assault in parks and open spaces) exceeds that which might be expected based on actual reported crime statistics (Burgess *et al.* 1988; Koskeh and Pain 2000). Women are statistically more likely to be attacked by someone known to them, at home for instance, and violent crime is far more frequently perpetrated against men, especially young black men, than against women. At the same time, it is important not to overlook the unpleasant realities for women of non-violent harassment, such as unwanted attention, name-calling, 'flashing' or groping. Whether real or perceived, fear of crime in the city is mutually reinforcing for men as well as women, through the local construction of dangerous and stigmatised people and places. In effect, a 'shadow of fear' (Sparks *et al.* 2001) combines with individual identification of self and other to constitute what are morally acceptable practices.

Case 5.6: Three-block fathers: mental maps of protective parenting in Chicago's South Side

This case draws on a short series of extracts from Kevin Roy's (2004) narrative research to illustrate the way mental map perceptions of danger and fear of crime circumscribe not only women and mothers' movements but also the involvement of men as fathers in family life.

Based on their perceptions of local neighbourhoods, fathers constructed mental maps of how particular physical spaces were dangerous or risky. Fathers assessed different neighbourhoods by asking a common question: 'Who's over there?' They considered whether an unknown area was dangerous based in part on what they knew of any families in the area. Fathers made reference to their own perceptions of daily-updated mental maps of safe and risky areas. Jelani, 23, staked his sense of control over his daily movements on such a mental map:

> My greatest fear is walking out of my house someday and getting shot or robbed. I mainly fear things that I can't control. I don't hang out in front of anyone's house. I try not to go outside when it's late. I don't go anywhere without a certain plan.

The most difficult challenge was to ensure both their own and their family members' well-being and physical safety. Malcolm protected his children with precautions in travelling on certain main boulevards in a bus after church on Sunday:

> We witnessed three people getting murdered in broad daylight on a Sunday. I had to put my kids on the floor (of the bus).

The large majority of men turned to their own mothers to help manage relationships with their children. Men's co-residence with their mothers and extended family (confined in safe 'three-block' areas) often allayed the concerns of maternal kin about dangerous neighbourhoods and unknown individuals. When Oscar moved into his own mother's new household in an unfamiliar neighbourhood he became attuned to his role as a 'three-block' father. He also began to shift his focus from his own well-being to concern for his daughter:

> I walk around and see how far I can go and see what is around – there is a park, so I can take my daughter there. Where I actually live, a three-block radius is OK. And I know where to go at a certain time, at night. I know not to go to this store. It's alright for me. I know how to conduct myself in certain situations.

Source: Roy (2004).

Comment

It is interesting how the quotes above (all from Roy 2004: 536–540) draw attention to strong parallels between the restricted movements of 'absent' fathers seeking to distance their parenting spaces from gang activity and the self-imposed curfew usually associated with women's fear of crime in the city. Feminist urban analysis has traditionally focused on women and a fear of the city itself (Domosh and Seager 2001: 99), whereby a generalised fear of violence or sexual assault accompanies the 'anonymity of the crowd' or the normative connotation of a woman stepping out alone in public or at night as transgressing respectable feminine behaviour. In this respect the design and maintenance of the built environment (lighting, security) play a significant part in shaping local perceptions of who 'belongs' or has safe access to an area at particular times of the day and night. Figure 5.4 shows how a derelict park and poorly lit subway may contribute to a heightened fear of crime and effectively limit movement through the area. Landscapes can invite or exclude access.

Figure 5.4 ***Fear of crime exacerbated by poorly designed public open space, Chicago***
Source: Helen Jarvis

Summary

This chapter has introduced the concept of everyday infrastructures, extending the understanding of this notion beyond the concrete articulations of roads and railways – to encompass institutional and emotional structures of constraint. Cases have been presented to show how these infrastructures reflect gendered assumptions at multiple scales of lived experience – from the body (in Yasmin's story), through social interactions in the built environment to structures of power and regulation at all levels of government bureaucracy (in Nicaraguan schools) and cultural norms of behaviour (in cultures of good parenting). Bound up with these multiple scales and infrastructures of daily life, we also stressed the significance to individual life-chances of a whole host of resource assets, entitlements and capabilities – beyond income and capital.

In Yasmin's story we highlighted the function of cultural values in a holistic understanding of resource entitlements and life-chances. It is not sufficient, for instance, to attribute highly differentiated experiences of social mobility and educational attainment to income-based housing class or a 'postcode lottery' of public service distribution. Resources such as time, transport, personnel, knowledge and values also form part of the equation. A report published by Demos, *The Good Life* (1998: 71), alludes to the 'pervasive' impact of unequal resource distribution where this is manifest in 'envy and a sense of injustice', through differences in knowledge and values as well as resources of income, time, transport and personnel. In neo-liberal Anglo-American economies there is concern that greater emphasis on self-determined consumer-citizenship and so-called 'choice' in public services, such as transport and education, creates new barriers and reinforces existing divisions between those who have choice resources (knowledge, time, instrumental competitive values, capital and income leverage) and those who are not in a position to compete for rationed public amenities or who value an ethic of care or defend the right to pursue non-Western, non-consumer values.

This chapter has engaged the reader in some hotly contested issues in planning and social theory, such as the extent to which the environment 'determines' behaviour (whereby 'good' behaviour can be cultivated and deviant behaviour removed) by design. Our aim was to shift the debate away from a preoccupation with environmental determinism, whereby social objectives (reduced crime, 'balanced' communities, walking and cycling) are tackled through design solutions alone, to a more integrated understanding of complex material, institutional, cultural and emotional factors. This led us to call for more inclusive urban design and governance, preferring the language of 'integration' to that of balance, where the latter implies you can have 'too much' or 'too little' of a community element (or facet of life) such as 'work' or 'family'.

From a feminist everyday life perspective we showed how the profile and training of architects and planners frequently results in a form of urban planning practice that is by the persistence of a narrow androcentric rubric 'disabling' rather than 'enabling'. The history of human settlement and development is littered with evidence of planners searching for utopia and interpreting this in terms of a fixed 'ideal' rather than in plural qualities of equally valid competing ideals. As David Pinder (2002) observes, in most visions of the 'good' city, the spatial aspects of master-planning are privileged under the assumption that social transformation will follow. This has led to criticism that enduring attachment to utopian ideals corresponds with a socially damaging environmental determinism, to the neglect of the lived reality of ordinary spaces and social interactions. From this perspective cities are sexist. As Clara Greed points out, 'many of the social issues of concern today are not spatially confined to one area. Whilst so-called ethnic minority areas might be physically identifiable, gender and disability issues occur everywhere and cannot be contained by special area policies' (1999: 5). Arguably then, a first step toward realising the goal of cities which are inclusive to all is to insist on socially progressive urban planning and a reorganisation of everyday life around the principles of social justice and an ethic of care.

Learning activity

For this learning activity, students are required to work in small groups of between two and four to create between them the questions and answers for a 'current affairs quiz'. Any individual in the group can set questions which probe a current theme or topical issue concerning urban gender relations. Students are encouraged to draw on specific news stories, images and headlines for inspiration, but they also need to research the academic literature to provide additional context for their chosen theme. The objective is to specifically locate the chosen subject or issue in time and space and to push the line of questioning (the quiz) along these axes. Internet sources can be used to identify contemporary issues but the quality of this data then needs to be evaluated by triangulation with other peer-reviewed sources.

Notes

1 Kalpana Sharma is Deputy Editor of *The Hindu* in Mumbai.

Further reading

Bakker, I. and Silvey, R. (eds) (2008) *Beyond States and Markets: The Challenge of Social Reproduction*. New York: Routledge.

Coleman, A. (1985) *Utopia on Trial: Visions and Realities in Planned Housing*. London: Hilary Shipman.

Gibson-Graham, J. K. (2006) *The End of Capitalism (As We Knew It): A Feminist Critique of Political Economy* (second edition). Minneapolis and London: University of Minneapolis Press.

Gilroy, R. and Booth, C. (1999) 'Building an infrastructure for everyday lives'. *European Planning Studies*, 7.3: 307–324.

Greed, C. and Roberts, M. (2001) *Approaching Urban Design: The Design Process*. London: Longman.

Jarvis, H. (2005) 'Moving to London time: household co-ordination and the infrastructure of everyday life'. *Time and Society*, 14.1: 133–154.

Kaika, M. (2004) 'Interrogating the geographies of the familiar: domesticating nature and constructing the autonomy of the modern home'. *International Journal of Urban and Regional Development*, 28.2: 265–286.

Roberts, M. (1991) *Living in a Man-Made World: Gender Assumptions in Modern Housing Design*. London: Routledge.

Web resource

http://www.gendersite.org/: gendersite is a UK-based online resource for gender and the built environment which reflects an innovative cross-disciplinary collaboration between gender studies, architecture and planning which combines academic research with equalities organisations in the voluntary sector.

6 Migration, movement and mobility

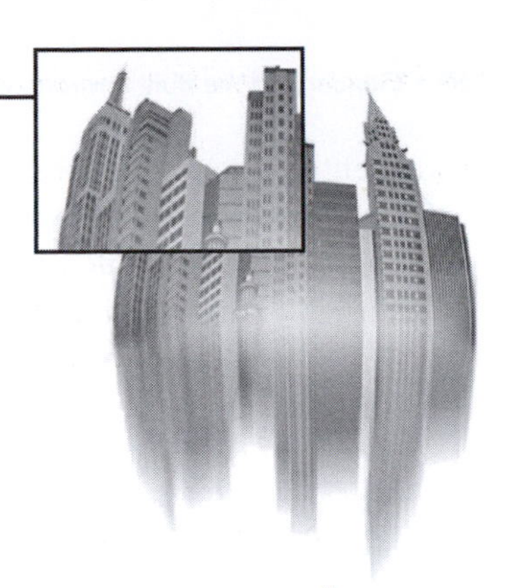

Learning objectives

- to be familiar with different types of population movement: migration, commuting, relocation and routine mobility
- to look at the relationships between poverty, power and gendered identity in people's experiences of mobility and confinement
- to show how assumptions of private 'automobility' impact on the shape of cities and experience of social exclusion
- to refine the above by looking at key intersections of cities, gender and mobility, comparing selected cases from the global north and the global south

Introduction

This chapter identifies a variety of population movements with the aim of better understanding highly gendered push and pull factors. From the outset we make the distinction between migration and mobility and this defines the two discrete parts of the ensuing discussion. We also identify different but interconnecting temporal scales of mobility and their local effects: daily local circulation, daily distance commuting, weekly commuting, occasional distance commuting, seasonal or task-specific labour migration and temporary, semi-permanent and permanent long-distance migration. In the first part we consider the nature of routine circulation, focusing on the divisive and damaging effects of the 'automobility trap' in urban design. Then we consider the nature of residential mobility and relocation. In the second part we look at the nature of migration, especially transnational migration, relating individual case studies back to themes of scale, power and interdependence raised in Chapter 4. Through these trends we again demonstrate the multiple, dynamic intersection of gendered identifications; the local ramifications of global processes. Our aim is to highlight the androcentric ways in which migration and mobility are typically understood and represented, suggesting ethnographic alternatives that better account for individual lived realities.

To develop these arguments we introduce case studies across this spectrum of mobility and migration, representing diverse household contexts, highlighting the intersections of gender, race, class and ethnicity which shape these push and pull factors and ultimately the integration, exclusion or segregation of cities around the world.

Movement and mobility: the puzzle of push and pull

The terms movement and mobility frequently assume positive connotations – of purposeful action, freedom to roam, open access and so forth. Yet simply 'getting around' is neither easy nor assured. At times there is a negative pressure to move which makes the experience stressful, involuntary, insecure and potentially dangerous. This is not only with respect to migrants seeking refuge from poverty or peril. As Domosh and Seager (2001) point out, just 'getting from one place to another takes time, money, confidence, and often machinery of some kind – and it can also take sheer endurance and will' (Domosh and Seager 2001: 110). Neither is it as easy as it once was to differentiate flows of circulation and relocation by distance travelled. In one context a daily journey of 30 miles or more could necessitate a house move and one-off household relocation. In another, greater distances than this are routinely undertaken as a daily commute; a life-absorbing process which through mere acceptance creates ever-increasing assumptions about 'reasonable' journey distance[1]. In this chapter, however, we generally make the distinction between long-distance migration and routine (daily) circulation. That said, the convenience of scale (or distance) movement differentiation is especially problematic in less developed countries, where the appropriate measure is not to and from the city but rather a more intense pressure of survival. It is at least as relevant to the lives of the majority of urban dwellers to consider the time and effort it takes to fetch water and fuel on foot in numerous repeated trips, as it is to view mobility in terms of distances travelled to waged labour.

Nevertheless, with the 'rolling out' of a dominant neo-liberalism, the polluting and congesting excesses of private motorised transport today flourish in many unlikely places. For instance, there has been a dramatic shift to the promotion of road-building and private motorised transport in India since 2000, from which time the Indian public's love affair with the motor car can be traced. A recent physical embodiment of the desire to close the gap between the requirements of modernity and income poverty in India is the unveiling of the Tata Nano, the cheapest car in the world (estimated cost $2,500) (BBC 2008). This trend is associated with a status-oriented lifestyle predicated on widespread use of scooters, mopeds, motorcycles and ultimately cars. These developments open up the possibility of a huge potential car-driving constituency (and not just in India – in China a rapidly increasing middle class shows every sign of being just as enamoured of consumerist trappings as its Western counterparts). Country after country is developing an 'automobility culture' whereby 2030 may witness 1 billion cars worldwide (Motavalli 2000; Urry 2004: 25). It is estimated that 40 million Indians can afford a car but infrastructural constraints limit the current total to about 7.5 million cars on the country's roads. By 2009 India is set to account for 8 per cent of global motor industry growth (Lloyds 2008). Rapid globalisation and modernisation in developing world cities are acting as a catalyst for a newly emerging, increasingly mobile middle class. Again these subject positions are gendered as well as being stratified by class, race and ethnicity. Figure 6.1 illustrates the local congestion and hazards relative to infrastructural constraints imposed by this rapidly developing 'automobility culture' – in this case on the streets of Bangkok.

Figure 6.1 ***Automobility in Bangkok: congested roads***
Source: Helen Jarvis

Paradoxically too, although transport planners report a clear link between travel patterns and social exclusion, whereby the distance individuals are willing or able to travel (their 'travel horizons') increase with income (Buchanan 2003), there is evidence too that some poor marginalised groups (notably young male migrants) undertake some of the most arduous daily journeys. This is a paradox, however, whose analysis and clarification have been hindered by the lack of a gendered understanding of the relational aspects of gender, class and spatial mobility in urban studies, in rich northern countries, between global north and south, and within the global south (but see McDowell 2003).

In the US context, we find evidence that long-distance commuting derives from a mismatch between affordable housing and available jobs, as well as the chosen living environment. Those on a minimum or modest wage among migrant day-workers and the middle classes alike make the difficult choice between sharing dense, frequently overcrowded and (for the poorest) substandard dwellings in close proximity to their jobs, or travelling long distances each day, hitching a ride or van-pooling with others, from peripheral family housing in remote bedroom settlements (Jarvis 2005a), or the rapidly growing peri-urban low-density settlements referred to as 'exurbia' (Brookings Institution 2006). This pattern of 'mobility' extends the working day by several hours and is a common livelihood strategy for low-income families in places as diverse as Cape Town, South Africa and Los Angeles, USA (Bénit and Morange 2006; Ellis *et al.* 2004).

These situations aside, it is widely observed that voluntary travel increases with affluence, whereas enforced travel increases with poverty. The increase in choice-derived mobility arises not only with respect to the frequency and distance travelled for pleasure, in tourism, but also as a means of 'shopping around' beyond the immediate neighbourhood for schools, specialist medical treatment, niche market goods and the like. In this respect Linda McDowell (2004: 146) observes that creating quasi-markets in public services (such as health and education) transforms citizens into consumers. Market-like 'brand' definition promotes competition between schools and hospitals and, with this, the time- and energy-intensive imperative for consumer service users to search hard for the market leader. By contrast, many poor households in otherwise affluent countries, forced to cope without stable waged employment, bear a number of different stigma attached to a world shrunk to pedestrian reach, particularly iconic gendered epithets such as 'single mums', 'welfare mums' and 'benefit cheats'. Some indications of these stigma and their effects are illustrated in Case 6.1 in relation to household resources compared for two samples of parents living in the city of Newcastle upon Tyne.

Case 6.1 Gender, generation, mobility and care

All too frequently travel behaviour is narrowly conceived of as an unproblematic 'trade-off' based on personal preference. It assumes a market-led and individualistic view of choice which neglects the reality that decisions are contested at the household level in relation to culturally specific aspirations and logistical constraints. As Nancy Folbre observes, 'choice is a funny thing, affected by both moral values and social pressures' (2001: 6). We see this in the way the practice of driving children short distances has come to reflect a dominant moral norm (protecting children from traffic and strangers) symbolic of 'good parenting'. Arguably too, school run travel behaviour is influenced by the altered composition and employment structure of contemporary households and pressures to pack more activities (of consumption and social reproduction as well as employment) into each day – the 'speeding up' of daily life. The opportunities and constraints of daily scheduling are heavily circumscribed by multiple economies and resources associated with gender and class divisions.

Recent research in Newcastle, England, explored the geography of pupil enrolment in relation to household resources, including household structure (number of income-earners), locating the home location of pupils attending each primary school in the city. This allowed for the comparison of actual catchment areas between different schools, as a calculation of the shortest path for each pupil and average distance travelled by pupils to reach their school.

Most children between the ages of 5 and 16 in the British state system will attend a school within convenient travelling distance from home (Dobson and Stillwell 2000). Yet within this geography there is considerable difference between an easily walked 0.5 km and a journey six times this distance or involving a number of busy roads. The results of the Newcastle study showed a significant mismatch between the schools parents chose to send their offspring to and the schools children would have attended if they had attended the school closest to their home. This mismatch reflected housing class differences but it also told a more nuanced story of gender and class inequalities within and between households.

The initial mapping exercise informed the selection of two contrasting schools: Woodland, which attracts a largely middle-class constituency of owner-occupiers; and Town, which serves a largely working-class population of households in social rented accommodation. In stark contrast, children who attend Town Primary live on average 0.4 km (maximum 2.9 km) away from the school, while

Woodland is attended by children who live on average 1.1 km away (maximum 6.7 km travel).

Table 6.1 indicates the differences in household resources associated with entitlement/capability, which clearly circumscribe the resulting differences in time–space prisms for chaperoning parents from the two sample populations. Another set of data which is not included here provided equivalent detail of social networks of support ('people' resources) available to each sample population. This highlights the meshing together of preference and possibility, where preferences are understood in relation to what is possible, as a function of the resources and capabilities available in a given situation (see Jarvis and Alvanides 2008 for full details).

Table 6.1 ***Typical profile of 'material' resources available to the two sample populations, based on an individual norm (baseline characteristics).***

Resource inventory: 'material' assets available to the household unit	*Woodland First School sample 'norm'*	*Town Primary sample 'norm'*
Number of income earners	1.5	0
Housing tenure	Owner-occupier (with mortgage)	Tenant (social rented)
Car(s) available for private use	2	0
Adult bicycle(s)	Yes	No
Kids bicycle(s)	Yes	Yes
Home PC with internet	Yes	Some (without internet)
Work PC with internet	Yes	No

Source: Jarvis and Alvanides (2008); authors survey of household resources.

Figure 6.2 illustrates by way of a time–space 'prism' (or action-space) the stark difference in spatial mobility and everyday coordination constraint which results from these unequal household resources. Mrs Tango is typical of the Town sample: she is a single parent who does not have access to a car or current paid employment (though she does undertake occasional 'cash in hand' cleaning jobs that are easily accessed by bus from her home). Mrs Tango's son attends the school closest to her social rented flat/apartment and the rhythm of her day is punctuated by the twice-daily walk with her son to his school and back, incorporating into these trips any necessary food shopping and bill paying. On the day shown, Mrs Tango spends the afternoon helping out in the classroom as an unpaid volunteer.

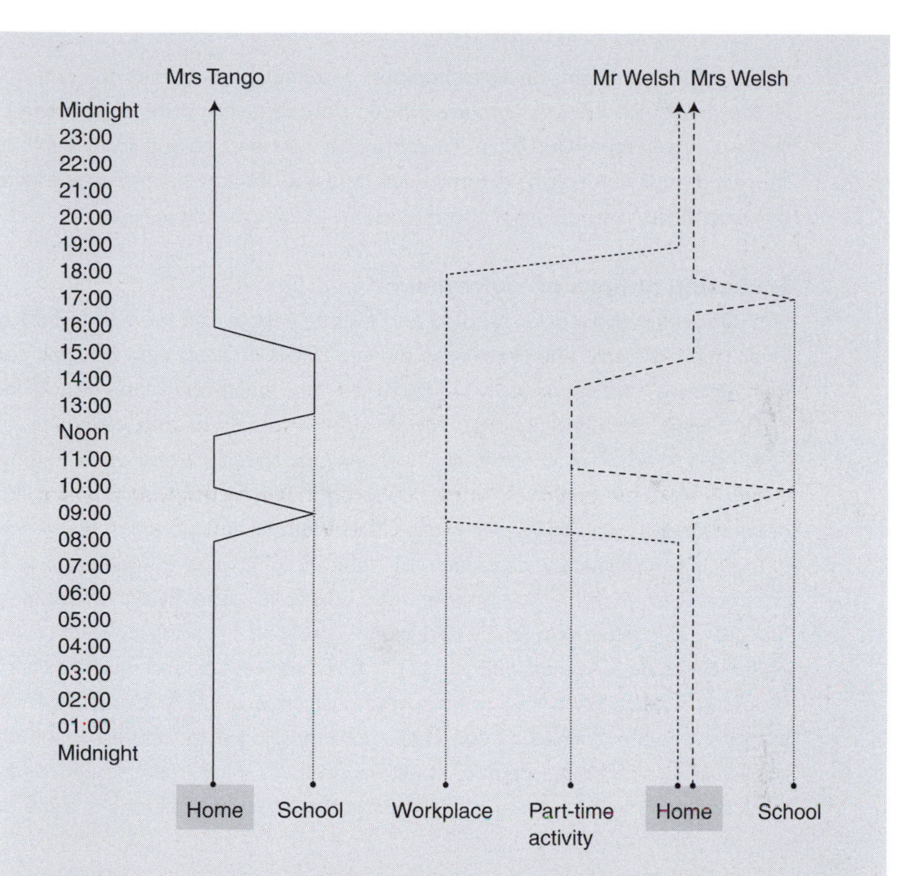

Figure 6.2 ***Time–space prisms for two Newcastle upon Tyne households – showing mobility constraints associated with gender and class divisions***

Mr and Mrs Welsh are fairly typical of the Woodland sample: they have 1.5 incomes and 2 cars between them. Mr Woodland commutes by car from home to distant employment each day and is unavailable for the school run, while Mrs Welsh undertakes part-time local paid employment that she fits between the twice-daily trips to the school by car. Both mothers in this case study share in common a strong moral sensibility that it is a mother's role to be there in person at the school gate for her children. Both have access to wider networks of family and friends than is represented by a household snapshot of people resources, but the scope this provides to lessen the constraints of the school run is circumscribed by this moral imperative to be a vigilant mother.

Overall comparisons between mothers chaperoning children to each of the two case study schools showed significant differences in the respective prism (or action-space), comparing stretched-out, repeated movements for car-driving mothers with the heavily circumscribed, parochial footprint of working-class mothers travelling to and from the school on foot and having to fit short, local shopping trips and cash in hand jobs into the limited time–space available between these fixed points of chaperone.

Extending notions of 'entrapment'

In a different study of 800 disabled and elderly people and more than 500 care-givers in a California homecare programme, 10 per cent reported that they got out 'almost never' and the majority left home less than once every three days (Decker 2006, www.americancity.org). Many elderly, disabled and care-givers feel trapped in their homes, cut off from social networks, hospitals and work. Among the vulnerable groups most likely to be trapped at home by transportation constraints are elderly and disabled women living alone. Without transportation assistance the existence of community amenities for vulnerable groups is meaningless. This circumscribed mobility (by gender, age, disability and living arrangements) reduces vital contact with family and friends, prevents full participation in civic life and imposes dependence with respect to basic necessities and amenities such as shopping, attending church, withdrawing cash from an ATM. Some destinations are only feasibly reached by car. There are many logistical barriers to the use of public transport for older people, those with mobility impairment, low incomes, or those unable to stand for long periods or transfer between platforms or climb steps.

Driving and car ownership

Although it is something of a generalisation, it is fairly accurate to portray everyday circulation in the global north as conforming to a dominant discourse of automobility in which 'driving and car ownership – is what normal people do' (Böhm *et al.* 2006: 8). This discourse is strongly linked to another, of individual self-determination and self-expression and of car ownership as a physical performance of liberal democratic ideals.

In most urban industrialised countries, full political, economic and cultural participation can be substantially restricted without a car. At the same time, the largest constituencies of public transport users (who may have restricted access to car ownership or use) are women, the poor and the elderly. In the 1980s and 1990s women's urban planning initiatives – for example, in London and in Toronto – focused attention on these transportation realities, redefining transportation as a 'women's issue' (Domosh and Seager 2001: 125). This research exposed, for instance, the direct budgetary and policy trade-offs between private versus public transportation systems. Cutbacks in public transportation services and a long-established privileging of car-based systems have subsequently resulted in profound gender- and class-specific consequences.

Yet, as a tangible instrument of emancipation, the motor car holds ambivalent status for feminist scholars. On the one hand Domosh and Seager observe that 'the car has been an especially powerful vehicle of women's liberation, both literally and metaphorically' (2001: 123). On the other hand, because of this, women's relationships with cars have been contested and controversial. The apparent freedom permitted by car use consists of a series of trade-offs that may themselves constitute an 'automobility trap' – for instance, the development of various forms of personal and public rapid transit systems encourages the spatial dispersion of sources of work, income and public services in a way that forces households and families to make use of these systems. The gendered division of household tasks can make women more dependent on and vulnerable to problems of transport use, and because of this, with essential use (particularly of cars) has come dependence and through dependence vulnerability; distance from work, school, hospital cultivates reliance on transport mechanisms whose failure can cause damage to the family/household. Finally, rather than freeing up time for the average family (and in particular female-headed households) in advanced industrial countries, automobility has added the social, temporal and financial costs of travel to time spent on essential income-generation at the same time that mass use of these systems has increasingly restricted any mobility advantages (traffic gridlock, overcrowding on bus/rail systems), while advertisers continue to depict the mythical freedom of the open road. Box 6.1 expands on this concept of automobility with reference to John Urry's (2004) definition and characteristics. We go on in this chapter to expand upon this concept (notably with respect to the global south), where current debate has not so far taken full account of the gendered implications of the rolling out of this dominant cultural system.

Box 6.1: Automobility

John Urry (2004) argues that the 'systematic consequences' of automobile domination go far beyond the constitutive technologies usually attributed to globalisation, such as the cinema, television and the computer. He identifies six components to the 'system' of automobility. Accordingly, automobility is:

1 the quintessential manufactured object produced by the leading industrial sectors and the iconic firms within twentieth-century capitalism;

2 the major item of individual consumption after housing which provides status to its owner/user through its sign-values (such as speed, security, safety, sexual desire, career success, freedom, family, masculinity) (Miller 2001);
3 an extraordinarily powerful complex of technical and social interlinkages with other industries, car parts and accessories; petrol refineries and distribution; road-building and maintenance; hotels, roadside services and motels; car sales and repair workshops; suburban house-building; retailing and leisure complexes; advertising and marketing; urban design and planning; and various oil-rich countries (Freund 1993);
4 the predominant global form of 'quasi-private' *mobility* that subordinates other mobilities of walking, cycling, travelling by rail and so on, and reorganises how people negotiate the opportunities for, and constraints upon, work, family life, childhood, leisure and pleasure (Whitelegg 1997);
5 the dominant culture that sustains major discourses of what constitutes the good life and what is necessary for an appropriate citizenship (e.g. via literary and artistic images);
6 the single most important cause of environmental resource-use.

(Urry 2004: 25–26)

Sudhir Challa Rajan sums up the socially divisive and environmentally damaging consequences of automobility by noting that

> in the course of a breathtaking century of cultural and social transformation, the car has almost indiscernibly turned into an ordinary part of daily life in contemporary Western life and acquired the quality of a human endowment and need, much like a home or clothing, (permeating) the daily behaviour of people, the purpose of institutions, and the structure of cities.
>
> (Rajan 2006: 116)

Gender and the automobility trap

The 'system' of automobility can be further analysed to expose gendered entrapment through the underlying assumptions:

1 The 'mismatch' (spatial dispersion) of homes, jobs, schools, shops and public services compels households and families to own and maintain a car to make use of these systems.
2 With essential use has come dependence and through dependence vulnerability (to breakdown, escalating fuel costs, road traffic hazards both on the road and as a pedestrian).

3 Rather than freeing up time for the average family in a rich northern country, automobility has added the costs of longer and more stressful journeys to time spent in employment, while at the same time restricting any mobility advantages through the treadmill of increased generalised movement (traffic gridlock, overcrowding on bus/rail systems).

Source: adapted from Sheller and Urry (2000: 737–757).

A common finding in transport research is that men in the global north on average travel longer distances and spend a greater amount of time on travel in their daily lives than do women. While mothers of school-age children undertake a high number of short journeys, women across all ages in the USA drive on average 60 per cent of the miles driven by their male counterparts (Rosenbloom 1993: 232). A particularly stark manifestation of this gendered difference in spatial movement relates to the emergence of the 'hypermobile cosmocrat' whose lifestyle is qualitatively at odds with (but frequently predicated upon) the automobility trap of the middle-class 'soccer mom' (Markovich and Hendler 2006). Case 6.2 expands upon the evidence of winners and losers in a hypermobile world, drawing on a report by the UK think-tank the Industrial Society (now the Work Foundation) (Doyle and Nathan 2001).

These gendered differences go beyond aggregate statistics on rates of car ownership and distances travelled to include non-representational bias, such as historical anxieties about women as drivers, as bodies 'out of control'. These include harsh social judgements of 'danger', 'wantonness' and a 'threat to family stability'. While the barriers to women driving are far fewer today, the persistent existence of sexist humour and statistical evidence that attests to women's more limited access to a car in couples where there is only one car shared between two reveals persistent discrimination. Domosh and Seager go on to observe that 'women are still outsiders at garages, auto shows, racetracks, and car dealerships – all consummate "men's spaces"' (2001: 125). In another example, John Urry notes the distinct way that women tend to inhabit cars relative to men. He observes that the proliferation of motor cars both in Europe and the USA from the 1950s exacerbated an already gendered 'cosiness of family life' by extending the time–space of male breadwinners relative to female homemakers:

> While working men became enmeshed in the stresses of daily commuter traffic into and out of urban centres, suburban 'housewives' had to juggle family time around multiple, often conflicting, schedules of mobility epitomized by the 'school run' and mom-as-chauffeur. Once family life is centred within the

> moving car, social responsibilities tend to push women, who now drive in very significant numbers, towards 'safer' cars and 'family' models while men often indulge in individualistic fantasies of fast sports car or the impractical 'classic car'… The distribution of company cars has also benefited men more than women, due to continuing horizontal and vertical segregation in the job market, which keeps most women out of positions with access to such 'perks'.
>
> (Urry 2006: 25–26)

Remarkably little has changed in these respects over the last 60 years, despite the erosion of the male-breadwinner norm. Indeed, the complexity of family chaperoning through a multi-task 'chain of trips' (to school, to work, to shop, back to school and after-school activities) can be said to be one mechanism for the reproduction of the gendered automobility trap discussed above, where the potential reach of everyday engagement (friendships, employment, access to services) shrinks in relation to the frequency and complexity of trips in and around the home neighbourhood. (We illustrated this in Case 6.1.) Compare, for instance, the image of the hypermobile cosmocrat conjured up in Case 6.2 with the everyday image of the 'woman attached' in Figure 6.3 with respect to mobility circumscribed by caring responsibilities and the physical limitations on movement these entail (burdened with pushchairs and prams, unable to navigate stairs or steep hills) (Tivers 1988).

Figure 6.3 ***Physical limits on mobility for mothers caring for young children, Newcastle upon Tyne public transport***
Source: Helen Jarvis

Case 6.2: The hypermobile cosmocrat

John Adams coined the term 'hypermobility' in 1999 to portray a trend of significantly increased individual mobility in affluent economies (especially by car but also by aeroplane both in terms of distances travelled and frequency of trips). This term has been widely adopted, mostly in discussions about the consequences of this for environmental degradation and loss of land to roads. Less frequently observed is the implicit understanding that individual practices of hypermobility reflect to a considerable extent manifestations of 'new masculinity'. Enter the cosmocrat. According to Micklethwait and Wooldridge (2000) the cosmocrat represents a new elite whose power lies in their ideas together with connections and chutzpah. These cosmocrats comprise mostly 'corporate man … the loyal retainers of sprawling multinationals'. The group also includes:

- 'Davos man'. CEOs, top politicians and assorted guru
- International entrepreneurs
- 'Sabbatical man': high-flying academics
- 'Agency man': agents of international governments and NGOs
- 'Carville man': travelling spin doctors and policy wonks
- 'Digital citizens' and others working in the new economy

The authors estimate that globally there are around 20m cosmocrats, around 8m of whom live in the US; they also estimate the numbers will have doubled by 2010. Above all, cosmocrats are highly mobile: much of their lives is spent on the move. Michael Bonsignore, Chairman of Honeywell, travels 200,000 miles a year. A typical round trip through Europe and China includes visits to nine cities in eleven days (Micklethwait and Wooldridge 2000).

Hypermobility is strongly associated with the 'world city' idea introduced in Chapter 4: a population of NY–LONdoners has been identified, who live, work and shop in the two cities as if they were one. For instance, Ron Kastner, theatrical producer and owner of a business doing specialised printing for investment banks, flies between London and New York five times a month, dividing his time between the East Village and Belsize Park. David Eastman, VP Wireless at Agency.com, often spends every few days flying between the two cities and is on first-name terms with the Virgin Atlantic business-class cabin crew (Chan and McGuire 2000).

A direct impact of hypermobility on the fabric of cities can be witnessed in the proliferation of 'instant offices', wi-fi internet availability on trains, boutique hotels (where the décor of the well-appointed apartment and the hotel room have morphed into one) and airlines promoting increased seat sizes and services for their business class traveller.

Source: Nathan and Doyle (2001: 14–15, 17).

Comment

A number of scholars have highlighted the implications for gender of these 'macho' modes of 'carefree mobility'. In a study of high-tech engineers in Cambridge, England, Doreen Massey (1995) made the observation that intense levels of fugue-like concentration, not to say corporate loyalty, typically presume that there is a 'wife in the wings' – picking up the cosmocrat's dry-cleaning along with his kids. In a rather more prosaic but poignant study, the journalist Melissa Benn exposes the plight of 'married lone parents' for whom the hard-headed practical realities of male spouse employment and the tendency for 'motherhood to increase a woman's conservatism' (Maushart 1997: 30) translate readily into the default position of 'the wife's sacrifice'. One of the interviewees in Benn's study explains, with a mix of bitterness and superiority:

> In the really top jobs, they want blood. They want you morning, noon and night. And that's how the world works. You might be able to do that right-on sharing stuff if you both work for the local council. But it's just not going to happen in the real commercial world. If one person is going to put in those kinds of hours, someone else has got to put in the hours to look after the children.
>
> (Benn 2005: 15)

In most of the research on heterosexual married couples it is suggested that the presence of children raises the stakes, making compromise more acceptable and gender inequalities easier to rationalise. But long working hours and demanding, less secure employment interacts with increasingly diverse household structures. For instance, Gillian Dunne (1997) has conducted research on gender divisions of labour in lesbian-couple families with children and identified relatively egalitarian arrangements. Moreover, a small but significant new demographic group to emerge in recent years is the 'single mother by choice' (Mattes 1994; Hertz 2006). Here the evidence is less clear whether family-unfriendly automobility can be tolerated or whether strategies of resistance are likely to emerge in the future. All this highlights the complex interdependence of new (increasingly polarised) mobility paradigms with changing household structures and gender divisions – in keeping with evidence of multiple and fluid masculinities and femininities.

Residential mobility

An enduring interest for academics, policy makers and planners has been to reflect (sometimes to model and predict) how the profile and composition of households impacts upon the shape of cities with respect to the connections between housing, employment, transport, migration and mobility. As Randolph observes of the potential for a spatial mismatch in this regard, 'it is households who consume housing, but individuals who participate in the labour market' (1991: 37). Close attention has therefore been paid to the relationship between household structure, notably gender divisions of labour, and decisions relating to housing choice and residential mobility. The crucial point here is that long-distance relocation is likely to entail the coordination or sacrifice of individual attachments to a job, place or social network. Popular reference is made to a 'spouse sacrifice' in this regard. Family studies scholars and children's geographers have similarly drawn attention to disruptions to children's attachment to place where they are invariably 'tied movers' in the relocation of one or other parent – or indeed, after divorce, consigned to a life of shuttling between two parental homes (Coltrane and Adams 2003; Hetherington and Kelly 2002; Furstenberg and Nord 1985).

According to early work by Mincer (1978), the economically inactive wife in a male-breadwinner household is a 'tied mover' who moves for the sake of her husband's career, although changes in the globalising economies of global north and south alike over the last twenty years now mean that such economically inactive women are increasingly rare. More recently it has been observed that in dual-income couples the wife's financial contribution from paid employment may inhibit such a move, making the husband a 'tied stayer' (Smits *et al.* 2003: 603). There is ongoing speculation as to whether women have gained any significant 'decision power' alongside greater labour market participation (Bielby and Bielby 1989; Bonney and Love 1991; Breugel 1996; Cooke 2003). The general view is that the normalisation of two earners in couple households contributes to both inertia and persistent gender inequalities whereby 'wives are more likely to be tied movers in migrating families while husbands, if they are tied at all, are more likely to be tied stayers than tied movers' (Mincer 1978: 754). Hardill (2002: 8) observes that women in dual-career households, even childless women, are more likely to be the 'trailing spouse'.

The relative propensity for people to move house and relocate varies by household employment structure (number of earners), household transition (divorce, adult children leaving home) and stage in the life course (such as with retirement). For example, Jarvis (1997) found that dual-career couples with children were less geographically mobile (using cross-sectional and longitudinal data for 2001 and 1991) compared with couples with children where only the male partner was in full-time employment. This variation in household rates of mobility is a major factor in the segregation of communities by age, class, race and sexuality (Hall *et al.* 1997). Moves ranging from a young person leaving the parental home, through family formation, marriage breakdown, the 'empty nest' (adult children leaving home) and various stages of retirement and ageing, all serve to shape highly segmented housing markets.

Attention has recently turned to the way that households and those with particular lifestyles are attracted to a particular place because of its climate, landscape or reputation for cultural diversity. This has long been observed with respect to the concentration of 'snow bird' retirees in the US 'sun belt' (Frey *et al.* 2000; Williams *et al.* 1997). It is similarly observed in clusters of same-sex households in cosmopolitan cities and neighbourhoods (such as San Francisco's Castro district) which boast a liberal politics and openness to diversity (Knopp 1995; Florida 2002; Black *et al.* 2000). In Britain, Duncan and Smith (2001: 484) find a correspondence between places known to provide an escape from the 'rat race' (such as Brighton or Hebden Bridge) and people forming 'alternative' family arrangements. Arguably, then, geographies of partnering and parenting do not rest with individual 'choice' but are instead informed by social relations in geographically situated households as part of neighbourhoods and community networks. This is another illustration of households representing a dynamic and fluid 'process' rather than a fixed mode of classification.

Adding further to the diversity of contemporary mobility experiences are the new and varied ways by which households extend beyond a single residential location. Indeed, a number of scholars argue that this spectrum undermines the meaning of labels such as 'household' and 'the family' (see Buzar *et al.* 2005). We find evidence of couples and families variously 'living apart together' (LAT) (where two people define themselves as a couple but maintain two separate homes and households) and 'living together apart' (LTA) (where one adult in a couple or family household lives away most weekdays in a second home, usually to engage in employment which is too far away to commute to on a daily basis). Case 6.3 offers a snapshot of the new trends of 'long-distance commuters'. In two-parent families, if one parent lives apart from their family for all or part of the week because their job is based hundreds, sometimes thousands of miles away, it is statistically likely to be the father who maintains 'absent' family relations (Benn 2005). Many dual-career families have in effect two homes, a family home and a 'pied-à-terre' metropolitan bedsit for the parachute parent who works long hours or works or travels much of the time away from the family home (Hardill 2002).

Other household transitions associated with distinct patterns of mobility have been identified, including childless young heterosexual couples, gay couples, 'empty nesters' (with grown children who have left home), mixed households of

Case 6.3: Living together apart

Although it's very hard to access robust data on LTA families, there is significant evidence that many more families in affluent countries live apart for some or all of the week today than did 20 years ago. As with all forms of migration and mobility there are push and pull factors. The push is that jobs are less secure and careers more likely to comprise a 'portfolio' of frequently changing clients, projects and skill requirements. Fixing a home on a single 'job for life' is no longer practical. At the same time, a significant pull factor is the possibility of accommodating longer-distance commuting without major sacrifice: faster and more comfortable and cheaper modes of transport, including cheap flights, and the possibility of sustaining relationships through regular phone conversations, email and internet wirelessly while on the move. According to Anne Green of Warwick University, who has studied the rapid increase in long-distance commuting:

> While 20 years ago it would have been unthinkable for a wife and children not to up sticks if a husband got a new job, these days things can be different. Women's employment has changed things: more and more families are dual-career, so one partner getting a job in another part of the country doesn't necessarily mean the other one's career can automatically be grafted there too…You also find that people (in the managerial classes) tend to value education particularly strongly, and if their children are happily settled in good schools they're often very loath to transfer them.
>
> (Moorhead 2007; Green *et al.* 1999).

Families move out for a better quality of life (bigger homes, gardens, schools) but the main breadwinner (typically a male professional or manager) commutes back to the city daily or weekly. Britons spend twice as much time commuting as the European average. According to research by Professor Glenn Lyons and Dr Kiron Chatterjee of the University of the West of England, rail commuters who are educated to degree level or above spend 50 per cent longer travelling to work than others. By 2016, it is reckoned there will be 1.5 million 'supercommuters' (mostly male) working in the UK but living overseas, using Heathrow, Gatwick and Stansted as commuter terminals (Wylie 2006). Eurostar estimates that half of the passengers who use its London to Paris service are commuters.

Part of the 'urban exodus' from big cities is in fact illusory. Some people commute hundreds of miles to work and some even live abroad at least part of the time. Often the argument made is that this lifestyle is adopted for reasons of improved quality of life (often in relation to children's education and access to green spaces). But, paradoxically, the consequences can impact negatively on quality of life in terms of exhaustion and disruption to social networks, as illustrated by a growing army of London 'weekly boarders' – people who rent tiny apartments, bed down on friends' floors or take up hotel room part-ownership options in London, Monday to Friday.

Novel services have grown up around the peculiarities and practicalities of super-commuting, as illustrated by bedroom exchanges advertised on the website 'Monday to Friday' (http://www.mondaytofriday.com/info_aboutus.php).

divorced/cohabiting/remarried parents with children from multiple relationships and various kinds of group households (students, migrant labour, religious communities) (Knox and McCarthy 2005: 301). This diversity of household types implies a similar diversity of mobility needs, access and vulnerability, and in this sense a spectrum of mobility can effectively be drawn between 'hypermobility' and confinement.

Globalisation, automobility and development

In the global south, the greater part of the rapid urbanisation has lacked the 'pull' forces of historical industrialisation and economic growth that has been the northern experience. As a consequence the primary driving force – poverty – has led to an entirely different 'automobility' environment. In many cities in Africa, Latin America, the Caribbean and Asia, a combination of poverty and the structural adjustment programmes during the 1980s and 1990s (see Chapter 3) has produced a series of effects that impact on and are impacted by multidimensional aspects of a work-driven system of automobility (or self-reliance). These set in train a vicious cycle of domestic disinvestment, employment lost to the closure of public sector companies and increased reliance on a growing informal sector which bears down increasingly on household survival strategies (Lingam 2005). In and around the city (particularly in terms of migration, which we discuss on p. 177) women in the global south have been mobilised in historically unprecedented numbers whereby individual self-determination and labour diversification have become survival skills: Case 6.4 illustrates this in the case of a household survey taken in the city of Masaya, Nicaragua (Cloke 2002).

Automobility for women in the south therefore entails the ability to move inside complex networks involving multiple locations and sets of actors constituting increasingly global chains of traders and 'workers', formal or informal, international or local, inter- or intra-city. Whether caused by adjustment processes or

Case 6.4: Gendered work in and around Masaya, Nicaragua

The results of a household survey taken in the city of Masaya, Nicaragua (Cloke 2002), illustrate the effects of automobility and labour diversity in a poor, less-globalised country in terms of the types of work done by sex and location. Nicaragua began its first structural adjustment programme with the IMF in 1991, following a period of revolutionary government from 1979 to 1990 which was accompanied by a prolonged and destructive civil war. Prior to 1990 and particularly during the period of the revolutionary regime, formal employment for women (particularly in the state sector) had risen sharply, as much as anything else because of the drain on male labour caused by the civil war. The structural adjustment programmes that ensued from 1991 caused a massive downsizing of state employment, concentrated in the health and education sectors – at least 70 per cent of those made unemployed were women (White 2001).

By the end of the 1990s although informal sector micro-businesses provided 70 per cent of all jobs for women and in 1998 54 per cent of micro-businesses were managed by women, of roughly 154,000 women engaged in commercial activity not a single one was in charge of a large, formal business (INPYME/INEC/PROMICRO-OIT 2001). The survey confirms a heavy concentration of women in informal commercial activities, although there is a considerable degree of gender overlap; it is also interesting to note that, in a country the size of England and Wales with a population at that time of about five million and relatively few cars, women's commercial activities were still spread all over the country and even outside. Transport for the poorest sections of society in Nicaragua consisted of an extensive but haphazard bus network and an informal truck-based hitchhiking system, both utilising a primitive and badly maintained road network, making travel for women who maintained primary responsibility for most of the domestic labour a major undertaking (see Table 6.2).

The greater diversity of jobs for men in the private sector is apparent here (even though in this survey the number of women interviewed outnumbered the number of men by a ratio of 1.5:1), particularly with respect to all kinds construction work, blue-collar skilled work (electrician, plumber, etc.) and their spatial dispersal over the whole country and internationally. The mobility overlap occurs, however, in jobs that are household businesses and in which the labour and travel are more evenly spread – bakeries and the sale and purchase of

Table 6.2 *Geographical distribution of types of formal and informal work done by male and female inhabitants of Masaya, Nicaragua*

Sex	*Proximity to home*	*Diverse activities reflecting multiple economies*
Female	Masaya/environs	Raising chickens/eggs for sale, *nacatamale*/tortilla making, goats for milk and cheese, agricultural labourer/own *finca*, flower-growing, selling firewood, making/selling furniture, making/selling baskets, clothes-washing/ironing, telephone call-placer, petrol-station worker, market dispatcher, private tutor, shop assistant, industrial designer, executive secretary/secretary, accountant, homeopathic medicine seller, bicycle repairs, mill worker, factory worker
	Departmental	Fruit selling/buying, clothes-making/seamstress, small general shop/stall, makes/buys/sells bread/cakes, cleaner/domestic employee, beautician/hairstylist, university lecturer, school teacher, lawyer, government official, meteorologist, doctor, clinical pharmacist, bank worker
	National/ international	Selling basic grains/beans, buys/sells shoes, itinerant goods/food vendor, craftwork
Male	Masaya/environs	Agricultural labourer/own *finca*, breeding fighting cocks, ironmonger, selling firewood, plumber, basket-making, carting/delivery, itinerant labourer, government worker, factory worker, fence-post maker, gardener, hotel manager, school teacher, barber, bicycle repairs, tyre repairs, private tuition, metallurgist/machinist, paint shop owner, dentist, owner of bookshop/stationery shop, waiter, market porter, fireman, billiards/bar proprietor
	Departmental	Driver/taxi-driver, soldering/welding, small general shop, accountant, sand plant worker, doctor, public notary, registrar, lawyer, electrician/electrical engineer, TV/radio repairs, National Assembly deputy, antique dealer, athletics umpire, clinical physician, goldsmith/jeweller, tailor, video rental business, mobile disco, security guard, bus conductor
	National/ international	Buying/selling fruit and produce, buying/selling shoes, coffee harvester, brick-laying, construction worker/ supervisor, craftwork, bakes/buys/sells bread, itinerant goods/food vendor, computer shop owner, economic adviser/consultant, making/selling furniture, fishing

bread, for instance, the buying and selling of fruit and vegetables, work as itinerant selling. Shoe-making, the selling of shoes and furniture-making are very important to the economy of Masaya and tend to be run as family businesses; within the household, moreover, there tend to be fairly strict divides as to who does what.

Source: Cloke (2002).

not, the increasingly mobile labour of these women acts as an important motor for globalisation (Sassen 2006). Saskia Sassen observes that women's 'survival circuits' are in effect crucial to the formation of novel economic arrangements in globalising cities, through 'emergent alterative political economies' (2006: 35; 2002). At the same time, within these networks and everyday infrastructures (linking back to Chapter 5) it has to be recognised that routine practices of walking, riding a bicycle or simply travelling by city bus are all effectively circumscribed by gendered assumptions and power relations. Some of the mundane but powerful ways in which women's and children's movements are modified and curtailed are illustrated in Case 6.5.

Household perspectives on migration and separation

In Chapter 5 we briefly introduced different types of migration to demonstrate how gendered analysis gives us a deeper understanding of the relationships of power within and between households, particularly with respect to processes that drive and sustain migratory types. Here we develop an understanding of gender, supported by the 'soft' networks (contacts, kinship, social and economic information) discussed in Chapter 4, as a tool by which to deconstruct migration and what is involved in its representation, performativity and practice.

The starting point of any gender analysis of migration has to be that economic push and pull factors are not gender neutral (Boyd and Grieco 2003). First, in macro-economic terms economic development affects gendered roles differently, promoting or hindering the migration of women or men by distance to a different extent. For instance, in Costa Rica, Silvia Chant (1991) found that reproductive labour played a more important role in determining household migration decisions than did pushes and pulls associated with wage labour (cited in Silvey 2004: 493; see also Mattingly 2001). Second, in receiving countries the economy is also gendered (for instance, the migration of women domestic workers to North America, the Middle East and Europe, and male agricultural

Case 6.5: Young women's experience of harassment on city buses in India

> A woman walks down a city street. A man whom she does not know makes an obscene noise or gesture. She counters with a retort or ignores him and walks on. This is a common enough sequence of events. It happens every day of the year... Superficially; this is a simple, ordinary encounter... But beneath the surface is a complexity of feeling, thought and intentions which has huge personal and political significance.
>
> (Bowman 1993: 517)

Women in India are increasing their use of public space and moving more freely about cities, some adopting Western dress or adapting traditional modes of dress (such as the salwar kamiz) in ways that allow greater bodily freedom of movement. At the same time, gender discrimination such as that depicted above persists in public urban space. Indeed, women's groups in India have noted a sharp increase in reports of sexual harassment and public displays of male domination in rapidly developing Indian cities. Although the threat tends to involve ritualised sexism rather than actual bodily harm, verbal harassment and groping effectively circumscribe women's movements through intimidation and fear for their safety (and dignity). Incidences of assault are particularly widespread on crowded city buses during rush hours (Vasudha 2005; Anand 2005; see also Baxi 2006, and, for the UK context, Pain 2001) (see Figure 6.4).

The widespread abuse of women on public transport in Indian cities functions as a strikingly visible and effective symbol of male dominance – limiting the integration of women into once male-demarcated space. Because women are vulnerable to constant sexual harassment on the public buses they report far greater concern about overcrowding than do men (Peters 2002: 15). By way of response, some women's groups call for special women-only bus services, while others would simply like to see more buses made available so as to reduce the overcrowding which cultivates harassment (Peters 2002: 15). A common complaint is that a largely male police presence does not enforce women's right to safe movement in public. Jaising argues that 'despite the fact that serious sexual harassment is rampant in India this largely goes unreported because women fear that complaining will attract negative attention' (1995: 53). Indian women themselves are divided on the problem of sexual assault and the solutions required. Some see the problem as resulting from young women emulating Western fashions and ideals and pushing the boundaries of strict norms of public decency. Widespread experience of harassment suggests that patriarchy remains deeply entrenched in the society and these injustices are dismissed as 'trivial teasing'.

Figure 6.4 ***Common assault: crowded city buses in Guwahati, Assam***
Source: Rituparna Bhattacharyya Sarma

Of course the ubiquity of such abuses serves the purpose of opposition to women's liberation well. Women are frequently deterred from travelling on public transport and moving freely about the city to attend college or pursue paid work outside the home where routine journeys feel dangerous, unpleasant or undermine their self-esteem and feminine identity. At the same time, the social institution of *pardha* (sometimes also spelled purdah) presents a more visible if equally significant barrier to women's occupation of public space and use of public transport. The custom (typical of Muslim societies) of gender-segregating access to public areas, including buses, can mean that the only viable transport open to women is the slow/inferior cycle rickshaw. Pro-growth government officials typically believe that rickshaws 'slow traffic' and as such should 'make room for cars' – a view which consigns women users to the margins (quite literally the gutter) (Peters 2002: 15).

workers to the southern USA). Emphasising the need to incorporate gender has also influenced network theory. The household decision-making process about who travels where to find work through 'soft', gendered survival networks, therefore, is complicated not just according to scale but in the way these are spatially gendered. A household in Mexico has to make a decision (filtered through the structures constituted by local cultural determinants and locally specific gendered hierarchies) about whether it will be better to send available female members to the *maquila* zone in the north, to try and get female or male members over the border into the USA, or to see if males or females can earn a living on the streets of Mexico City or perhaps in one of the tourist areas such as Tijuana or Acapulco.

Traditional sex roles and stereotypes notwithstanding, according to the Economic Commission for Latin America and the Caribbean (ECLAC) the available data indicate that women constitute just under half of migrants globally. Within that figure, they comprise most of the internal migrants in Latin America and the Caribbean and are the majority of migrants constituting the intra-regional flows (Martínez and Reboiras 2001; United Nations Population Division 2001). Immigration to North America, the most popular destination for migrants globally, stood at 12.5 million immigrants in 1960, but by 2005 it was 44.5 million (UN 2005); the same year, female migrants made up 54.6 per cent of all new permanent migrants in the USA (Jefferys and Rytina 2006: 5) and the US Department of Labor calculated in 2002 that women migrants at that time made up about 40 per cent of illegal immigrants (Passel *et al.* 2004: 2). For such women, levels of domestic violence and sexual and physical abuse are notably higher than among native-born women in the US, and levels of access to healthcare and health insurance are all lower, adding to the precariousness and vulnerability of the migrant experience in receiving countries, as well as the gendered reactions to movement, entry and settlement. In addition, as precariousness and vulnerability have increased in countries of origin after the debt crisis of the 1980s and the adjustment programmes of the 1990s, the make-up of the female component of internal migration has changed – in the 1960s and 1970s single women were the largest component, but after the 1980s increasing flows of married women with young children have been forced to migrate (Pessar, 2005).

Lastly, the problem with the orthodox economic way in which migration tends to be treated with reference to absolute numbers is that it exacerbates the impression of permanence in this form of automobility and feeds into the oppressive, diluvium political discourses about global migration. There are, however, rhythms and patterns of migration and return; two-thirds of current migration to the US is temporary labour, whose share of the whole doubled between 1994 and 2001 (IMF 2004) and in addition, this increase has been demand driven, which is to say that labour shortages in the US itself have driven the provision of more and more extensive visa-types to make up the shortfall. There is also considerable evidence to suggest that, whereas the number of undocumented migrants returning to their country of origin from the US within a year has halved over the last 20 years, this is in the main because increasingly hostile migration policies have made the border more difficult to cross, implying that such anti-migrant policies have the opposite effect from that intended. Susan Mains (2002) observes, for instance, that the US Border Patrol situates its disciplinary practices and new surveillance technology (including infra-red night vision cameras) in gender stereotypes, replacing the individual migrant story (crossing the border legally or casually) with abstract representations of the 'Mexican masculinity' and the 'problem' of 'aliens' (Mains 2002: 193). In short, migration has to be understood as a social and political process which is 'part and parcel of the various gender politics of migrants' lives and broader political and economic processes' (Silvey 2004: 495).

Transnational migration

The trend of people living in cities is irreversible, with the number of people in cities set to double by 2030, and migration (local, national and international) and the resulting diaspora are major factors driving this urbanisation process. Urbanisation often brings about changes in the roles that men and women play in society, for example, with respect to a shift in the mode of employment (such as from formal to informal) or household management and leadership in the everyday politics of urban migration. In this respect, Nira Yuval-Davis (1999) stresses the importance of examining in gendered and non-ethnocentric ways, recognising this as a multi-layered construct through which different layers – local, ethnic, national, state, cross-border and supra-state–affect can reconstitute the relationship and position of each layer. This is illustrated in the variety of ways that first-generation immigrant Muslim women in Oslo, Norway, are observed to interpret and practise gender relations within the framework of Islam (Predelli 2004: 474). This reinforces the idea that the conditions of diaspora affect women's and men's status and position in family, society and religion – often in contradictory ways such that new social contexts influence gendered practices alongside long-established norms and values.

We need to consider the interacting dynamics of gender and the urbanisation process, particularly in the light of the processes of neo-liberalisation and globalisation discussed above; in the decades up to 2000 women increased greatly in numbers as a proportion of migrants globally, and Asian women were the fastest-growing category of foreign workers, increasing by 800,000 a year up to 1998 (ILO 1998). There are a number of different reasons for this, including the economic determinants of a globalising labour force that have feminised workforces within poorer countries, increasing the rate of break-up of family units in those countries and therefore the necessity for female household heads to find work outside their own countries. Other reasons include the increased demand for female labour by MNCs outsourcing to areas of cheap labour supply (such as

China) and yet others to do with growth in the global trafficking of women and children as part of the sex industry. How different cultures view the economic participation of women and men also has a significant impact on the gender make-up of the flows of migration and, ultimately, on the sex ratio of urban populations, as we see in Case 6.6.

Finally it must be noted that transnational migration due to economic motivations also contributes to the diversity of family forms previously noted, giving rise to households characterised by the spread of household functions across borders. These transnational households must rely on livelihood strategies such that financial caretaking takes place from afar, while 'other-mothers' provide local care to members left behind (Schmalzbauer 2004: 1,320) (see also Case 3.4, p. 79).

Summary

In this chapter we have identified some of the many different types of mobility and migration alongside the multi-scalar implications of these intersecting flows. We have stressed that mobility functions as an expression of vulnerability and insecurity as often as it does freedom of opportunity. This was illustrated by case studies ranging along a spectrum of hypermobility to confinement as well as by the paradox that labelling and discrimination result from spatial entrapment (as in the case of UK school enrolment and differential household resources). Stigma and exploitation were also shown to result from involuntary migration bound up with sex trafficking and indentured employment.

We observed a rising cultural and economic ubiquity of private motorised transport contributing to a mobility imperative or 'automobility trap' of increasing spatial dispersion which has profoundly gendered, double-edged ramifications. In the USA, research on deprivation, poverty and insecurity often points out that the 'clapped-out' car is the last possession to be relinquished – long after a house has been lost to debt – with the homeless effectively clinging onto the American Dream (at least to the possibility of being able to reach places of employment) from a makeshift home inside their car.

We introduced a number of different case studies from around the world to illustrate contrasting experiences of localised circulation, access to particular modes of transport, and the implications within and between households of confinement, hypermobility and transnational family migration. These served to demonstrate the critical intersections of gender, generation, poverty and uneven development as well as the multiple, dynamic intersections of gendered identifications – of

Case 6.6: Modern slavery?

China

China's rapid economic boom in the twenty-first century exploits the availability of significant flows of poor migrant workers, forced to look for wage labour in the booming urban metropolitan areas of Beijing and in the Pearl River Delta. Amnesty International reports an estimated 150 million Chinese people moving long distance, typically without dependants, from poor rural communities to urban labour markets and certain servitude, since 1995. Classed as 'temporary workers', these migrants have uncertain legal status and as such are forced to take on bonded labour (frequently as construction workers), working 12 hours a day, seven days a week, denied full rights to housing, education and medical care (Waits 2007). Unsanitary and overcrowded housing conditions for these migrant workers are reminiscent of those which mobilised urban philanthropy in nineteenth-century Europe under the auspices of the sanitary movement. Moreover, in 2002 the *China Business Review* (Zhang Ye 2002) reported that while the overall ratio of men to women migrants nationally in China was about 2:1, in the Pearl River Delta the ratio is reversed; the new export processing zones have created intensively feminised spaces driving the massive economic growth that China has experienced recently, but under male-dominated systems of supervision.

The sex sector

Yet this is the twenty-first century, and more sinister than exploitative industrial labour are the hazards of modern slavery, not least of which is growth in sex slavery. Monzini (2005) presents a disturbing portrait of the trafficking of women and girls for the purpose of street prostitution and sex tourism centres. The scale and profile of sex work vary geographically but the sector is estimated to have a global turnover of $5,000– $7,000 billion, greater than the combined military budget for the whole world (European Parliament 2004, cited in Monzini 2005: 24). A number of discrete trends of growth are associated with newly expanded markets in Europe (largely supplied by the post-socialist and Balkan states) as well as other 'developed' and 'less developed' countries. On the one hand, hundreds of thousands of women from poverty-stricken rural regions are being driven to supply 'new' prostitution markets in China as well as countries hit by war and the arrival of troops from abroad (Cambodia and Kosovo being good examples). On the other hand, the internet has made it possible to advertise more secure and sophisticated contact systems for 'escort services'. Sex tourism has also expanded, especially in Southeast Asia during the 1990s, catering for international customers who provide a major source of foreign currency. In Thailand, annual income from the sex sector is between $22.5 and $27 billion, or 10–14 per cent of gross domestic product (Monzini 2005: 26).

Seiko Hanochi (2003) provides a powerful account of Japan's sex sector, which is its most lucrative service industry. Ethnographic research reveals a complex social hierarchy differentiating local from foreign exotic dancers. The latter are charged exhorbitant sums for food and shelter and by this means forced into risky prostitution through financial dependence on gangmaster pimps. Bound up with the increased flow of international migrants (predominantly women fulfilling the social reproduction needs of receiving countries) is the way social reproduction is increasingly marketised, commodified and linked to remote, unresponsive and increasingly illegitimate (often criminal) structures (Bakker and Gill 2003: 13). Close attention to the various macro, meso and micro sites of sex sector globalisation highlights another paradox, that sex districts function both as the loci of intense competition and as communities of resistance through shared interest (Hanochi 2003: 155). Indeed, movements to advance the rights of sex workers are spreading around the world and in 1997 the First National Conference of Sex Workers was held in Calcutta, attended by 3,000 delegates. While modest efforts are underway within the sector to eliminate the stigma and discrimination accompanying the practices of prostitution, international action is needed to address the profound exploitation and inequality that transnational trafficking and domestic prostitution represent.

partnering, parenting, household structure, social stratification, race, ethnicity and vulnerability to harassment in public. We closely engaged with the everyday realities of social reproduction, shifting attention away from headline statistics to qualitative, ethnographic approaches to mobility, migration and the co-constitutive impact of these transitions on the urban social landscape.

Learning activity

For this activity you will trace out mobility options in your community based on access to the public transport available. Assign groups to represent different actors in the community (elderly, lone parent, mobility impaired) and have each consider the different places people need to travel to conduct a range of economic, social and bureaucratic functions. Then investigate the means available to this group – how easy or difficult is it to travel, considering the need to conduct multiple tasks in one day? What does this say about who is included and excluded from daily life in your community?

Notes

1 By 2003 the UK had an average commute of 45 minutes per day, the longest in Europe (twice as long as in Italy). According to data from a national car-drivers' organisation, commuting passenger miles in the UK increased by 6 per cent between 1993 and 2003 and the average distance travelled went up by 17 per cent, to 8.5 miles (RAC Foundation 2003).

Further reading

Dowling, R. (2000) 'Cultures of mothering and car use in suburban Sydney: a preliminary investigation'. *Geoforum*, 31: 345–353.

Katz, C. (2001) 'Vagabond capitalism and the necessity of social reproduction'. *Antipode*, 33.4: 709–728.

Mahler, S. and Pessar, P. (2001) 'Gendered geographies of power: analysing gender across transnational spaces'. *Identities: Global Studies in Culture and Power*, 7.4: 441–459.

Mattingly, D. (2001) 'The home and the world: domestic service and international networks of caring labour'. *Annals of the Association of American Geographers*, 91.2: 370–386.

Mitchell, K., Marston, S. A. and Katz, C. (eds) (2004) *Life's Work: Geographies of Social Reproduction*. Oxford: Blackwell.

Pratt, G. (1999) 'From registered nurse to registered nanny: discursive geographies of Filipina domestic workers in Vancouver, B.C.'. *Economic Geography*, 75: 215–236.

Silvey, R. (2004) 'Power, difference and mobility: feminist advances in migration studies'. *Progress in Human Geography*, 28.4: 490–506.

7 Homes, jobs, communities and networks

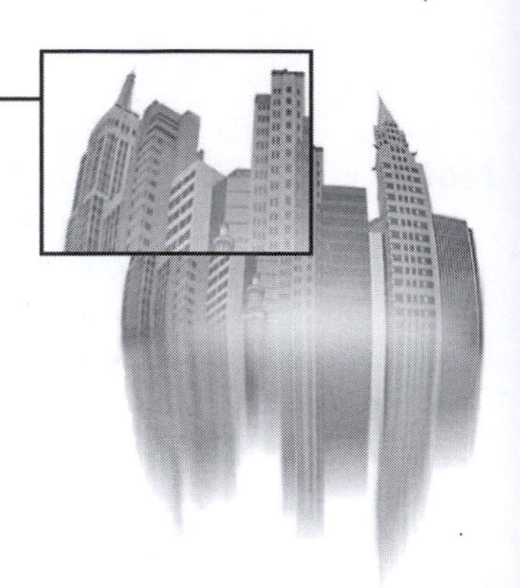

Learning objectives

- to look at persistent gender divisions in the home and workplace
- to refine the above by showing how gender and power relations transcend these sites through network interdependence
- to be familiar with concepts of identity, performance and emotional labour
- to identify networks of reciprocity and exchange from specific case studies

Introduction

This chapter explores the fluid and dynamic intersections of homes, jobs, communities and social networks with the aim of better understanding the co-constitution of cities and gender. It is tempting from an urban ethnographic, everyday life perspective to reduce the dynamic complexity of people–place interactions to fragments of activity which are played out in discrete sites over the course of the day – in the street, at the marketplace or shop, at the school gate, on public buses and in factories and offices. Indeed, this provides a straightforward and engaging way of highlighting specific discriminatory practices, such as unequal pay, the so-called 'glass ceiling', and sexual harassment and homophobia on public transport, for instance. We adopt this approach ourselves in a number of case studies. At the same time, this 'static' approach fails to adequately reflect the messy reality of the overlapping spheres of identity which arguably constitute the essence of the everyday. The purpose of this chapter then is to shift attention away from individual sites and scales of analysis in order to trace the circuits, networks and cultures of reciprocity and learning which intersect and transcend discrete realms, such as 'home', 'work' and 'family' in the reproduction of urban daily life.

We begin with the classic critique of the idealised notion of the 'home as a haven' but then segue into new concerns for the corrosive effects of a housing-led system of welfare and gendered expectations of home-based care in the community in advanced neo-liberal economies today. The selected case studies make the connections between a persistent, socially and geographically uneven shortfall in affordable housing and persistent gendered norms of 'good' (respectable, prudent, feminine) household management. This discussion again points to the home as representing something 'more than material', through multiple economies, as myriad attributes of status, wealth, parenting, retirement and strategies to extend or let out individual rooms. Key to understanding the interdependence of home, work and everyday life are the concepts of identity formation and emotion-work, which we introduce alongside the gendered work of caring and the performance of gender in the new economy. While these discussions primarily focus on gender identities and performances located in the conditions of advanced neo-liberal economies (notably Britain and the USA), we remind readers of the uneven local effects of global restructuring, as well as support networks which can be transnational in reach, by drawing comparisons with the global south.

Gendered homes, inside and out

Arguably, there is a powerful normative if not positive association between the home and an idealised and subordinated feminine realm. There is a long tradition in feminist critique of the sexist city (Greed 1994; Darke *et al.* 2000) which emphasises a gendered binary division, relating to the physical separation of homes from sites of employment (such as with suburbanisation) and distinctions between 'indoor' 'wifework' and 'macho' 'outdoor' home construction and maintenance, typically involving machinery (Nelson and Smith 1999: 125). Notions of the 'ideal home' persist, for instance, in the gender stereotypes which sell magazines on cookery, cars, home decoration, needlecrafts, home improvement projects and power tools.

While the social and material conditions of housing, employment and family life have changed fundamentally in recent decades, as we go on to illustrate, idealised notions of 'home' continue to exert a powerful influence on the social construction and symbolic representation of gender. As Linda McDowell observes (1999: 71), the 'reality' and the symbolic meaning of the home (or, as we go on to stress, household networks which pass through multiple manifestations of an idealised home) combine to produce the construction of a particular version of a home in different ways in different societies. In other words, gender role identities are largely reproduced at the micro-level of the home locale. This suggests that in order to appreciate the relevance of 'the home' in a globalised, mobile world, we need to first understand the concept of identity formation and belonging. The key term 'identity' and what it has come to signify are explored briefly in Box 7.1. This helps explain the construction of a masculine 'outside' world of employment, alongside a feminised realm of 'indoor work' including that of caring for others and monitoring their emotional well-being.

Box 7.1: Identity

The term 'identity' has been very widely used in feminist scholarship since the 1990s. It is a complex and highly contested term which is central to the recognition and articulation of difference. Strictly defined, identity means the collective aspects and state of 'being' a specified person (or thing) – but in the social sciences it has come to mean the way individuals recognise themselves (how they present themselves in relation to others in particular contexts) rather than as a definable identity-category. The starting point is usually to differentiate 'self' from 'others'. It is also quite common to find identity used in relation to a chosen or assumed 'style' (of dress, home or all-round 'lifestyle'). These static and voluntaristic (choice-based) notions of identity have been criticised by post-structural feminists (see Chapter 3) who recognise in identity formation the reproduction of plural and dynamic 'narratives of the self' which have direct reference to social structures and power relations which may be constraining as well as enabling. Feminist geographers also observe that 'who we are' is related in quite fundamental ways to 'where we are', in that identity is spatially as well as socially constituted. In short, questions of identity are hard to divorce from a wider cultural politics of representation (McDowell and Sharp 1999: 133).

Some of the ways in which this inside/outside gender segregation continues to be reproduced through repeated representations in urban daily life can be observed in Figures 7.1 and 7.2. Figure 7.1 shows two commercial window cleaners abseiling down a glass tower block in the city of Chicago. The Mexican men carrying out this low-waged, dangerous, physical work appear to display a very public performance of hyper-masculinity in their agile and acrobatic (though essentially repetitive) labour. The gender and status distinction between cleaning the outside windows of an office block and cleaning the inside windows of a home is strikingly made not only by the scale of the job and exposure to the elements but also by the mechanical quality of the equipment involved: ropes, pulleys, harnesses, helmets and boiler suits. In the gendered occupational hierarchy of post-industrial tertiary sector cleaning, this is arguably the male-typed equivalent of ship-building and coal-mining.

Figure 7.1 ***Scaling the heights on low wages, male window cleaners abseiling, Chicago***
Source: Helen Jarvis

Figure 7.2 shows another taken for granted reproduction of gendered stereotypes by the hackneyed association of 'secretarial college' with a frilly-skirted doll-like female on this item of street signage. Of course, the damning suggestion here is that the job of a secretary is not only essentially female but involves little more than standing around looking appealing. Photographed in 2006, this sign is of course intended to poke fun at a bygone era of 1950s femininity. Yet the sense of parody is largely overtaken by the realities of persistent gender segregation in contemporary society, observed at every turn on this and the surrounding city streets.

The post-modern home

Widespread recognition that gender identities are not fixed (by sex), but instead are continually reproduced as plural and dynamic 'narratives of the self', has accompanied parallel claims in urban studies of the emancipatory potential of the post-modern home and city. Yet there is a paradox here in the way aesthetic styles, identities and self-expression have been allowed to proliferate at the same time as urban daily life is being brought under increasing neo-liberal surveillance. This is evident not only with respect to street-level 'zero tolerance' and the privatisation

Figure 7.2 ***Sexist signs to convey emotion-work: secretarial college, Chicago***
Source: Helen Jarvis

of gated new home developments (Fyfe 2004) but also, more subtly, in the myth of gender democracy in the post-modern home.

The early 1980s witnessed growing distaste for the 'brutal aesthetic' associated with Modernity. The post-modern home was said to 'look forward to the past' in the way it combined a traditional vernacular of decorative ornament with colour, playfulness, pleasure and gratuitous comfort. Superficially the post-modern home appeared to represent a 'retrieval of feminine tastes' through the mass fabrication of previously hand-crafted designs, where convenience and hygiene were important considerations. Yet Domosh and Seager (2001: 28) stress that this rupture in the dominant aesthetic and rejection of 'macho modernity' did little to liberate those who were left to maintain the domestic realm (see pp. 201–2 on modern maid-work).

Instead a form of 'feminised post-modernism' can be identified in the more recent 'new urbanism', in which the design ideals of compact, mixed-use village-like urban arrangements are reproduced within proprietary settlements to emphasise family-oriented, liveable, human-scale neighbourhoods (Leccese and McCormick 1999). Whether in practice the new urbanism lives up to the intentions of inclusive design is highly debatable. This is the subject of ongoing research and debate. We return to consider the question of whether it is possible to create a gender-neutral, family-friendly or barrier-free city through design considerations alone in Chapter 8. Here, though, it is illuminating to reflect on the difference in both aesthetic qualities and purpose of two contrasting modes of post-modern housing with respect to waterfront apartments in Chicago (built in the late 1990s) (Figure 7.3) and village-like condominiums intended for a mixed community of households in Portland, Oregon (significantly remodelled in 2000) (Figure 7.4).

Homes fit for all?

The phrase and narrative 'homes for heroes', coined by then Prime Minister David Lloyd George immediately following World War I, crop up repeatedly in the history of home building in Britain. The phrase is politically laden, implying both the construction of new homes as a project of 'society building' and the moral association of home with duty and hard work. The general election of 1918 was momentous in several respects; it was the first in Britain in which women could vote and it put housing at the top of the political agenda for the first time because of the urgent need to accommodate the returning troops and their families. The 'homes

Figure 7.3 ***Variations on the post-modern home (1), waterfront apartments in Chicago***
Source: Helen Jarvis

Figure 7.4 ***Variations on the post-modern home (2), village-like condominiums in Portland, Oregon***
Source: Helen Jarvis

for heroes' campaign led to the 1919 Housing Act, which required local authorities to build homes with public subsidies in the first significant wave of social housing. It soon became apparent that the scale of construction needed was too costly for public subsidies alone, so the emphasis shifted to the private sector and the idea that lower-paid and working-class people should aspire to buy their own home.

The phrase 'homes for heroes' has resurfaced regularly since this time in media reports of a lack of adequate, affordable housing for 'key' workers. More particularly, the phrase has been mobilised with respect to reports of appalling conditions found in military quarters and the disproportionate number of ex-servicemen to be found among the homeless population (Higate 2000). David Cheal describes military organisations as 'total institutions' which 'make extreme demands upon individuals by a rigid power structure' (2002: 68). This rigid power structure is reflected in the way army wives are assigned military quarters which are cut off from other communities on the urban fringe; they see their husbands deployed without much notice and, as Cheal explains, 'it is understood by everyone that whilst on a tour of duty, military personnel should not be distracted by the mundane worries of their wives and children' (2002: 68). The politics of military housing are clearly androcentric; rank units of housing are distributed to the male solider (the hero) with scant regard for the feminised work of home-making. In turn, issues of poor housing and rigid power structures come to light in disproportionately high rates of divorce and domestic violence among men and women who are 'married to the military' (Tivers 1999; Hosek *et al.* 2002; Griffin and Morgan 1988).

A persistent shortfall in the quantity and quality of homes in Britain has contributed to unresolved debate concerning the role of the home and housing provision in the urban social economy. In 1942 the British government commissioned William Beveridge to report on ways to rebuild Britain after World War II. In this he named the five 'evil giants' of his time as 'want, disease, ignorance, squalor and idleness'. In particular, the standards of housing called for a massive programme of reconstruction and slum clearance but progress was slow. When the BBC broadcast the landmark docudrama *Cathy Come Home* in 1966 (see Case 7.1) the full horror of deep-rooted squalor and deprivation – conditions which should have improved over the intervening years – provoked a national sense of shame. While it is shocking to think about Cathy and others like her who were separated from their families and made homeless by poverty and social exclusion, our instinct might be to consign this story to history. Yet, for the urban poor in British cities today, where marginalised populations reflect intersections of gender, race, class and ethnicity, housing insecurity, rough sleeping, 'sofa surfing' and many other types of homelessness persist. More generally, a shift in government policy and cultural attitude which now views the home and housing as a private asset and store of wealth has led to a sharp increase in divisions in housing security and welfare, as we elaborate with respect to housing-led welfare.

It is also important to remember the stark differences between generally high standards of housing provision and the norm of home ownership in the global north and the precarious forms of shelter that are home to a variety of household situations in the south. Gender differences in access to housing and legal rights to property are the focus of an extensive part of the housing studies of the south, not least in Africa (see, for instance, Kalabamu *et al.* 2005; Moser 1987). One clear manifestation of unequal access to housing can be observed in the different housing careers and property rights that a man and a woman in a couple often face. To exemplify this, Rhema's story is presented in Case 7.2 (Cadstedt 2006: 107).

Housing-led welfare and care in the community

It is increasingly evident that the home is today expected to provide much more than shelter, privacy and a place for status display in advanced neo-liberal economies. Successive governments of Britain and the USA, for instance, have favoured home

Case 7.1: *Cathy Come Home*

In 2007 the UK homeless charity Shelter marked the date 40 years on from the day the shocking and controversial film *Cathy Come Home* was first screened, highlighting the plight of homeless families. The film highlighted failures in the housing system, lack of security and failures in urban living conditions for many marginalised households, problems which have not been resolved 40 years on. In Britain, where *Cathy* was filmed in the 1960s, we no longer find sex-segregated workhouses for the unemployed, but other indignities associated with homelessness, poor and insecure housing still persist. Indeed, some of the most 'successful' cities today have an acute shortage of affordable housing for individuals and families living on a modest or low income and the poor such that insecure housing threatens to undermine family stability and exacerbate rates of domestic violence, stress and debt.

In December 1966 the BBC broadcast the landmark docudrama *Cathy Come Home*, written by Jeremy Sandford and directed by Ken Loach. A quarter of Britain's population tuned in to see the appalling story of Cathy and her family. It reflected the miserable housing conditions of many inner city slums at the time and the large-scale human agony of the most vulnerable people in society who were hardest hit by the acute housing crisis of the 1960s. Full of youthful optimism, Cathy moves to the city, where she meets and falls in love with the mischievous Reg. Soon after marrying, when Cathy is pregnant with their first child, Reg is injured in an accident at work. Reg loses his job, they can no longer afford the rent, and so begins the family's slippery descent towards homelessness. Increasingly poor and desperate, Cathy, Reg and their three children are forced to move in with relatives, then into a caravan on the outskirts of the city and eventually to a derelict slum building. Cathy and the children end up in a hostel, while Reg is left to fend for himself. In the final, heart-rending scenes, Reg is forced to leave the family to look for work. Cathy is evicted from the hostel, and the children are taken away from her and placed in local authority care. The viewing public was outraged that this could happen in modern Britain. The screening of *Cathy Come Home* was arguably a pivotal moment in British history as it galvanised public opinion and voluntary sector determination to prevent bad housing from tearing families, and lives, apart.

Today the slums of the 1960s are gone, but 40 years on the housing crisis still exists, and it is widely recognised that housing conditions and residential location are key factors in the health, education and well-being of vulnerable groups in society.

Source: Extracts from Shelter (2006: 5).

Case 7.2: Home as dependence for Rhema – Mwanza City, Tanzania

Rhema was born in the 1970s in Mwanza City, in Nyakato, an outskirt settlement where her parents have a house. She stayed with them until she got married in the mid-1990s. After the marriage, she and her husband lived in his parents' housing in Igogo, an old, centrally located, unplanned settlement in Mwanza City. After a few years, when they had two children, they moved out. Rhema says her mother-in-law did not want to have her there any longer. They started to rent one room in a house in Mabatini; this is where they lived at the time of the interview. A friend of her husband knew the landlord. Her husband currently has a job in town, while she describes herself as a housewife. About two years ago, Rhema went back to her parents' home a few kilometres away because her husband had beaten her. She got money for the bus fare from an NGO working for women's rights in the area. However, her husband came and brought her to the hospital and from there to his parents' house. She is back living with her husband in the rented room and the children are still with his parents, but they plan to bring them home. A couple of years ago, they bought a plot up in the hills close to this area. They had planned to start constructing their house the year we met but due to the problems they have had, she does not think that they will manage to start before next year. Two years after I met Rhema, I returned to the house and the landlord told me that Rhema and her husband had moved in with relatives in Igogo. It can be assumed that she is back at her husband's parents' home. (Cadstadt 2006: 108)

Rhema's housing career has features in common with other female respondents' stories. To live with parents or other relatives until marriage and then be dependent on the husband for arranging the housing situation is a dominant pattern for the married women in the study. Her dependence on her husband is illustrated by the fact that she had to seek help from an organisation to get money to cover the bus fare to go to her parental home, a few kilometres away. (Cadstadt 2006: 109)

ownership above all other tenure and this has contributed to the cultural normalisation that individuals and families build up a store of wealth in their homes to 'draw down' in time of hardship or to fund personal care in old age or adult offspring education. It is generally true that home ownership is popular and represents a dominant 'tenure of choice' for around 70 per cent of British households. In the USA, Forrest and Murie (1995: 91) note the extraordinary extent to which buying a home is considered a rational investment, synonymous with 'having made it', associated with social cachet. They observe that 'the concept of private property is as fundamental as the belief in individual rights' (Forrest and Murie 1995: 87).

Of course the 'credit crunch'[1] of 2008/2009 has revealed the fundamental flaws in neo-liberal promotion of housing-led welfare and privatised risk. Belief that the home represents a flexible investment, much like an illusory cash dispenser, has contributed in no small measure to the problem of high-risk borrowing and negative equity. Banks can be held responsible for adverse lending, while changes in public spending on welfare and a market-led economy have led households to spend rather than to save their primary asset – the home. Consequently the illusion of social mobility built on home ownership coincides with unprecedented levels of household debt in Britain and the USA (Citizens Advice Bureau 2003; US Department of Housing and Urban Development 2004).

Bound up with this crisis in housing finance are the household welfare effects of a 'rolling back' of state services and a shift in the emphasis of funding and risk from the state to the household. Put simply, a neo-liberal welfare paradigm assumes that private markets deliver basic needs and wants more efficiently than the state. In Britain this process began symbolically with the mass sale of council housing under Margaret Thatcher's 'Right to Buy' agenda. In a home-ownership-led model of privatised welfare, a build-up of equity in the home replaces public insurance schemes as a source of 'pension plan' as well as a source of wealth from which to fund the care of elderly relatives, sponsor adult offspring through higher education and pay for specialist medical treatment and personal care in old age (Jarvis 2008). Of course for most people this source is more perceived than real and in any case is not infinitely elastic.

Compounding this financial pressure on the home is a parallel shift in health policy away from state-assisted residential care for elderly and disabled people needing high levels of personal care, toward a policy of Community Care (or care in the community). Under this policy the majority of the care burden (that is, the unpaid work as well as the cash costs associated with this) falls upon informal carers, who are susceptible to feeling isolated and experience poor mental and physical health themselves (Carpentier and Ducharme 2003). The scale of the problem is masked by the 'invisible' nature of home-based care – which is itself, of course, now on the increase. These neo-liberal policies exploit not only the gendered work of caring but also the material attributes of, and personal identification with, the home itself. This is because the home is a key site of care which is becoming institutionalised by the medical trappings of in-home nursing care (Milligan 2000; Dyck *et al.* 2005).

The gendered work of caring

In stark contrast with the idealised (largely mythical) image of the home as a haven of effortless domesticity, it has to be recognised that the home frequently harbours domestic violence, depression, ill health and anxieties over financial debts (Allen and Crow 1989; Darke *et al.* 1994). We are used to judging the contemporary home in the global north in terms of the comfort, cleanliness and convenience (Shove 2003) of its interior, and this in turn contributes to a rising treadmill of consumer expectation which has tended in turn to escalate standards of cleanliness and presentation. These are mutually reinforcing trends.

At the same time, in the more affluent cities, the magnitude and pace of technological developments noted in Chapter 5 have drawn attention to the perceived 'speeding up and intensification' of daily life. Both in urban studies and gender studies literature we find frequent reference to a 'time-squeeze' and associated 'care deficit'. Scholars generally agree that we are witnessing acceleration in the pace of life (although this can be more perceived than real), a rise in time-saving innovations, increasing stress and (gender) role overload (Southerton 2003), yet the relationship between time-squeeze and gender is not straightforward. On the one hand, labour-saving technologies in the home have reduced the time required for certain domestic chores such as laundry and specialist cooking (utilising microwave ovens, blenders, food processors, etc.) and in the case of ready-made meals and disposable nappies time and effort (if not the environment) have been saved by the substitution of household effort by a consumer product. On the other hand, standards of cleanliness and interior design have increased the frequency and intensity of some domestic chores (outfits changed several times a day, pale carpets and chrome surfaces, hotel-style presentation of guest bedrooms) (Cox 2006).

Time–budget analysis points to a persistent 'second shift' for women, who typically combine paid employment with the lion's share of household work. In the case of dual-income couples there is evidence that the battle of the sexes over who should do what in the home is side-stepped to a large extent by the practice of out-sourcing unpleasant or routine housework (Cox 2006). In this way individual women are variously the beneficiaries and the subjects of these transformations. In the absence of radical collective solutions to the realities of childcare, cooking and cleaning, and the failure of men in sufficient numbers to pick up a fair share of domestic reproduction work, some women are improving their wages and status relative to men in paid employment by out-sourcing their unpaid 'second shift' to low-wage cleaners, nannies, au-pairs and childminders – almost all of whom are women, typically migrant women of colour (Ehrenreich and Hochschild 2003).

We know from repeated 'who does what' surveys that women consistently perform the lion's share of unpaid domestic, caring and emotion-work, especially as mothers but also as daughters, lovers, neighbours and volunteers. The myriad activities of domestic and caring work extend far beyond the confines of the home: this translates visibly in the streets of towns and cities around the world into those who are 'attached' to (confined by) responsibility for young, old or mobility impaired dependants (Jarvis *et al.* 2001); but also less visibly in the 'juggling' and coordination of care-giving remotely from places of work. It also extends to the emotion-work of family members, neighbours and volunteers. Box 7.1 expands on the concept of emotion-work.

Box 7.2: Emotion-work

The terms 'emotion-work' and 'emotional labour' were first used by Arlie Russell Hochschild to describe the 'management of feeling to create a publicly observable facial or bodily display' (1979; 1984: 4). The terms built upon Erving Goffman's theory of dramaturgical social interaction (front stage, back stage, surface acting, deep acting, face-work). Nicky James went on to define emotional labour as 'the work involved in dealing with other people's feelings, a core component of which is the regulation of emotion' (1989: 15). The term 'relational practices' has also been applied to emotion-work as the 'collaborative work' of relationship building, which is typically rendered invisible and not recognised or rewarded as 'work' (Fletcher 2001: 1). Examples of emotion-work might include: a parent's efforts to keep her children happy; a nurse comforting a bereaved relative; or a trade-union leader rallying members to strike for fair wages. Historically the display of emotions has been denigrated as a feminine attribute. With economic restructuring and the proliferation of service sector employment, workers are increasingly expected to display (or convey) tightly scripted emotions as part of their job: cabin crew welcoming passengers onto the plane as if it were their own private living room; call centre operators smiling down the phone; shop staff wearing the shoes and clothing that are for sale in the store as living mannequins.

A recent pan-European survey suggests that even though more women than ever before are working for pay outside the home, and a caring and sharing image of 'new man' is propagated in the press, just one in ten partnered men across Europe admits to doing a fair share of the household chores (Dobson *et al.* 2007). Moreover, there is evidence that unequal household gender divisions account for high levels of psychic stress, depression and low well-being among economically inactive 'housewives' (Griffin *et al.* 2002). This remains the case 30 years after Marilyn French published her satirical feminist novel *The Women's Room* to highlight the mass of women's frustrations, resentments and a common feeling of being trapped in demoralising roles of domestic servitude (French 1997, first published 1977). In the USA today, 'even women who earn considerably more than their husbands seldom persuade their guys to put more hours into family work' (Ehrenreich and Hochschild 2003; Folbre 2001: 17). Consequently, as

Lynne Segal (1990) points out in the title of her book *Slow Motion*, while we have witnessed radical transformations and restructuring of gender roles and participation in employment, evidence of any fundamental change in dominant masculine identities and concomitant expectations of femininity has seriously lagged behind.

The 'hidden' nature of much unpaid relational work, especially emotion-work, is revealed in 'the female world of cards and holidays – the work of kinship' (di Leonardi 1987). As Carol Gilligan noted more than twenty years ago, 'women's place in man's life cycle has been that of nurturer, caretaker, and helpmate, the weaver of those networks of relations on which she in turn relies' (1982: 17). While the intervening decades have brought significant changes in affluent societies to the way nurture and care-taking are expressed, frequently through niche market consumption of products and services, the tasks of writing postcards to kin while on holiday, buying gifts for friends and relations and organising family meals and celebrations still largely fall to women (di Leonardi 1987; Clarke 2007). Case 7.3 expands on this notion of gendered emotion-work with respect to the impact of contemporary consumer culture on mothering in relation to a treadmill of expectations associated with the children's birthday party – in the UK context. Over the years, British children's parties have evolved as more elaborate affairs and at the same time they have also become increasingly commercialised. This raises the question of whether the burgeoning culture of children's parties signals the vibrancy of neighbouring activities and friendships or the commercialisation of an idealised (highly gendered) facet of domestic work and further evidence of the penetration of market forces into the non-economic worlds of nurturing and social (re)production (Clarke 2007: 265–266).

In a largely quantitative study of neighbouring activities in the USA, Campbell and Lee (1998) found that gender predicted 'good neighbour' volunteering activities more strongly than life course and participation in wage employment. This result was explained by the social construction of gender roles in the study areas, whereby the women interviewed felt emotionally and socially responsible for contributing to their immediate community and helping those among their neighbours who they felt needed outside help. This finding is repeated across a number of studies of different aspects of community volunteering and neighbouring. For example, in a detailed study of charity shop volunteers in Britain, Suzanne Horne and Avril Maddrell (2002: 73) identify gender and age as the most significant characteristics influencing volunteering. Moreover, they note how this retail landscape tends to remain feminised in large part through the 'snowballing' recruitment of volunteers through existing social networks of women volunteers.

Case 7.3: Contemporary consumer culture and the birthday party circuit

The ethnographic study cited here by Clarke (2007) involved 76 households spanning a range of ethnic and social backgrounds representative of a fairly typical London semi-urban ethnically diverse street. The methods and approach are similar to those used by Sandra Wallman in her classic (1984) urban domestic ethnography *Eight London Households.*

The typical contemporary child's home-birthday party commonly involves the careful selection of a theme; the making or buying of the ideal cake; the choice of miniature items for guests' 'gift-bags'; the co-ordination of age-appropriate games and activities; the purchasing or making of party clothes or fancy-dress outfits; the selection of party plates, cups, hats and tablecloths; the making [of] or shopping for desirable party treats and snacks; and, most importantly, the careful selection of just the right number and range of party guests. These selections are weighed heavily against the myriad of choices exercised by other parents at previous parties and the potential choices of future parties. Hence the children's party occupies, besides an obvious role as a marker of age, a heightened temporality as a genre of consumption (Clarke 2007: 265).

On Jay Road, an ordinary street in North London, Julie, the mother of a six-year-old girl, laments that a mother 'unfortunate' enough to have their child's birthday fall in August (between school in-take years) might have attended over twenty parties since the previous September due to the number of friends her child has accumulated. The sheer financial and organisational labour led many friends to be increasingly ambivalent about 'the snowball' effect of arranging and attending party after party and buying a seemingly endless round of suitable birthday presents. Belinda, a mother of three children living in a street adjaçent to Jay Road, observes that the 'party circuit' seems to have expanded in size and lavishness since she was a child brought up in the same neighbourhood.

While Julie and Belinda try to find ways to negotiate the sheer enormity of the commitment to children's parties with 'down-sizing' strategies, round the corner Camilla Knowles, mother of four young children, day-dreams about themes and novelties for her children's parties (Clarke 2007: 267–268). Camilla is well known amongst mothers in and around Jay Road as she attends many mothers' meetings and events and is dubbed 'super-mum' by her peers....Mothers on Jay Road marvel at Camilla's birthday parties and view them as the pinnacle of a mode of creative mothering in which attention to detail is manifest (Clarke 2007: 268).

Source: Clarke (2007).

Comment

On one level, children's birthday parties, their organisation, design and orchestration, can be seen to characterise the 'invisible labour' of the gendered work of caring. Yet, on another level, this study exposes a 'more than material' tension or contradiction between commercialised (costly, convenient, 'shortcut') expressions of love and 'authentic' home-made goods collectively idealised as a measure of 'commitment and housewifely skill' (Clarke 2007: 269). Elsewhere it is noted that efforts to value 'quality of life' more than income and consumer goods as something 'authentic' (or, put bluntly, home-made) is a gendered, socially stratified social movement.

This reinforces our understanding that gendered identities are produced and reproduced through the intersection of performers, performance and performativity. As Gregson and Rose (2000) explain with reference to contrasting accounts of performance by Erving Goffman and Judith Butler, performance (doing) and performativity (the social construction of doing things a certain way) are 'intrinsically connected through the saturation of performance with power' (Gregson and Rose 2000: 434). They go on to argue that 'space needs to be thought of as brought into being through performance and as a performative articulation of power' (Gregson and Rose 2000: 434). In this chapter we have illustrated these intersections of space, embodied action and discursive network in a range of domestic and neighbouring contexts of everyday life in Western cities. Now we expand this notion of the discursive network with reference back to the multiscalar processes introduced in Chapter 4.

Servicing the affluent home

The expansion of niche markets into all walks of life in affluent economies has resulted in not only rising living standards but also a treadmill of consumer expectation. This has led to increasing demand for migrant labour to provide new varieties and increasing forms of 'out-sourced' household domestic practices. Brenner and Theobald (2002: 65) point out that neo-liberalism cultivates a type of 'possessive individualism' which finds its expression not only in fortified, gated, semi-privatised public urban space but also in competitive consumption of 'keeping up with the Joneses'. This trend encourages self-realisation through new forms of consumption in and around the home, including (for the most affluent) recruitment of 'concierge' services and paid help in the home to undertake much of the additional work involved in the purchase and maintenance of new gadgets and fashion commodities. In turn this 'out-sourcing' of household work typically draws on, and drives, the flow of low-wage, unskilled, migrant labour to servicing sector jobs in the city (see Mitchell *et al.* 2004; Pratt 2004). At the extreme, the intimate emotion-work of selecting, buying and wrapping wedding and birthday gifts can be bought as a professional service.

A cursory glance through the advertising materials of some of the organisations that have capitalised on rising demand for cleaners from dual-income families reveals the paradox that remarkably persistent, traditional, sex-typed stereotypes overlay the phenomenal increase in professional women's participation in paid employment since the early 1980s. One of the largest domestic maid services, Molly Maid Inc., operating through thousands of franchised businesses in towns and cities across the USA and Britain, was recently valued at over £2 billion ($3 billion) (WhichFranchise 2008). The sales pitch claims that 'only Molly Maid can give your house The Pink Glove Treatment'. Every aspect of the brand is gendered in a stylised way. The name; the colour and logo (pastel pink with a frilly maid's cap); the marketing of a non-profit philanthropic 'Ms Molly Foundation' to assist victims and families affected by domestic violence – all key into gendered assumptions of female domesticity and decorative/subordinate femininity. Yet the intersections of gender, class, race and ethnicity remain omnipresent too. The franchise literature clearly differentiates between 'those women who clean and those who don't' with the message: 'our maid service franchise owners don't clean homes. We train you how to recruit, train, motivate, and manage your employees to do the cleaning' (mollymaid.com 2008). There are other very large residential cleaning services just like this too; feminist commentators have reported on the factory-like employment practices of corporate-run services such as Merry Maids and the Maids International in the US context (Ehreneich 2001; Ehrenreich and Hochschild 2003). Recent survey evidence suggests that somewhere between one in six and one in three households employs outside help to do some portion of its routine housework (Roberts and Hill 2005).

In New Zealand, a national home-maintenance franchise organisation has emerged to meet rising demand from new household types, especially older, single-person, dual-income and affluent professional households, for odd jobs to be done commercially where they are no longer undertaken informally as projects of 'do it yourself'. Again, the advertising material peddles narrow, anachronistic, sexual stereotypes, suggesting (as a caricature), for instance, that 'if your husband is not manly enough to put up those shelves – hire a real man, hire a hubby!' The advertising strapline on the website of Hire a Hubby is 'Hire-a-Hubby – for all those jobs your hubby can't or won't do because he's in the pub, or watching football!' (www.hireahubby.com/au, accessed 25 July 2008). Similar 'niche markets' are opening up for male stereotyped skills associated with yard work (lawn mowing, path clearing), household and car repairs, assembling flat-pack furniture and programming electrical appliances.

One UK 'hubby out-sourcing' business advertising franchise opportunities in the Midlands differentiates between two classes of hubby – tech hubby and general hubby. Here the on-line advertising goes: 'Be a hubby – have you got your own tools, and transport? Do you have your own PC?' (www.househubbies.com/, accessed 25 July 2008). The idea that it is possible (not to say comfortable, convenient and clean) to hire a husband and purchase the services of a maid by the hour illustrates the problematic intersection of persistent sexist discrimination with contemporary consumer commoditisation. The emergence of 'handy man' agencies along similar corporate lines to 'domestic cleaning' services – that is, paying minimum wage to poorly protected, often illegal migrant labour – raises the question of to what extent commoditised forms of male reproductive labour differ from those of paid domestic work performed by women (see Campkin and Cox 2007; Cox 2006; B. Anderson 2007).

Again, the intersections of poverty, vulnerability and embodied, gendered assumptions of who 'can' (or 'should') do which jobs where are critical here. How do we differentiate between jobs that are poorly paid and lacking basic protection? Along these lines, Barbara Ehrenreich poses the question: 'who cleans up in your house – a man, a woman … or a cleaner?' (Ehrenreich 2003; see also Shelley 2007 on the exploitation of migrant labour generally). This question goes to the very heart of the issue of multiple deprivation and stigmatisation: the cleaner is deemed of such low status that he or she is de-sexed, less than man or woman – barely a human machine. According to Ehrenreich (2003), marriage counsellors popularly recommend that dual-career couples hire a cleaner (a cleaning lady) to perpetuate the myth of gender democracy in their relationships and as an alternative to squabbling over whose turn it is to empty the dishwasher. These are profoundly gendered systems of occupational segregation at work – but they are global, intersectional and increasingly also dehumanising and exploitative for the poorest and most vulnerable gendered bodies of women *and* men.

The performance of gender in the new economy

As we saw in Chapter 3, international competition in labour market production has spawned new varieties of 'non-standard', 'flexible' contracts and working time arrangements. Restructuring has been so deep that it is no longer possible to capture working time and contractual arrangements for individual workers in a single category such as full-time or part-time, temporary or permanent employment (Casey *et al.* 1997: 19). Increasingly, the gendered status of employment is bound up with social capital and, as such, the fluid network relations connecting home, work, kinship, community and occupational affiliations.

From the mid-1980s feminist scholars began to shift their attention away from structures of occupational segregation and individual workers to consider the performance and cultural characteristics of the work itself. This shift in interest accompanied a general 'turn' to culture for explanations of why particular inequalities and normative stereotypes persist. Studies on gender and work (gender at work as a process) began to explore the ways in which gender-specific traits and characteristics are attributed to men and women through the work they do in particular settings. This approach highlights the way gendered identities are created and recreated at work, rather than being 'held' as fixed and unchanging attributes that individuals bring with them to the labour market (Stichter and Parpart 1988). Understood in this way, the gendered identities of both jobs and workers are negotiable and contestable. Jobs are not gender neutral; rather they are created as appropriate for either men or women, and the sets of social practices that constitute and maintain them are constructed so as to embody socially sanctioned but variable characteristics of masculinity and femininity (McDowell 1999: 135).

A particular challenge for the intertwining of urban studies and gender studies in this book is the reflexive and iterative transmission of workplace cultures through everyday household practices in the urban arena. Arguably from a household, everyday life perspective, certain workplace cultures, terms of employment and working time arrangements run counter to local norms of gender democracy in other spheres. For instance, many aspects of 'new economy' employment, such as collaborative development, business travel (remember the hypermobile cosmocrat in Chapter 6) and commercial activities that run 24 hours a day, seven days a week, run up against trends of gender democracy and efforts to 'balance' home and work (see Presser 2003) in more egalitarian ways. The origins of much software development, for instance, reside in all-night, caffeine-driven, 'buzz group' experimentation. This is not dissimilar to the 'studio culture' noted for architecture students in Chapter 5 – where this deadline-driven training becomes professionally rooted in promotion prospects, where clients are regularly entertained in corporate clubs and restaurants and tightly competitive deadlines are met by putting in excessive overtime on a regular basis. This work-focused identity has variously spawned practices of casualisation, up-skilling/deskilling, self-employment, chronic job insecurity, 'self-exploitation' (Ross 1997; Pratt 2000) and both the intensification and extension of work in an 'always on' existence (Jarvis and Pratt 2006).

These work-related pressures are perhaps never so apparent as in the 'hot spots and cool places' (Pratt 2002) of the new media and high-technology sectors. A paradigmatic high-tech hub is that of California's Silicon Valley, transformed from a gentle landscape of orchards to the powerhouse of the new economy in less than four decades. Case 7.4 explores the recent history of the Valley with respect to the paradoxical co-existence of a liberal politics and high-profile gender equality campaigns alongside the emergence of a culture of work which is antithetical to most aspects of caring and social reproduction work. In the commentary that follows we offer a snapshot of the 'macho' high-tech work culture prevalent in this context today. It highlights the practical and emotional spillover effects of the way information and communications technology allows paid work to invade the home, as well as the limited coordination of family members and domestic arrangements from places of work.

The powerful grip that a work-centred existence now has on a youthful, hyper-mobile elite in the new economy is captured in the following quotation from Ellen Ullman (1997), in her book *Close to the Machine*, on the alienating effects of working from home with new technology:

> The workplace [is] the last place where your position in the order of things is still known, where people must put up with you on a regular basis, over a long period of time, and you with them. Families scatter, marriages end, yet the office and the factory have hung on a bit longer as a staple human gathering place. Maybe this is why the decline of industrial work and the downsizing of corporations have produced such anxiety: the final village is dissolving, and those of us without real jobs – where will we meet each other now?
>
> (Ullman 1997: 145)

Once again, as the burgeoning literature on work and lifestyle in Silicon Valley attests, we see the intersections of gender, generation, race, class and social capital at work in the complex stratification of life-chances – the dynamic of winners and losers that shape, and are shaped by, gendered urban transformation.

Communities and networks

The term community tends to be used quite loosely in the urban studies literature and feminist scholars have been critical of its 'plastic nature' and the way it is used to evoke 'warm persuasion' in urban regeneration initiatives (Young 1995: 235). Whether or not it is conceptually robust in academic terms there is no avoiding this popular term in the media and across all levels of government rhetoric. Fraser and Lacey (1993) start to deconstruct the term by pointing to the

Case 7.4: The corrosion of home–work–life balance in Silicon Valley

In 2000 a woman, Amy Dean, headed the Santa Clara Central Labor Council and another woman, Carly Fiorina, headed Hewlett-Packard, the oldest and most prestigious firm to have developed in the Valley. When Fiorina was chosen for this position in the summer of 1999, she became the first women to run a FORTUNE 30 firm. By the same token, when Dean assumed the helm of the Labor Council in 1994 at the age of thirty one, she became the youngest such leader in any large council in the country, let alone the youngest woman leader.

These women were not the first in the area to take leadership roles. There was a historical precedent that goes back to the late nineteenth century. The work of women throughout the years eventually led to the area being dubbed the 'feminist capital' of the USA in the 1970s, when Janet Gray Hayes became the first woman mayor of a city larger than 500,000. This reputation was consolidated in 1980 when women won majorities on both the San Jose City Council and the Santa Clara County Board of Supervisors (Matthews 2003: 183). Paradoxically, even as an unprecedented number of women were being elected to office in the Valley, a culture was taking shape in the local high-tech arena that has been repeatedly described in terms of its macho qualities: work-obsessed, hard on family life, celebratory of the male nerd (see, for instance, Bronson 2000). How could such contradictory developments have occurred in the same place at the same time (Matthews 2003: 184)? One of the toughest aspects of the typical Valley high-tech professional job (which Matthews argues differs from previous defence contract and public sector professions in the Valley history) is that employees feel that they are on call twenty four hours a day (Matthews 2003: 187).

From orchards to high-tech to anti-growth environmentalism

Women's organisations flourished in the Valley. Scholars who have studied the emergence of anti-growth activities have noticed that there is a strong correlation between women finding political voice and women using that voice to protect the quality of life in their community. As one scholar in the Valley put it: 'to a great degree, the revolt against development was led by women ... the construction company, real estate developers, public work agencies, and building trades unions that formed the core of the dominant pro-growth coalition, by contrast, were overwhelmingly dominated by men' (Matthews 2003: 192). (Note the resonances here with the learning activity on Finger Wharf on pp. 213–15.)

To understand how a macho culture of work came in to being in this 'feminised capital' it is important to note how male the founding myths of Silicon Valley have been. Consider, for instance, the 1999 film *Pirate of Silicon Valley*. A 1990 Ford Foundation funded study of the impacts of Valley culture on men and fatherhood singled out Apple Computers as one of three firms for in-depth research. The research reported that 'the cult of the macho workculture is alive and well at Apple' (Matthews 2003: 210). This culture of work has taken a toll on family life, as demonstrated in a number of studies of work–life imbalance in Silicon Valley (Stacey 1998).

Source: Matthews (2003: 183–207).

Comment

While a significant gender gap remains in a number of professions (such as architecture, as we noted in Chapter 4), there are now more female than male solicitors and women are colonising specific sectors of the professions, such as general practice medicine. Yet as a number of feminist scholars observe, the light, bright workplaces of the new economy are no less heavily gendered than the assembly lines of the old economy (see Gill 2002; Perrons 2003).

The high-tech Silicon Valley economy represents a workforce ecosystem which is highly gendered in large part because technical prowess has been constructed as the key to status. Nationally in the US, only 19 per cent of the science, engineering and technology workforce are women (Commission on the Advancement of Women and Minorities in Science, Engineering and Technology Development 2000), 20 per cent in the Valley region (Collaborative Economics 2001: 5). There is a running joke that, not so long ago, university engineering departments were built without women's toilets (English-Lueck 2002).

Jan English-Lueck observes that differences in power play out in the small details of daily life:

> The way people use their technology acts as a signpost pointing toward the role of power. To whom do people talk, and how? When is it preferable to leave an e-mail, or a voice-mail message at 2.00pm., rather than to communicate with 'sneaker mail', a face-to-face conversation with a co-worker down the hall? How do people learn that just because an e-mail *can* be sent to the corporate president, it does not mean the message *should* be sent? Power relationships are expressed through technology within households as well. Some people are expected to leave their pagers on at all times, while other members

> of the family can turn the cell phone off. Children can be tethered to parental control by a pager. Power means having access to others, while limiting access to oneself.
>
> (English-Lueck 2002: 31; see also Darrah 2002)

way it variously refers to an entity (people who are connected by kinship and/or within a particular locality or through shared values or histories) and to an ideal, such as with the idealised village community promoted by the new urbanism and the urban village movement. Sometimes the term is used as if community and residential areas are synonymous. This is illustrated in the notion of creating 'balanced communities' by mixing home owners and tenants in new housing designed to cater for a spectrum of household and income types (Gans 1961; Minton 2002). In reality there are many examples of divided communities and multiple communities within an otherwise homogenous residential area, such as when ethnic and religious differences intersect with housing class or consumer culture and the like. In advanced consumer societies, for instance, people are as likely to identify themselves with others around them who share a taste and politics for alternative food networks (local, organic, allotment cultivation) as they are with others who attend the same church or trade union.

Household members effectively participate in spheres of activity inside and beyond the nominal boundaries of the household (the 'home sphere') – through employment, social movements, trade-union or political activity, by escorting children to school and entering into reciprocal ('do-it-yourself') exchanges with other households or extended family. These activities and social interactions both shape and are shaped by household ideology, through the mediation of gender, race, class and ethnicity. Arguably, households do not function on the head of a pin or in a social vacuum. A useful way of illustrating local, cultural situatedness is through the metaphor of the household 'as a node in a multilayered web or the locus for a number of networks of relations: economic, social and technological' (Wheelock and Oughton 1996: 156) which stretches through communities, social and kin networks and a multiplicity of formal and informal economic opportunities. This is similar to the way Jarvis presents the household as part of a wider social and material infrastructure (see Chapter 5) where processes of household decision-making are represented over time and space 'within a web of networks: of social and kin relations; of resource provision; and of information, knowledge and learning' (1999b: 226).

From this brief review we would argue in favour of focusing on household daily life rather more specifically than ambiguous notions of community. This approach allows us to highlight the existence of diverse economies whereby what economists and politicians usually refer to as 'the' economy (in terms of markets, wages and capitalist enterprises) is only one aspect of the economic activities in which people engage (Gibson-Graham 1996; Pavlovskaya 2004). In practice, diverse economies might be related to communities, where any surplus generated locally by women is typically reinvested in the community in ways that especially benefit social reproduction and a caring economy (Escobar and Harcourt 2005: 11–12); or diversity may reflect growth in 'alternative' networks – food cooperatives, time-banks, toy exchanges, baby-sitting circles, self-employment or philanthropic initiatives.

Previously we stressed the multi-scalar, co-constitution of urban settlement patterns and the local effects of labour market restructuring. Recent developments in theories of networked or mesh-worked 'glocalities' have prompted feminist scholars to look beyond homes and jobs as a clutch of buildings or a geographic 'place' to pay closer attention instead to the neighbourhood as a located bundle of resource attributes and amenities which serves larger areas, even regions (Horelli 2002: 9). In this respect, while it is still constructive to reflect on everyday activities, locations and journeys in terms of an infrastructure (or, as we stressed in Chapter 5, multiple non-material as well as material infrastructures), it is important to recognise 'hybrid networks of people, activities, services and technology' (Latour 1993: Horrelli 2002). The aim then is to examine the intersecting and interdependent people–place–network relations which constitute diverse gendered realities.

There is a clear precedent for thinking about the neighbourhood as a social dynamic in the work of the late Jane Jacobs and her seminal 'anti-planning' text *The Death and Life of Great American Cities*. The social and economic vitality that Jacobs (1961) eulogised in economically mixed and ethnically diverse traditional inner city neighbourhoods (such as Greenwich Village, New York City) crucially functioned through the 'problem of organized complexity' (or cultivated serendipity), which she considered part of 'an organic whole' (Jacobs 1961: 43). This interrelationship combined casual street interactions, 'eyes on the street' and a mixture of land uses with the pivotal concept of 'social capital': networks of trust, good will, fellowship and both civic and social engagement. Tracing the roots of social capital in the daily lives of urban dwellers and their relative 'neighbourliness' continues to have significant bearing in urban studies debates where the concern is to cultivate or 'engineer' social cohesion in otherwise divided communities and neighbourhoods. Consequently, much has been written concerning the definition of social capital: what it is/how it functions; why it is so important to community cohesion; whether it is under threat in the contemporary consumer age and what can be done to sustain it (see Bourdieu 1983; Coleman 1988; Putnam 2000; Field 2003).

According to John Field, the central thesis of social capital is that 'relationships matter' and consequently 'social networks are a valuable asset' (2003: 2). While Jane Jacobs observed the critical function of the street as a primary arena for these ties and interactions, others have pointed to a variety of sites and situations, including the home, family, kinship, workplaces and colleagues, trade-union federations and protest groups.

A number of researchers have considered the intersections of social capital with cultures of work and class but the relationships between networks and gender democracy are less well known (but see Dyck 1996; Uttal 1999). In early research on the role of class, Willmott (1987) observed that households with more resources across multiple economies have more choice over housing, transport and employment and that this relationship is co-constitutive, so that increased opportunities influence positive social capital and this way increased access to all manner of capital (material, social and cultural). More recent research focuses on the role of class and gender in friendship networks (Pahl 2000; Sayer 2005). Social network analysis is able to provide a flexible method of exploring community life, change over time, and networks of resource provision (Phillipson *et al.* 1999).

One of the most important ways that social capital functions with respect to the co-constitution of cities and gender is as 'lively conduits' of information and learning (Pratt 1998: 26; Jarvis *et al.* 2001: 88–90). This subtle sense by which gendered moral significances are transmitted through everyday interactions on the street and through community engagement is perhaps best illuminated by Susan Hanson and Geraldine Pratt (1995) in the seminal text *Gender, Space and Society*, which studies a single urban area, Worcester, Massachusetts, through a feminist geographic lens. Hanson and Pratt (1995) describe the relationship between socio-spatially determined networks, relative inertia and the choices available to households in terms of housing and jobs:

> Living in one place for a long time has important implications for how the housing and job markets function because residential stability nourishes the development of personal networks. Insofar as these networks of friends and acquaintances, relatives, neighbours, and co-workers are also tied to particular locations, they in turn foster rootedness. One way that personal contacts do this is by being lively conduits of information about housing and jobs.
>
> (Hanson and Pratt 1995: 190)

This intersecting, multi-scalar conceptualisation of locally embedded capital and networks and circuits of learning has subsequently been developed in a number of different studies, to help explain differences in partnering, parenting, neighbouring and civic engagement (see, for instance, Holloway 1998; Duncan and Edwards 1999).

More recently still, connections have been made between the 'soft' normative cultural networks of partnering and parenting and more organised, politically transformative activist networks. Case 7.5 introduces the background to activist mothering networks (such as Mothers Against Violence) which exemplify these new contexts for collective action (William and Roseneil 2004). We should also point out, for balance, that men's groups and activist fatherhood have also developed in the same context, though typically mobilising distinctly different gendered practices and strategies of political protest. One of the most famous examples of this, in the UK context, is the Fathers 4 Justice campaign, which evokes the now famous image of the activist father who in 2004 scaled the balcony of Buckingham Palace dressed as Batman to protest for a change in family law – to give him rights of access to his children in a typically bitter custody dispute (Sheldon and Collier 2006).

Summary

This chapter has traced the conceptual and empirical connections between homes, jobs, communities and networks. Rather than looking at these spheres in isolation, our aim was to highlight the multi-scalar intersections and interdependence of the co-constitution of cities and gender within and between these realms. First, we briefly reviewed the literature on the home and the workplace, critically examining the disciplinary isolation by which urban studies and gender studies traditionally endowed the private home and public street and workplace as discrete sites of gender discrimination. Moving towards an integrated analysis of cities and gender we suggested that discriminatory practices and cultures of constraint underpinning home, work, family and all aspects of 'life' are better explored through social networks of reciprocity, learning and exchange, or in communities of resistance (whether virtual, single issue or place based). Key to making these connections were the concepts of gendered identity and emotion-work, which we introduced in the context of changes in the meaning of home and transitions to a neo-liberal policy of community care.

Previously we noted that the notion of a masculine urban core and a feminised suburban periphery assumes a caricature of gendered urban space that rarely holds up to scrutiny. Trends of urbanisation, suburbanisation, counter-urbanisation and gentrification indicate a patchwork of land use types which defy a simple core versus periphery classification. More importantly, activities of production (paid work) and those of unpaid social reproduction, caring and emotion-work are not

Case 7.5: Activist mothering

Feminist scholars of women's community activism argue for the importance of examining family as well as neighbourhood institutions and social networks for the development and expression of political consciousness. Analysis of political practice illustrates the contradiction between women's performance of apparently traditional female roles and the revolutionary actions they adopt for the benefit of their families and communities. Activist mothering (illustrated by the organisations identified at the end of this case) is a good example of this.

Temma Kaplan (1982) uses the term 'female consciousness' to describe women who make political claims on the basis of their gender roles and participate in radical political action. In a similar vein, Maxine Molyneux (1986) differentiates between 'practical gender issues' and 'strategic gender issues' to capture the way women activists organise around their practical everyday needs for food, shelter, day care and housing versus organising around their gender-specific identities. Obviously, this distinction often breaks down in practice as analysis of women's community work demonstrates. The work on 'activist mothers' provides a new conceptualisation of the interacting nature of labour, politics and mothering – three aspects of social life usually analysed separately – from the point of view of women whose motherwork historically has been ignored or pathologised in sociological analyses – whereby the women studied challenge the false separation of productive work in the labour force, reproductive work in the family and politics (Naples 1998: 112).

Activist mothering includes self-conscious struggles against racism, sexism and poverty – challenging essentialist interpretations of mothering practices. Women of colour as activist mothers, especially those living in poor neighbourhoods, must fight against discrimination and the oppressive institutions that shape their daily lives and, consequently, as mothers they model strategies of resistance for their children. For example, Latin American and African-American women's struggle against racism infuses their mothering practices inside and outside their 'homeplace' (Acosta-Belén 1986; Naples 1998). Lessons carved out of the experiences of 'everyday racism' contribute to mothering practices that include 'handing down the knowledge of racism from generation to generation' (hooks 1990: 41).

Activist mothering organisation links:

- Mothers Against Guns: http://mothersagainstguns.org/
- Mothers Against Violence: http://www.mothersagainstviolence.org.uk/

- Mums' Army (UK magazine campaign): http://www.takeabreak.co.uk/mums-army

Source: Naples (1998: 112–114)

spatially separated but are instead dynamic and contingent upon multiple tasks and elaborate coordination work performed in tandem 'virtually' or through 'between places', on the move and as the result of vigilant clock-watching. At the same time, here we pointed to the enduring distinction in status between the sites and stereotypes of work and employment that are 'naturalised' as female as opposed to those endowed with masculine power and status. The roles in which women undertake waged work 'in public' are quite typically feminised and assigned low status; they are marked out hierarchically even if not spatially segregated through their concentration in caring and quasi-domestic occupations in locations close to their own domestic caring commitments.

As with other trends of spatial segregation the exact nature and extent of women's 'entrapment', whether through labour demand, labour supply, human capital 'preference', commuting or spouse effects, are highly contested (see, for instance, England 1993; Hanson and Pratt 1994; Cooke and Bailey 1996). It is unclear from current research, for instance, whether the sorts of dense, compact communities advocated by proponents of feminism and the urban village approach to resolve the problems of fragmented suburbs are actually successful in addressing problems of gender-constraining routine coordination (Markovich and Hendler 2006: 424). Resolving these questions and unravelling home–job–community–network interdependencies require that we look beyond static behavioural studies in these domains – to consider issues of governance, in order to assess the future possibility of more inclusive, liveable cities.

Learning activity

Read the following description of the changing fortunes of a former industrial waterfront in Woolloomooloo, Sydney, Australia; consider how this case study exposes the intersections of city and gender in relation to intersecting identities of gender, class and affiliation (trade unionism), land use and new social movements mobilised around land use and growth control. Some of the questions you might ask of this case are included below but you might also think up more of your own. The following website links may be useful for background research.

- http://travel.webshots.com/album/53477509cWEaDy
- http://en.wikipedia.org/wiki/Finger_Wharf
- http://www.cityofsydney.nsw.gov.au/development/documents/2030/CouncilMeeting031207/071203_PDTC_ITEM02_ATTACHMENTN.PDF

For most of the twentieth century the Finger Wharf at Woolloomooloo was a bustling workplace. For the male 'wharfies' who loaded and unloaded cargoes here, working conditions were often harsh, dirty and dangerous. Most labour was casual – a job today did not necessarily mean a job tomorrow. Daily work was handed out at the wharf gates by the 'bull system', which favoured the biggest men and those least likely to complain. Physical deformities, broken bones, torn muscles and diseased lungs were common among workers on the wharves. The Waterside Workers' Federation (a trade union) fought to improve wharf labourers' conditions. Opposition from employers and government was often violent but through the tenacity of the trade union the bull system ended and conditions slowly improved. By the mid-1950s, wharfies had attendance pay, sick leave, annual leave and first aid equipment, medical leave, washrooms and clean places to eat at work.

During the 1970s, new container ports, cruise liner facilities and airports in other places around the city of Sydney gradually replaced the wharf's function. The work and workers moved with them. For nearly a decade, this enormous building lay derelict and decaying. In 1987, the State Government decided to demolish the Finger Wharf.

After more than 70 years at the heart of Sydney's cargo and passenger handling industry, the Finger Wharf at Woolloomooloo had become a local landmark – a symbol of community struggle for both economic development and social justice. The State Government's decision to replace the Wharf with a new marina caused a storm of community protest. Supporters of the Wharf, including both former wharfie families and 'slow-growth' environmentalists, formed the Friends of the Finger Wharf. This group supported a Green Ban on demolition, promoted environmentally sensitive retention of the Wharf and encouraged public participation in local planning decisions. Powerful property developers and politicians still pushed for the Wharf to be torn down. The minister assisting the premier condemned it as 'an eyesore'. In January 1991 some 120 placard-waving protesters converged on the wharf to demonstrate against its demolition. The wharf was temporarily saved but the government was not prepared to conserve it and it was left to rot. After another year of public meetings, protests and debate, the government called for redevelopment proposals that would preserve Finger Wharf and in May 1993 the contractors moved onto the site.

Today the wharf represents a quintessential example of waterfront regeneration (gentrification). It incorporates private up-market 'gated' apartments for sale and for lease, a boardwalk for exclusive restaurants and wine bars, as well as a private marina.

Ask yourself the following questions:

1 How would you describe the land use and community profile of the Finger Wharf today? (Think about who lives and works here and how the space is used by day and by night.)
2 What evidence do you find of changing gender divisions, inequality and uneven development over the course of history in this case?
3 How might the different protagonists (wharfies, local residents, environmental groups, politicians, property developers) reflect on the relative 'success' of participating in this process?

Note

1 Credit crunch is a term used to describe a sudden reduction in the general availability of loans (or 'credit'), or a sudden increase in the cost of obtaining loans from the banks.

Further reading

Halford, S. and Leonard, P. (2006) *Negotiating Gendered Identities at Work*. Basingstoke: Palgrave.

Hanson, S. and Pratt, G. (1995) *Gender, Work and Space*. London: Routledge.

Hayden, D. (1980) 'What would a non-sexist city be like? Speculation on housing, urban design, and human work'. *Signs*, 5.3: 170–187.

McDowell, L. (1997) *Capital Culture: Gender at Work in the City*. Oxford: Blackwell.

Matthews, G. (2003) *Silicon Valley, Women and the Californian Dream: Gender, Class and Opportunity in the Twentieth Century*. Stanford: Stanford University Press.

Pratt, G. (2004) *Working Feminism*. Edinburgh: Edinburgh University Press.

Sennett, R. (1998) *The Corrosion of Character: The Personal Consequences of Work in the New Capitalism*. New York: W. W. Norton.

Walkerdine, V. (2003) 'Reclassifying upward mobility: femininity and the neo-liberal subject'. *Gender and Education*, 15.3: 237–248.

Part III

Representation and Regulation

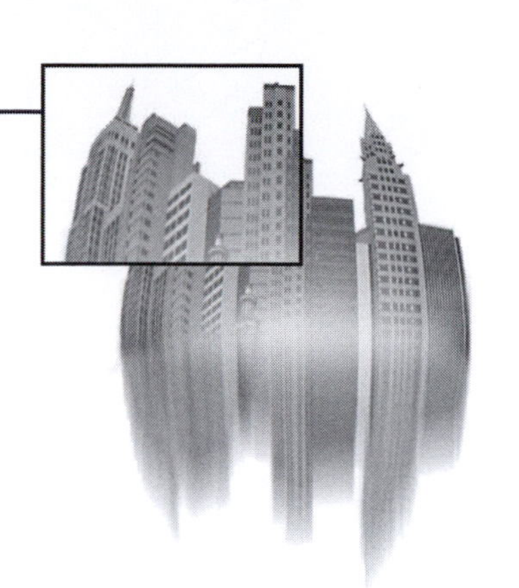

8 Planning and social welfare

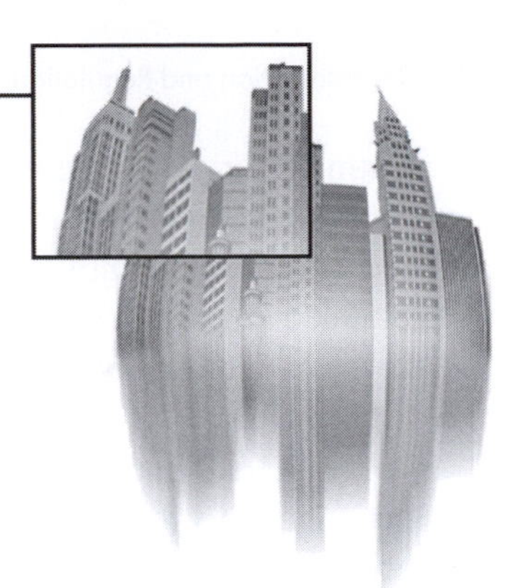

Learning objectives

- to gain familiarity with concepts of governance and gender mainstreaming
- to recognise diverse and intersecting functions of governance, resistance, surveillance, participation and control
- to critically examine how cities are planned and managed as multi-layered gender systems
- to appreciate the unintended consequences of institutional ideologies (including sexism) in the representation, regulation and performance of planning and social welfare

Introduction

This chapter explores the way cities of the global north and south are planned and managed, alongside the dynamic gender systems by which societies are segregated and organised. By asking who plans, how, what and for whom, this chapter builds upon earlier observation of androcentrism discussed in Chapter 5. Here we lift our gaze from the everyday infrastructures of homes, jobs, transport and emotion-work to consider the role and influence of multi-layered governing institutions. Our discussions are structured in two parts to reflect two principal routes by which traditional gender norms may be challenged: the first by working to change the apparatus of the state and associated governance (from within); the second by grassroots resistance – against the state.

In the first part we differentiate between government and governance, making specific reference to the social and cultural construction of gender systems and gender contracts. We illustrate the way these systems can be observed to function in direct and subtle ways; in self-help housing in Africa, in neo-liberal policy in

Latin America and in efforts to incorporate gender mainstreaming in a more progressive model of urban governance in Porto Alegre, Brazil. We also examine the way gender identities and performances which do not fit prescribed norms are branded as deviant or criminalised by street-level bureaucrats. This is illustrated in the case of outdoor prostitution in London.

In the second part we consider the role of resistance movements and collective action against the prevailing gender order. We develop our argument by considering a number of alternative planning visions. Bound up with this discussion are unresolved debates concerning the feminist political goals of gender mainstreaming. We also point to the corroding effects of competitiveness and growth as fundamental drivers of non-representative urban governance, issues to which we return in Chapter 10 with respect to future prospects for urban social justice and, more immediately, the part to be played by transformative classroom teaching and practitioner engagement.

The 'master plan'

Historically the planning and development of cities have been derived from a supremely masculinised belief system that places a heavy emphasis on the male-dominated sectors of economic activity in the city and elsewhere. This legacy of struggle over space and representation, between the masculine City-as-Citadel and the feminine City-as-Garden, dates back to the Platonic ideals of Grecian philosophy and, even further, to the ancient city-forms of the Middle East. Akkerman describes the idealisation of geometric linear layouts in a rigid and hierarchical form as a 'Platonic-Cartesian faith in geometry and mechanics' (2006: 242). In this orderly social geometry, line and angle clash with the messiness of daily life in a continuous (and continuously unsuccessful) effort to force fluid socio-cultural complexity into neat compartments and orderly processes. Moreover, in the global south, in countries which have inherited many of these urban planning and regulatory preconceptions whilst entirely lacking the centralised state mechanisms, the public wealth or the learned experience of economic evolution through industrialization, such ideals have caused even more problems.

This partial, masculinised view has resulted in a preponderance of the severest impacts being experienced in the more feminised sectors of the socio-economy globally, by virtue of their being rendered invisible or unimportant. This belief system is particularly notable in the language and ideas behind the 'master plan' and its association with competitiveness and growth as the dominant purpose of development. In this context, those with the power to commission and execute the 'master plan' believe in a form of environmental determinism which we go on to show is not borne out by the evidence.

Although words like masterpiece, master plan and master bedroom are widely used as if they are gender neutral, it is hard to ignore the way the historical association of 'master' with maleness and male authority permanently keys gender into sexist language (see Romaine 1999). We go on to illustrate in this chapter how both top-down and street-level bureaucracy reflect an enduring legacy of sexist language, structures and practices. Viewed this way, what shape the built environment and the systems (and absence) of welfare provision are not simply decisions from 'the' state (top-down) but instead are played out in the multiple gendered sites and dilemmas of front-line planners and welfare workers who interpret public policy in daily practice (see Lipsky 1980).

Gender mainstreaming

These failings have not gone unchallenged. Gender mainstreaming, participatory democracy and sustainable livelihoods have all developed rapidly over the last 20 years as analytical narratives through which to reposition marginalised groups and communities in an effort to remedy persistent exclusions and bias. According to Horrelli (2002) 'gender mainstreaming' variously represents a policy, an approach, a philosophy, a mechanism, a strategy, and a method. It is a transformative process, meaning 'the (re)organisation, improvement, development, and evaluation of policy processes so that a gender equality perspective is incorporated in all policies at all levels and all stages, by the actors normally involved in policy making and in practice' (Council of Europe 1998: 6). The 'mainstream' that is being influenced includes less tangible cultural norms and assumptions (attitudes and minds) of regional 'gender systems' or 'contracts' (which we define on pp. 223–4) as well as governing institutions. We saw evidence of efforts to instigate gender mainstreaming in the Nordic project New Everyday Life (Forskargruppen 1987) and the EuroFEM toolkit introduced in Chapter 4. A more detailed outline of this concept is included in Box 8.1.

Box 8.1: Gender mainstreaming

The concept of gender mainstreaming emerged during the United Nations Third World Conference on Women in Nairobi in 1985. It harks back to the 1960s when Betty Friedan campaigned for the National Organisation of Women (NOW) with the manifesto 'to take the action needed to bring women into the mainstream of American society, now, full equality for women, in fully equal participation with men' (Friedan 1976: 119; quoted in Oakley 2002: 55). In a policy-making context the aim is to routinely incorporate a gender perspective in all research and proposals for reform, to prevent 'solutions' being arrived at which maintain or exacerbate persistent inequalities. An internationally shared ambition to 'promote gender equality and empower women' is the third of seven Millenium Development Goals (MDGs).

Gender mainstreaming in theory

Distinctions need to be drawn between gender mainstreaming as a practice (such as a 'test' of gender awareness in government policy initiatives) and as a form of theory or as a project of political transformation. For Sylvia Walby (2005: 321) gender mainstreaming as a form of theory 'encapsulates many of the tensions and dilemmas in feminist theory and practice over the past decade and provides a new focus for debates on how to move them on' (see, for instance, Walby 2001; Woodward 2003). She identifies six major issues:

1 how to address the tension between 'gender equality' and the 'mainstream';
2 whether the vision of gender equality invoked by the mainstreaming process draws on notions of 'sameness', 'difference' or 'transformation' (or inclusion, reversal or displacement; see Verlook 2005: 344);
3 whether the vision of gender equality can be distinguished from the strategy to get there, or whether these are two dimensions of the same process;
4 the relationships of gender mainstreaming with other complex inequalities, especially those associated with ethnicity and class, but also disability, faith, sexual orientation and age (see pp. 91–2 on intersectionality);
5 the relationship between 'expertise' and 'democracy', and the rethinking of the concept and practice of democracy to include gender relations;
6 the implications of the transnational nature of the development of gender mainstreaming, including the influence of international regimes and the development of human rights discourse.

In short, making sense of the contested issues bound up with the theory of gender mainstreaming requires that the (local) meaning of gender equality is addressed as well as the project of gender mainstreaming (Walby 2005: 321).

We argue that if urban planning and social policy are ever meaningfully to reflect the ideals of gender mainstreaming they need first to recognise that cities and households function through a system of gender contracts which are locally, culturally and temporally constructed and reproduced. We turn now to explore these multi-layered and intersecting systems of hierarchy and segregation.

Gender systems, gender contracts and urban governance

Recent developments in feminist theory affirm that the social norms, rules, regulations and principles that organise and structure urban societies are not gender-neutral entities but are instead inscribed with specific norms for masculine and feminine roles (Young 2005: 3). There is, in effect, a hierarchical mesh-worked intersection of regulating systems, namely national welfare regimes, gendered governance, gender systems and gender contracts. In the urban studies literature it is frequently implied that national welfare regimes exert the most powerful influence. We question this thesis and instead draw attention to the myriad gender contracts which together constitute gendered governance (see Box 8.1). Within this framework, gendered governance can be viewed as the macro-political aggregate of combined state and civic spheres of influence. Within this, the state-wide gender system can be thought of as organising dominant narratives such as heteronormativity, patriarchy and individualism, for example (Mortensen 2006: 124). Finally, and of particular significance here, gender contracts can be thought of as the 'invisible power relationships that determine roles, responsibilities, privileges, status, sexuality and behaviour of men and women within households, communities, the market and the state' (Young 2005: 3).

From an everyday life perspective we can identify myriad local gender contracts underpinning a single dominant gender system and it is through the continual engagement of this system with local contracts that the prevailing gender order is preserved or transformed (Kalabamu 2005). The feminist historian Yvonne Hirdman (1991) is credited with theorising the relationship between state-wide gender systems and their constituent, culturally constructed gender contracts. Hirdman's (1991) notion of gender systems has been extremely influential in Sweden, where a distinct dual-breadwinner/dual-carer gender system is widely regarded in feminist policy circles as the system of governance most closely resembling a gender egalitarian, social democratic society. This Scandinavian system is supported by gender contracts which link through household gender divisions, employment law, childcare subsidies, as well as progressive media representations of masculine and feminine identities.

Box 8.2: Governance

Governance broadly refers to the sum of the many ways individuals and institutions, public and private, plan and manage the common affairs of the city. The term is widely used in urban studies and development studies but is less often found in feminist and gender studies literature.

In the context of urban social geography, Knox and Pinch (2006) mobilise the term to differentiate between the apparatus of the state and that of civil society in the variety of methods by which societies are governed, or regulated. In this way, 'the term is used to indicate the shift away from direct government control of the economy and society via hierarchical bureaucracies towards indirect control via diverse non-governmental organizations' (Knox and Pinch 2006: 319).

Within development studies literature the term is used to emphasise the continuing dynamic process through which conflicting or diverse interests are negotiated. For instance, the UN-Habitat Global Campaign for Good Urban Governance seeks to engage informal groups and the social capital of citizens alongside formal governing institutions to empower marginalised groups through participatory democracy. In this context the goal of good urban governance is to construct the inclusive city through cooperation and consensus (UN-Habitat 2003).

Arguably, the way that the term governance is used in much of the literature on cities perpetuates androcentric norms, largely because the governing institutions considered are rooted in a narrow focus on formal economic growth. By contrast, Jo Beall (1996) argues that urban governance must be gender sensitive if it is to be equitable, sustainable and effective. In this she points to the crucial importance of participation and civic engagement. We wish to go further by arguing that if urban planning and social policy are ever to be equitable, then greater understanding is first needed of the gender contracts that constitute the prevailing systems of government *and* governance.

The local specificity and effects of gender contracts are clearly illustrated in the case of self-help housing in Lobatse, Botswana, in Case 8.1. In this case it is especially notable that the macro-level imposition of a housing 'master plan', alongside imported building techniques and planning codes, competes with traditional gender divisions and practices of housing provision found elsewhere in Botswana. This highlights both the malleability of gender contracts (in this case, who is involved in housing construction and how this provision is valued) and at the same time the enduring, though negotiated, male control of the most highly valued resources. This case shows that everyday practices and livelihood strategies are subject to negotiations that take place within and between different layers of governing systems (or gender order). Vulnerability, therefore, can be understood as the situation of not being able to negotiate these systems or being unable to resist the authority of 'experts' who may be in a position to ration or set conditions to access essential resources. One reason why the women of Lobatse lack the skills (and access to land) necessary to build homes in the self-help housing project is that municipal bureaucrats and regulatory institutions (architects, engineers, project manager and funding bodies) have keyed the parameters of permissible construction into dominant Western, male, 'modern' expectations of separate gender roles. This 'expert' discourse is so powerful that the women have 'internalised their inability to build modern homes'.

National welfare regimes

In the developed world, national welfare regimes are commonly identified by the extent to which the state, the market or the individual and their family make provision for times of hardship (in terms of healthcare, pensions, unemployment, childcare, parental leave and the like) (for international comparisons, see Gornick and Meyers 2003; O'Connor *et al.* 1999). Since it is rare to find any national or city-regional context in which the state, market or family solely provides for individual well-being, it can be difficult to unravel the relative importance of each of these layers of influence. A common approach is to fix on institutional differences, such as levels of taxation and statutory social security provision in circumstances of unemployment, assuming in turn that these shape individual choices and behaviour (Hakim 2000: 172). Yet there is a paradox in the way the stronger welfare state may not necessarily foster gender democracy (Folbre 1994: 161). Understanding this puzzle requires an understanding of the way welfare regimes compete with cultural norms such as the powerful normative assumptions of 'proper parenting', ideal family arrangements (two heterosexual parents) and 'respectable femininity'. This is illustrated by the politicisation and media vilification that has frequently been the lot of the 'single mum' in the UK and the USA.

With differentially empowered groups, the key to fundamental change has still to be conducted through negotiation with the state, as much the arbiter of last resort for the vast majority as a source of transfer payments. This is despite the way the 'rolling back' of the state in recent decades has reduced the scale of welfare state provision. The extent of welfare reduction is such that previously universal social insurance benefits, such as housing assistance and unemployment payments, are now restricted to the most needy, typically stigmatised and subject to the conditions that recipients meet certain behavioural norms (such as actively seeking employment, sending children to early years programmes and school, and not causing a 'nuisance' in the neighbourhood). Consequently, while the scale of state welfare intervention (and what some would call the 'nanny state') has shrunk, the bureaucratic surveillance of conduct has increased in many direct and subtle ways in affluent neo-liberal economies. Moreover, enduring assumptions that separate a competitive, growth-oriented public sphere from a feminised, care-burdened domestic sphere result in a poor fit between the structure of planning and welfare systems and

Case 8.1: Changing gender contracts in self-help housing

While gender contracts in traditional Tswana societies arose out of men's control of resources critical to subsistence livelihood – land and cattle – as well as dominance of, and exclusion of women from, the public sphere, analysis of data in this study shows that the negotiation and definition of the emerging gender contracts in self-help housing in Lobatse were based on at least three contract realities: (i) male-dominated state apparatus controlled access to land for self-help housing; (ii) more men than women had better paid, verifiable and regular income; (iii) women lacked skills to build homes permitted by the state in self-help housing areas.

Contract 1: *House construction is 'masculine work'*. A principal gender contract, from which other contracts in self-help housing appear to be derived, is the widespread acceptance that only men could undertake house building and related construction work. Although some women had previously built traditional houses in their home villages and unplanned areas in Lobatse, female respondents said they could not build homes under self-help housing schemes. They had internalised their inability to build modern homes.

Through the enforcement of urban development standards, building codes and regulations, state apparatus had mystified self-help housing and, therefore, defined what is socially acceptable housing. Only people, mostly men, equipped with modern building skills and abilities to read and interpret maps, engineering drawings and architectural plans could effectively participate in housing construction.

Contract 2: *Men are responsible for housing provision*. The second contract is a logical follow up to the first one in that respondents expected men, the builders, to provide houses for their families. These findings contradict the traditional gender contract in Botswana where women were responsible for housing provision while men were expected to provide expensive and socially valued items such as cars and cattle.

Contract 3: *Women furnish and take care of houses*. An overwhelming majority of the respondents said they expected women to provide furniture and domestic appliances. These expectations are consistent with the traditional gender reproductive contract where women were responsible for domestic chores.

Contract 4: *Men make financial decisions in self-help housing.*

Contract 5: *Women play supportive roles in self-help housing.*

Source: Kalabamu (2005: 257–258).

men's and women's lived realities on the ground. For example, Nancy Fraser (2000: 1) points to evidence that just 7 per cent of American families with children maintain the sort of family defined as 'normal' by the prevailing welfare system, that of a father and a mother together, with the mother at home and the father employed (Fraser 2000: 1; Marotta 2002).

There is a tendency nevertheless for advocates of more egalitarian gender systems to highlight national comparisons between the low taxation, weak social security of the USA and the high taxation, generous welfare provision of, say, Sweden. Who would not be struck by the 65 weeks (15 months) of paid leave parents in Sweden are entitled to share (at 80 per cent of earnings for 12 months) compared with the 12 weeks of *unpaid* family and medical leave parents (namely mothers) in the USA *might* be entitled to use to care for a new-born child if their work history and employer satisfy the eligibility criteria (Jarvis 2005a: 54; Gornick and Meyers 2003). Arguably, national differences in overall welfare provision can have a significant impact on individual men, women, children and families, as Nancy Folbre (2001: 202) observes with respect to her conceptualisation of families as 'little welfare states'. Yet they do not tell the whole story.

Critics point to the questionable assumption that a private reality (cultural tastes, identity) can be read off from systems of social welfare. Life is far richer than is implied by a single scale such as the state: there are many intersecting gender contracts which serve to variously reinforce and transform the prevailing system. Macro-level comparisons which 'read off' gender inequalities from differences in state planning and welfare policies are criticised for neglecting difference and diversity. On the one hand, there are those who focus on the opportunities women and men have to express individual preferences, such as the extent to which they engage with parenting or employment. For instance, Catherine Hakim (2000) views the 'choice' for mothers to work outside the home as being something highly individual, in which the state and national identity barely register. She draws attention to the neglect of individual differences in women's market resources and investment in human capital (education and skills) relative to what she calls 'home-centred' versus 'work-centred' preferences. Her argument is that such work–lifestyle preferences are at least as important as social and economic factors in determining women's employment patterns (Hakim 2000: 174). Critics point out that this paradigm models choice as both 'free' and 'rational' where in practice it is shaped by 'social ties and socially negotiated moral rationalities' (Duncan 2005: 52).

On the other hand, Bob Connell (1995) points to the different milieux in which gendered subjectivities are constructed, such as the workplace, home and kin networks. These cut across national systems to create local cultural differences which may be stronger than shared national 'headline' trends associated with government policy or provision. Similarly, Birgit Pfau-Effinger (2004, 1993), and others researching local differences in women's employment in Europe, argues that behaviour is steered neither by the state nor by the individual in isolation. Pfau-Effinger (1993) places social ties and moral responsibilities centre-stage, understanding people to build moral identities and reputations (as the 'good enough' mother, the loyal worker or supportive spouse), thus representing a person with particular competencies. People negotiate these identities and reputations in relation to others. Thus when relationships are changed, for example, through divorce, people may respond by renegotiating their identity (Smart and Neale 1998). This evidence of multi-scalar difference brings us back to a crucial recognition of fluid and dynamic gender identities and the gendered body.

Reflecting on local cultural specificity brings us back to the lived reality of vulnerable livelihoods in situations where the concept of a state welfare regime is of limited relevance. Forms of urban governance and planning that seek to engage with disempowered communities and groups in the cities of the vast mass of excluded urban poor in the south need to recognise the cooperative, semi-cooperative and non-cooperative communal behaviour which results from everyday coping strategies, as well as the unequal structures of power within which they are conducted. The 'moral economy' that Scott (1976) referred to, for instance, by which social risks are insured informally within collective, reciprocal arrangements, frequently represents a volatile and unhealthy cocktail of identity interfaces and power negotiations (for rich and poor alike) through which belonging and exclusion are negotiated and enforced. A good example of this is suggested by the unofficial village and caste *panchayats* in India, which attempt to oversee and regulate the accelerating intermixing of caste, religion and sex occurring even in the rural hinterland, up to and including the exercise of the power of life and death. Poverty and vulnerability, then, are also 'anti-plan' and continuously clash with the orderly regulation of the city envisaged as indispensable to modernity – they are a constant reproach to the ambitions of urban regulators and a totalising vision of strategic urban planning.

Street-level governance: embodied surveillance

So far the discussion indicates clearly that cities, societies and household decisions are not determined by master planners and national welfare regimes in any monolithic sense. In reality, cities are planned and managed through multi-layered gender systems, not least of which are the street-level, grassroots and everyday practices of surveillance which we turn to consider in this section. In practice, dealing with the effects of 'fire, disorder, robbery, disease, war, natural disaster, terrorist violence and pestilence' falls less to 'the plan' than it does to individual performances and practices of emergency workers, welfare providers and 'the taken-for-granted heartbeat of the city', as dramatically revealed during great urban disasters like 9/11, the flood in New Orleans in 2005, and the Madrid and London train bombings in 2004 and 2005 (Stenson 2007: 32). This returns us to the everyday life approach theorised in the earlier chapters. As with Lipsky (1980), we recognise the crucial role of street-level bureaucrats and we agree with Lyon (2007: 83) that bureaucracies as mundane as human resource management function as effective forms of governance and surveillance.

Ehrenreich (2001) provides a ready example of this in her searing exposé of US minimum-wage conditions of employment. She cites the case of the computerised personality test (not to mention the routine drug test) which stands between a hard-up job-seeker and a minimum-wage job at a large multinational such as Wal-Mart. It is easy to see how the personality test is used as a measure of zero tolerance toward non-conformity in bodily performance, clothing and social values (Ehrenreich 2001: 124–125). Elsewhere there are myriad ways that people's character, their 'worth' or perceived 'threat' are read off and judged (even if unconsciously) on the basis of body shape and appearance, whether in a job interview, by security guards or the police or when someone weighs up whether to give cash to a homeless pan-handler (Dixon *et al.* 2006; Simonsen 2003; Colls 2002; Longhurst 2001; Domosh 2000; Grosz 1994).

As Townley (1994: 16) observes, the gatekeepers of employment epitomise gendered subjectivisation in the same way that front-line welfare workers effectively translate entitlement and 'merit' (as someone worthy of help) in situations where male/corporate agendas rule. While acknowledging the seminal work of Lipsky (1980) on street-level bureaucrats, it is clear that welfare outcomes (such as access to childcare or emergency accommodation) cannot be simply 'read off' from formal policy discourses: the uneven development of street-level services intersects with enduring practices and normative ideologies (such as proper parenting and respectable femininity) to effectively stigmatise, discriminate and ration the delivery of welfare. In short, management techniques (such as the survey), new technologies (such as closed circuit television) and human resource managers (and front-line welfare providers) combine to make activities and individuals knowable and governable (Lyon 2007: 84).

Another example of street-level governance is the way that cash-strapped municipal leaders leverage private finance for urban revitalisation by promoting the city as an 'unmissable opportunity for developers' and by offering assurances that urban 'disorder' will be policed out of the area by a zero-tolerance approach such as that popularly called 'broken windows'. The phrase 'broken windows' was coined in a 1982 edition of the *Atlantic Monthly*:

> Consider a building with a few broken windows. If the windows are not repaired, the tendency is for vandals to break a few more windows. Eventually, they may even break into the building, and if it's unoccupied, perhaps become squatters or light fires inside. Or consider a sidewalk. Some litter accumulates. Soon, more litter accumulates. Eventually, people even start leaving bags of trash from take-out restaurants there or breaking into cars.
>
> (Wilson and Kelling 1982)

An intersectional reading of this text exposes a barely disguised horror of difference as 'deviance'. The implication is that if nothing is done in response to the broken window this signals to those who would otherwise 'invest' in the neighbourhood (middle-class home-owners, employers) that no one cares about 'unacceptable conduct' (LeGates and Stout 2007: 256).

Hubbard *et al.* (2007) point to the way this has led to the extension of private property (and the employment of private security firms) as a means of 'civilising' the streets (by removing rough sleepers or anyone appearing to pose a threat, such as groups of youths dressed as 'goths' or in hoodies or anyone shouting or singing or drinking in public). More specifically, the authors point to the incompatibility of street prostitution with urban gentrification. Here there is a tension between the tacit encouragement of some forms of corporatised adult entertainment by the local state (Hubbard 2004) and the labelling of street prostitutes as 'other' and a threat to the reimagined metropolis – to be physically removed from gentrifying neighbourhoods (Hubbard *et al.* 2007: 203–205).

The local effects and unintended consequences of a zero-tolerance approach can be seen in Case 8.2 with respect to street prostitution in London. Governments across continental Europe have experimented with different approaches to 'clean up', 'tolerate', 'de-stigmatise' or 'criminalise' prostitution, such as confining it 'indoors' or to specified zones of the city. In Case 8.2 researchers from the London-based Poppy Project highlight the way the problem has typically been defined through the supply of prostitutes (they use this term in preference to the

Case 8.2: The Poppy Project (www.eaves4women.co.uk/POPPY_Project)

The Poppy Project was set up in 2003. It is funded by the UK Office of Criminal Justice Reform to provide accommodation and support to women who have been trafficked into prostitution. It has 35 bed spaces throughout London. It also commissions research and undertakes advocacy work on the issues of prostitution generally.

Women in street prostitution are routinely criminalised. Prostitution is not an offence, but soliciting and loitering for the purposes of prostitution is. Women in street prostitution are routinely arrested and charged with such offences, and experience a 'revolving door' of going to court, being fined, and then going back onto the streets soon after to make enough money to pay the fine. (Bindell and Atkins 2007: 12–13)

Whilst police and local policy responses have been 'nuisance' driven, relatively little research has sought to document impacts on the quality of life for those living in areas of street prostitution, rather than buyers, and the impacts on their everyday life. These include female residents being propositioned by kerb crawlers, increased traffic, used condoms and needles in the street, and a general sense of fear and threat. An incompatibility is apparent here, with those involved with street prostitution choosing residential areas, since it increases their sense of safety, whilst local community members resent the intrusion and disruption involved (May *et al.* 2001). Tolerance zones have been one attempt to address this conflict, but there is a growing body of research which shows they solve none of the problems inherent in street prostitution. The location of zones is always a subject of contention – either because those who live or work close to them object, or the intended beneficiaries find them unsafe and unpleasant. (Bindell and Atkins 2007: 12)

Punishment of women in street prostitution is usually disproportionate to that of kerb crawlers. Women in street prostitution are also much more likely to be issued with an Anti-Social Behaviour Order (ASBO). Introduced in 1999 as part of the Crime and Disorder Act, these orders were designed to help police and local authorities tackle young people's creating havoc on housing estates, as well as problem neighbours. ASBOs are served under civil law but breaching them is a criminal offence. Therefore, women who persistently breach ASBOs could be given up to five years in prison.

Kerb crawlers tend either to be ignored by police, or given the option of signing a voluntary Acceptable Behaviour Contract (ABC), agreeing not to enter the area where the offence took place, and not to pay or attempt to pay for sexual services.

According to a senior police office (Inspector Andy Bennett, Avon & Somerset Police Force, August 2006), 'the publicity surrounding men who have appeared in court or received an ASBO can be very damaging to their reputation' (Bindell and Atkins 2007: 13).

Comment

This case is shot through with evidence of local, cultural-specific gender contracts: the belief that it is inevitable that men will turn to the purchase of sexual services; that men who buy sex should not be stigmatised by this act (they are respectable family men), while women who sell sexual services do not deserve this dignity; women who sell sexual services are a 'nuisance' to be cleaned up or swept out of sight. In turn, these contracts, or beliefs, feed into the hyper-surveillance and criminalisation of these women's behaviour – by the police force (like the military a 'total institution' which has a powerful influence on the attitudes and response of individual front-line officers on the beat); by multiple agencies working to provide a range of social and medical services, social workers, counselling services; and by residents who feel their own gendered subjectivities (ideas about respectable femininity) under threat.

The Poppy Project Report uses the term 'prostitute' rather than the commonly used academic term 'sex worker' because 'very few women in the sex industry use the term "sex worker" to describe themselves. This term dignifies and de-stigmatises the sex industry and the buyers/exploiters, and not the women; it undermines efforts to locate prostitution as a form of sexual violence' (Bindel and Atkins 2007: 5).

Street prostitution is a dangerous and unhealthy business. Women in both on and off-street prostitution experience numerous physical and mental health problems. In a study of prostitutes in four counties, 62 per cent of the women interviewed reported being raped; 73 per cent had been assaulted; and 72 per cent were currently or formerly homeless. Three out of four women in prostitution became involved aged 21 or younger and one in two aged 18 or younger (Child and Women Abuse Study Unit 2003).

academic term 'sex worker') rather than male demand to purchase sex. They contrast evidence of prostitutes driven onto the streets by human trafficking and drug addiction with a survey claiming that 10 per cent of men in London aged 16–44 reported having paid for sex in the past five years (Home Office 2004).

We noted in Chapter 1 that deviance and unacceptable conduct are popularly identified with an erosion of 'family values' such as with increased lone motherhood, teenage pregnancy, absent fathers, undisciplined youth, lack of respect for property and the like. Crime prevention policy interventions such as 'broken windows' implicitly assume that a 'ghetto culture' functions through 'deviant norms' and an absence of 'middle class role models' (Murray 1996; Wilson 1987, 1996, cited in Kearns and Parkes 2005: 34). The application of intersectional theory allows us to counter the narrow moral order of broken windows with an understanding of multiple axes of deprivation and the reality that neighbourhoods divided by gender, class and racialised differences in US and UK cities are becoming *more* distinct, literally bounded by gates and barriers, as part of a strategy of the white middle classes protecting their interests and properties from a range of urban 'others' (McDowell 2006: 838). Evidence in Case 8.2 that some communities mobilise to exclude prostitutes while other residents exhibit greater tolerance prompts us to now pay closer attention to different forms of resistance, collective action and ways that different interest groups 'reclaim' the streets.

Resistance movements and collective action

As the global population increases, more people of different ethnicities, religions and socio-political beliefs have come to live together in urban areas than ever before, so that inter-communal struggles have multiplied in number and intensified, especially struggles for and about urban space. In each of these struggles, however, the conflicted space represents something more important than place and location; it becomes emblematic of identity, if not the right to belong.

Resistance assumes many guises, as we go on to show. The act of going on strike for fair pay and conditions remains a painfully necessary course of action in many quarters. The feminised textile and garment industry, for instance, harbours some of the worst sweatshop conditions and the hidden, highly exploitative practice of paying unprotected home-workers a poverty-level piece rate. Figure 8.1 shows placard-waving garment industry workers out on strike outside the trendy merchandising marts of San Francisco in the late 1990s. Other very different manifestations of resistance include private and individual acts of consumption and product boycott, modes of dress and public displays of political identity ranging from street marches to donning wristbands, badges and T-shirts to publicise a cause. Figure 8.2 illustrates this individual expression of resistance with respect to a rail of pink pro-abortion T-shirts hanging in a San Francisco shop window.

Organising in the south

Whatever the apparent differences in group types and make-up, there are growing numbers of organisations and federations representing marginalised groups and

Figure 8.1 ***Placard waving garment industry workers, San Francisco***
Source: Helen Jarvis

there is a growing knowledge among these organisations of existing norms, rights and legal structures and a growing confidence in asserting and lobbying for equal rights (UN-Habitat 2000). Despite the diversity of different types of groups, the principal focus of their interests and therefore their lobbying tends to be similar. In the global south the common themes are equality of access to land and credit for building, for instance, as well as security of tenure (including a widespread need for legalisation of property and land titles), particularly for female-headed households. Sylvia Chant observes a spiral of disadvantage for this household type, for instance, in that 'when female heads are unable to buy land and housing and have no option but to rent or share accommodation, this may restrict the range of informal domestic-based income-generating activities they are commonly forced to engage in due to lack of formal employment and/or help with childcare' (2007b: 41).

In Mumbai, women's groups are developing new forms of politics to challenge existing power structures which draw on their existing strengths – based on savings, federating and setting precedents (Appadurai 2001a: 33). Others such as Grassroots Enterprise Management Services (GEMS) are non-financial services that help women identify livelihood options through micro-enterprise development of micro-enterprises (Viswanath 2001). These groups and meetings serve important purposes beyond instrumental assistance. Perhaps

Figure 8.2 ***Logo resistance: women's liberation T-shirt, San Francisco***
Source: Helen Jarvis

most importantly they bring to the attention of the local authorities knowledge of women's issues and a more representative local population. While the authorities are nominally responsible for the people in a given area, in fact without the lobbying of grassroots organisations they have little or no idea who these people are, where they are or how many of them there are. In this way, the alliance of groups is permanently engaged in a 'politics of knowledge' (Appadurai 2001a).

A practical way to unravel these intersecting and competing interests is to consider the transformations that have occurred in World Bank urban policies over the past four decades. Up until the 1970s the World Bank tended to view cities as receptacles for rural migrants and as distractions from rural development policies (Zanetta 2001: 520) in which the sites and services approach sufficed. This changed in the 1980s alongside a realisation of the complexity of the situation, and cities began to be envisaged more in terms of their systemic components, in which public policy and institutions played an important role.

It was not really until the 1990s that cities were recognised as important centres of production in their own right, in which the role of public policy should be to facilitate production and the functioning of markets, a vision which combined neatly with the macro-economic priorities of structural adjustment. The millennium saw the beginnings of a more holistic approach to cities in which sustainability and livelihoods was a priority and in which the destructive mistakes of the previous two decades began to be recognised through efforts to address inequality and poverty alleviation (see Chapter 9). But all of these policy changes were slow reactions to fast-changing circumstances which had already moved beyond the analytical capacity of the World Bank, particularly because of a series of global economic crises, in the 1970s and 1980s in particular, that acted to change the policy environment.

Because of the debt crisis in the 1980s and later the problems associated with the structural adjustment policies that resulted from neo-liberal analysis of that situation, increasingly critical voices were raised against development-as-project generally and structural adjustment in particular. The growth in numbers, types and financial power of non-governmental organisations (NGOs) throughout the same period lowered the analytical gaze to the role of women, environmental destruction and sustainability and looked increasingly to grassroots mobilisation as a source of innovation, rather than the increasingly lacklustre meta-narratives of the World Bank. These critiques undoubtedly affected the ways in which the World Bank and the International Bank for Reconstruction and Development (IBRD) wrote marginalised groups of women and urban projects into their visualisation of development. Related to this, Yunus (1994) observed that this inclusion was always subordinate to the overriding priority of unrestrained economic growth without which there could be no development-as-modernity.

Alternative and inclusive planning visions

Come what may, as a response to the exclusionary nature of both the neo-liberal 'anti-plan' and resistance to the feelings of helplessness engendered by increasingly rapid globalisation, over the last 20 years increasing numbers of initiatives have sprung up in the global north and south, both as acts of resistance and to articulate an alternative vision of planning and social welfare. Case 8.3 highlights this proliferation of alternative visions, by identifying in snapshot fashion a small selection of the groups and organisations to have emerged internationally in recent years. Figure 8.3 illustrates the high-quality production of materials increasingly typical of activist groups which have access to relatively inexpensive new media publishing. While the group illustrated (Moms Rising) represents a left-leaning middle-class US organisation, the point remains that new media publishing technologies have transformed the opportunities for activist groups to disseminate alternative manifestos for change. (Also refer back to Case 7.5 on p. 212.)

Case 8.3: Alternative planning visions

Very few women are in a position to contribute in the 'growth machine' power partnerships of contemporary urban governance. Major players draw from 'elites' in government, consultants, academics, media and business. These coalitions of power effectively shape city spaces. This allows them at the same time to build their own power and influence, perpetuating not only the androcentric bias noted in Chapter 1 but also the interests of a privileged elite and the interests of capital efficiency over social reproductive sufficiency. Is there scope for urbanism that promotes not the ideal city but a good city for everyday life, a city for all – the vulnerable and excluded? What role should women play in planning cities of the future? What prospects are there for an inclusive urbanism? The following represent a sample of initiatives from across the world which are responding to these questions in a variety of ways.

Toronto's Alternative Planning Group

Established in 1998, the Alternative Planning Group (APG) is a partnership of four social planning organisations representing women in the context of marginalised ethnic minority groups. A number of social planning initiatives focus on anti-racism, access and equity. The group raises awareness of unequal access to public services and is increasingly vociferous in the transformation of social planning in Toronto (http://www.hispaniccouncil.net/id17.html).

Sprawl-busters: home town America fights back!

'Home town America fights back!' is the declaration of a web-based new social movement which promotes resistance to planning applications for edge of town 'big box' chain stores. Dubbed the 'anti-Wal-Mart movement', the goal of sprawl-busters is to develop an alternative power coalition drawn from community activists, many of them parents and self-employed home-workers, informed by unpaid consultants – to run successful campaigns against large-scale megastore development and as a means of protecting local independent shops and traditional neighbourhood shopping centres (www.sprawl-busters.com).

See also Reclaim the Streets, http://rts.gn.apc.org/.

The Global Women's Strike

A diverse group makes up the Global Women's Strike, formed to win economic and social recognition for unwaged caring work in different settings around the world. International coordinator Selma James (architect of the 1970s Wages for Housework campaign) observes that

> caring for others is accomplished by a dazzling array of skills in an endless variety of circumstances. As well as cooking, shopping, cleaning, laundering, planting, tending, harvesting for others, women comfort and guide, nurse and teach, arrange and advise, discipline and encourage, fight for and pacify. Taxing and exhausting under any circumstances, this service work, this emotional housework, is done outside and inside the home.
>
> (Global Women's Strike website, 2008)

The Global Women's Strike and the campaign Wages for Housework is as relevant in affluent countries like the USA and UK as it is in Bolivia and the developing world more generally: it is mobilising women to fight for social justice and a caring economy. The claim is that as 70 per cent of those who live in conditions of poverty in the world are women, it is women who should spearhead the direction of economic changes.

One success story of the Global Women's Strike is the women-dominated grassroots Bolivarian Revolution to 'guarantee equality and equity between men and women in the exercise of their right to work [including recognition that] work in the home is an economic activity that creates added values and produces social welfare and wealth' (Article 88 of the Bolivarian Constitution, 1999). In February 2006 President Hugo Chavez decided that female heads of households would receive minimum wage payments and Social Security in recognition of their housework. This is a remarkable vindication of more than 30 years of feminist campaigning around the world on the principle of 'wages for housework'. The difference to 'housework' as understood in Bolivia as compared say with the USA (where the typical housewife is financially dependent on a male breadwinner) is that entitlement to minimum wages recognises a whole host of unpaid essential community care work whereby women are instrumental in the provision of preventive healthcare, midwife services, organising community kitchens to feed those with no income (Fischer-Hoffman, http://mrzine.monthlyreview.org/cfh150206.html, accessed 16/02/06).

In this way Article 88 is radically different from Welfare in the United States. The latter is built on punitive 'workfare' measures which require women to pay for motherhood either out of the 'surplus' of wage employment or as a second shift on top of recognised work outside the home. According to Fischer-Hoffman (2006), 'the basic idea is that children who are born poor are innocent and deserve some basic support, but once they grow up and become mothers (or fathers) themselves, they have no "excuses" for being poor'; there is a compulsion to put waged work before the unpaid care of children and community (www.globalwomenstrike.net).

See also Reclaim the Night, www.reclaimthenight.org.

Figure 8.3 ***Mom's Rising: activist mothering publicity materials***
Source: Helen Jarvis

Changes at least in the way urban authorities and planners have begun to respond to these kinds of resistance initiatives can be seen in phenomena such as the Good Urban Governance Campaign in India (GOI 2001; Jain 2003). Faced with a situation where India needed to build 33 million houses in five years, 16.75 million of them in urban areas, the government of India launched a campaign recognising the importance of aspects of urban governance such as sustainability; the decentralization of power and resources to the closest relevant level; increasing efficiency; transparency and accountability among those responsible for service delivery; and, perhaps most importantly, an ongoing process of civic engagement with all the responsible and affected parties to ensure maximum participation. As yet there is limited evidence by way of assessment to show how this campaign is working, but it does indicate the potential in this area.

Even in the twenty-first century, there are relatively few cities in which truly progressive ideals of urban planning and regulation have actually been put into practice. One group of cities where progressive ideals have been actively pursued is in the Porto Alegre, Brazil, under the control of the Partido do Trabalhadores. This is illustrated in Case 8.4. Here too there is continuing need for critical reflection on the practical consequences of these ideals and measures by which to evaluate planning policies seeking to promote a non-sexist, inclusive city.

Case 8.4: Women, participation and urban governance: the case of Porto Alegre, Brazil

Although experiments in more inclusive regimes of urban governance are becoming more common in the cities of the global south, there are few examples of direct democracy as extensive and well known as that of the participatory budgeting systems in the cities of Brazil, including Sao Paolo until 2004. The cities of Porto Alegre and Belo Horizonte are the most well-known examples, but by 2006 participatory budgeting systems had expanded to cover 170 cities across Brazil (Avritzer 2006), emerging from the growth in number and type of urban assemblies in cities across Brazil in the 1970s and 1980s. The spread of this new kind of governance undoubtedly reflected the success it had in providing basic amenities (water, sewage, pavements, etc.) to poor communities in the cities where it was implemented. In Porto Alegre:

> The percentage of households with access to the sewage network rose from 46 per cent in 1989 to 85 per cent in 1996, and access to running water rose from 80 per cent to 98 per cent during the same period.
>
> (Novy and Leubolt 2005: 2028)

In Porto Alegre, 16 administrative regions had regional and thematic assemblies which voted on regional issues, votes being decided by a majority of residents. Residents with a direct interest in a given theme (health, transport, etc.) could participate in that thematic assembly and there was no limit to the number of themes in which a resident could participate; a woman could participate as a woman, as well as a mother with health, education and transport concerns, for instance. The number of participants in each assembly decided how many representatives went forward for the next phase, the intermediary assembly (Avritzer 2002).

In the intermediary assemblies the delegates chose five priority themes from 12 outlined in the budget and they then chose project allocations for each priority theme. Priority ranking was chosen by given criteria, the amount of previous access of the population to the public good in question, the size of population of the region in question and what the regional population chose. The allocations were then referred back to sub-regional assemblies and there was a process of ratifying the budget and electing councillors for a budgetary council responsible for executing the wishes of the population. The whole process gives the impression of being very complicated but took place between April and July of each year:

> The importance of these assemblies stems from the fact that they create a public method of decision of the works by the population, supplying an alternative to the particularisation and the obscure form of decision-making that are common practice in municipal administrations in Brazil. The regional assemblies introduce some important elements: a democratic and public mode of negotiation among the population. Democratic because the participation of the population is the criteria through which delegates and counsellors are appointed, transferring to this same population a decision-making power that it did not previously possess. Public because the necessity criteria are known, the population of the region is known and the meetings where the decisions of ranking are taken are open and speech is free. In this manner, participation and discussion with clear criteria have substituted the fragmented mode of resource distribution that existed previously.
>
> (Avritzer 2002: 8)

For our purposes what is particularly interesting about Porto Alegre (as one of several examples) is the way in which high levels of participation of women have developed over the years. The percentage of women participating in the process in Porto Alegre grew from 46.7 per cent in 1993 to 56.4 per cent in 2002 (as illustrated in Table 8.1), at which point women were obviously the majority of participants. This would be extremely unusual anywhere but even more so in a patriarchal culture like that of Brazil and indeed Latin America in general (this

Table 8.1 ***Participation in budgetary assemblies in Porto Alegre***

	2002	*2000*	*1998*	*1995*	*1993*	*Percentage of total population*
Up to 2 MS	39.5	24.9	30.9	n.a.	n.a.	11.4
2–4 MS	29.9	29.3	26.1	n.a.	n.a.	19.4
4–8 MS (more than 4 MS)	18.4	22.7	21.1	n.a.	n.a.	64.1
8–12 MS	5.1	10.0	9.7	n.a.	n.a.	
More than 12 MS	6.8	13.1	12.2	n.a.	n.a	
No answer on salary	–	–	–	–	–	5.1
Women	56.4	57.3	51.4	46.8	46.7	53.3
Black	28.1	20.9	n.a.	n.a.	n.a.	10.9

Note: MS = minimum salaries.

Source: Novy and Leubolt (2005: 2029); from *CIDADE* (2003), *CIDADE* and PMPA (2002: 25) and Marquetti (2002).

high level of participation has to be qualified, however; at the higher levels of the process (such as councillor) the percentage of women diminish). There is in addition a high level of participation by lower-income earners (although there are considerable variations in participation rates from city to city) and those with only elementary education, as well as high levels of participation by women at the assembly levels. The fragmented and elite-dominated models of budgetary allocation that had dominated in many Brazilian cities until the conversion to participatory budgeting has been inverted, and this in turn elicited high levels of participation by a poor majority which, until that time, had been excluded from urban power structures.

Among the marginalised groups in the cities of the global south, substantial improvements could be made through relatively simple changes in the way urban infrastructure is planned, such as efficient sources of energy (especially new forms of fuel for cooking and heating), joined-up thinking on transport, water and sanitation systems (UNMP/TFEGE 2005: 66); or, to take one frequently glaring omission in the juridical environment, enacting, implementing and upholding legislation to guarantee the property and land rights of women and girls. If it was true in 1985 that: 'the term urban management is beginning to take on a new richer meaning; it no longer refers only to the systems of control but rather, to sets of behavioural relationships, the process through which the myriad activities of the inhabitants interact with each other and with the governance of the city' (Churchill 1985: v), then the neo-liberal project would appear to have constituted a direct breach. In practice, control was taken out of the hands of urban residents and effectively externalised from the nation-state as the strategic concerns of privatisation and liberalisation took precedence over working with these behavioural relationships. As we have seen, however, neo-liberalism and structural adjustment in their turn play an important part in generating a thriving transnational activism, arising not just from the built urban environment but as a direct result of the exclusionary power structures within which it is embedded.

What the alternative visions we have offered above and in previous chapters have in common is that they all attempt to engage with some version of Geddes' 'synoptic vision' (Geddes 1915: v) – by taking as a central premise the interconnectedness of the whole (of which gender mainstreaming provides a contextual background). Thus we come back once again to the idea of a gendered, holistic urban management in which 'a performative approach [to gender] resists locating emotions or gender in either embodied city dwellers or urban environments, insisting instead that they necessarily infuse and reside in both' (Bondi 2006). According to McGill, a more progressive urban planning process (whether in the global north or the global south) should incorporate a number of inclusive characteristics:

- it should embrace all the players in the city-building process, not just be a dispenser of resources;
- it should harness the driving force of urban development, not attempt to govern it artificially, through a master plan;
- it should be horizontally integrated to overcome sub-sectoral myopia;
- it should be vertically integrated (to overcome town planning's Achilles heel of being divorced from budgeted implementation);
- it should be capable of responding to opportunities that present themselves (the implied release of the innate capacity of the community or informal sector, through NGO participation) (McGill 1998: 466).

Going back to our previous suggestion that the city is dressed in the raiment of conflict, however, all progress towards social equity, gendered planning and sustainability (now and in the future) seems almost inevitably destined to carry on foundering in the stormy seas of the socio-political. Scott Campbell (1996) suggests that planners are forever caught in a triangle of priorities, equity and social justice, environmental protection and economic development, which result in three kinds of conflict: development conflict, resource conflict and property conflict. What we might add to this is that conflict is not in itself a bad thing, but the supposition that conflict can be resolved through centralised strategic planning is always a bad thing. Centralised authorities are not (as many would like to pretend) natural and objective disseminators of the best existing truth; instead they are themselves hostages to the imbalances of power inevitable in the socio-political environment responsible for their creation. We would therefore suggest that the way forward for gender mainstreamed cities of the future lies in processes that seek to devolve power-through-planning to the lowest feasible level while at the same time constantly re-examining processes and mechanisms by which that power-through-planning might be realised in a participatory, horizontally active, vertically challenging micro-democracy.

Summary

This chapter points to a multi-layered, multi-scalar, gendered system of planning and social welfare alongside culturally constructed gender contracts. Within this framework we identified multiple and fluid family and household arrangements, systems of planning and public policy ranging from 'master plans' to participatory democracy and national welfare regimes through city-regional governance to street-level bureaucrats and front-line emergency services. We extended notions of governance beyond formal planning and public policy to recognise the way prevailing gender systems function through intersecting social and cultural norms and ties, multiple economies (of paid and unpaid work, gifts, volunteering and exchange), household resource capabilities and livelihood practices. Case studies were introduced to illustrate the diverse and intersecting functions of governance, resistance, surveillance and collective engagement which ultimately contribute to patterns of social inclusion and exclusion.

Notwithstanding this critical analysis, there are some fundamental design and planning innovations discussed above and in previous chapters that can make a substantial difference in the inclusiveness of the city; approaches towards inclusive urban design where problems such as social exclusion and homelessness are addressed as part of the planning process from the beginning, rather than as a kind of embarrassing afterthought. Design apartheid needs to be discarded and arguments for ideas such as 'family-friendly' and 'older-person-friendly' city initiatives (introduced in Chapter 4) taken into serious consideration. The barriers to all of the more inclusive design initiatives which we have discussed either here or in previous chapters lie in the structures of urban governance (the growth machine) rather than with architects and planners themselves; inclusive planning and participatory governance get closer to representing the politics of everyday life.

An intersectional approach mobilises a powerful critique of the privileging of one social category or definition of moral worth over another, whether this privileging seeks to explain persistent inequality or to prescribe urban and social policy interventions. Transitional patterns of identity and their concomitant performances can be brought to the city, experienced in an urban childhood or enforced through ethnic or gendered structures, that change over time and become imprinted on the urban built façade. For planning to be more inclusive it needs to work with those patterns and the self-referential power structures with which they intertwine. This is recognised in the plea by feminist scholars for inclusive, participatory planning – building on greater awareness of the embodied, performed, emotionally experienced qualities of the city (see Greed 2005; Healy 1997). Accordingly, Liz Bondi (2006: 1) states:

> Consideration of gender as a facet of embodied identity is essential to understanding urban experience, helping to forge commonalities among, and differences between, city dwellers whether living and/or working in close proximity or far apart ... Approaching gender as performative emphasises the non-cognitive, non-rational character of routine everyday practices. It is therefore sensitive to, and offers a useful theoretical framework within which to explore, emotional dimensions of urban life.

This is not to say that cities (location decisions, travel patterns and so on) are entirely 'man-made' (see Chapter 5). On the contrary, exposing this binary renders it visible as a construct to work against, to unsettle and ultimately deconstruct. As Lynda Johnston (2005: 124–125) points out, feminist research has increasingly deployed the binary of male/female bodies and spaces as a means of subversion and resistance, resulting in a 'paradoxical space' by which bodies simultaneously occupy the centre and the margins, understood not just as men or women but as sexed bodies that are contingent upon time and place (Johnston 2005: 125; Rose 1993). A similar argument has been made for intersections of gender, class, ethnicity and race in cites of the south which have been rendered parochial and marginal by hegemonic, ethnocentric ideologies of world city status (Robinson 2006).

One inevitable consequence of this battle between the feminised 'incoherent' and masculinist 'coherence', between 'sanity' and those whose reality is deemed insane, is that the city is in every corner, point and straight line shrouded in conflict. This can be public, highly symbolic conflict such as the increased use of regulations and ASBOs to keep the homeless out of city centres in the UK, the clearance by bulldozers of squatter settlements in Mumbai or the 2007 ban on rickshaws in Delhi by the Delhi High Court; or it can be private and personal conflict, such as the alienation and vulnerability women may experience in a planned public space where the use of trees and bushes creates a menacing atmosphere after dark, or the daily struggle of a disabled person to get to work by accessing a public transport system not constructed to take into account the reality of disability.

Arguably, despite rising levels of grassroots resistance to persistent urban inequalities we find evidence at every layer of planning and welfare of the damaging effects on intersecting inequality of neo-liberal emphasis on competitiveness. The challenge remains, therefore, for a transformative feminist politics to subvert androcentrism and establish an ethic of care as a project of gender mainstreaming.

Learning activity

Read about and explore some of the on-line organizations and movements concerned with developing more inclusive/non-sexist cities (such as World Urban Forum). Critically examine the features commonly associated with gender democracy and the suggested ways to eliminate gender apartheid in urban design. Draw up a 'manifesto for change' based on a summary and consensus of the main features identified.

See, for example, http://www.powercampnational.ca/en/resources/resource/best_practices_for_girl_and_young_women_friendly_cities_towards_an_international_dialogue.

Further reading

Folbre, N. (2001) *The Invisible Heart: Economics and Family Values*. New York: The New Press.

Hirdman, Y. (1991) 'The gender system'. In T. Andreasen (ed.) *Moving On: New Perspectives on the Women's Movement*. Aarhus: Aarhus University Press.

Markusen, A. R. (2005) 'City spatial structure, women's household work and national urban policy'. In S. S. Fainstein and L. J. Servon (eds) *Gender and Planning: A Reader*. New Brunswick, NJ: Rutgers University Press, 169–190.

Sandercock, L. and Forsyth, A. (2005) 'A gender agenda: new directions for planning theory'. In S. S. Fainstein and L. J. Servon (eds) *Gender and Planning: A Reader*. New Brunswick, NJ: Rutgers University Press, 67–85.

Young, I. M. (2005) 'Justice and the politics of difference'. In S. S. Fainstein and L. J. Servon (eds) *Gender and Planning: A Reader*. New Brunswick, NJ: Rutgers University Press, 86–103.

9 Urban poverty, livelihood and vulnerability

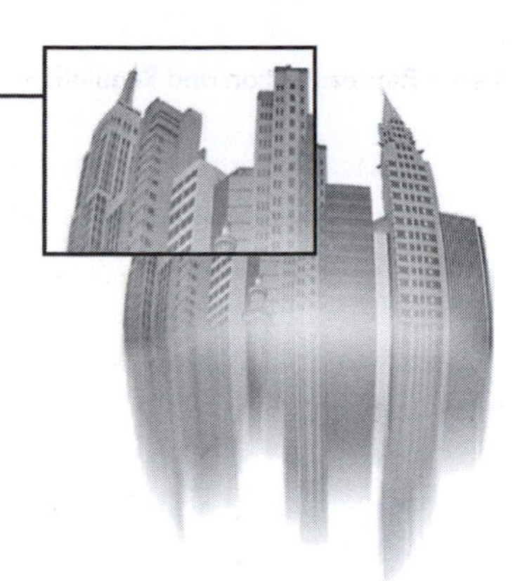

Learning objectives

- to differentiate between the concepts of poverty, vulnerability and exclusion and to understand in what ways urban poverty and rural poverty differ
- to understand what livelihoods are and how this concept contributes to our understanding of the micro-level actions of individuals and households within the wider social, political and economic context
- to be able to critically apply the concepts associated with a sustainable livelihoods framework and with multiple economies to analyse urban poverty from a gendered perspective in different contexts of the global south and the global north

Introduction

Urban poverty and livelihood insecurity are global problems, meaning both that they affect people in all countries of the world and that they require solutions that draw from global experience and expertise. This chapter considers in turn the experiences of urban poverty and livelihood insecurity across the global south and within the global north. The scale and depth of the problem are far greater in the global south and for this reason we largely consider the conceptualisation and problems of poverty from a southern perspective. At the same time, we believe it is also constructive to compare this context with 'new' expressions of urban poverty found in advanced neo-liberal and post-socialist cities. Here, poverty increasingly involves not only multiple deprivations but also social isolation and a spatial concentration of poor households, especially lone-parent households, in stigmatised neighbourhoods and communities. Moreover, contrary to the common association of poverty with unemployment, the new urban poverty increasingly encompasses new sub-classifications of the 'working poor' (Smith *et al.* 2008; Buck *et al.* 2002; Castells and Portes 1989).

This chapter builds on the concept of multiple economies and different types of assets that we first introduced in Chapter 1, adding to this an understanding of multiple and diverse livelihoods. We find repeated evidence that 'wages are not enough' to protect individuals, families and households from poverty and the risk of losing their home (Smith and Wallerstein 1992: 254). In practice, non-market activities and assets, many of them essentially feminised, have a significant impact on individual and collective well-being. From a southern perspective we consider the contribution of a sustainable livelihood framework, which effectively takes a multiple livelihoods approach from theory and translates it into development policies and outcomes. In the context of advanced neoliberal economies we briefly consider the insights multiple economies and livelihoods can provide for policies seeking to combat social exclusion where this highlights the negative effects of diminishing collective welfare outlined in Chapter 8.

Common to the wide-ranging case studies we discuss in this chapter is evidence of a growing polarisation of formal and informal economies; a growth of informality at the low end of the job market, in consumption and employment; increased reliance on social capital (networks of reciprocity and exchange); and a spatial concentration of poverty, multiple deprivation and social isolation – which tends to militate against successful poverty reduction. Our discussion of these issues exposes the clear gender implications to both deeply entrenched and 'new' expressions of poverty and vulnerability. In conclusion we argue that these concepts and intersecting economies have to be understood in locally specific contexts to take full account of the co-constitution of social inequalities, gendered norms and uneven development.

Conceptualising urban poverty in the global south

The conceptualisation of poverty has been the focus of much debate and consideration in recent years and rightly so (Sen 1983, 1987; Chambers 1989; Townsend 1971, 1985). A number of scholars have called for everyone engaged in development to be more reflexive about what it is they are attempting to eradicate and how the different ways of defining poverty have implications for the means of reducing its incidence. This attests to the significant shift in thinking that has occurred in recent years, from early representations of poverty which narrowly focused on measures of income and consumption alone to composite social indicators and recognition of participatory assessments based on the poor's own perceptions of what poverty is (Narayan *et al.* 2000; Chambers 1997). A well-established example of a composite measure of well-being which combines income and social indictors is the UNDP's Human Development Index (HDI).

This is not to say that income- and consumption-based definitions are no longer supported. On the contrary, major multilateral development agencies, among others, rely on head count measures of poverty, such as the proportion living on less than the equivalent of US$1 per day, which are based on consumption outcomes. Halving poverty measured in this way remains the indicator used to assess progress on achieving the first of the Millennium Development Goals (MDG1). Current projections are that there will still be more than 700 million people living on less than US$1 a day in 2015 (Chen and Ravallion 2007: 1). Moreover, this aggregate indication takes no account of structural divisions by gender, generation or any other form of social difference.

There is a general consensus that poverty not only functions through a lack of income or low consumption, but is multidimensional and associated with multiple deprivations (housing, income/consumption, basic services, rights, voice and governance) (Sen 1999; World Bank 2001a). It is in devising measures of poverty that this conceptual complexity can become lost, with poverty measures often falling back to reductionist income or consumption levels, as noted above for MDG1. This suggests that the barriers to moving poverty measures beyond income and consumption are technical rather than conceptual as there is little resistance in development discourse to conceptualising poverty as being multidimensional. The same cannot be said of efforts to incorporate structural constraints within the analysis of the many factors causing poverty. The acceptance of structural factors, such as ethnicity, gender inequality, remoteness and the like, varies especially widely within a southern perspective, because this acceptance has considerable implications for the transformations required to achieve sustained reductions in poverty over the long term. Many mainstream development agencies, particularly multilateral and bilateral lending and donor organisations, for instance, tend to focus on issues of access to and delivery of services (such as water and sanitation) instead of considering how to support people to better negotiate access, to broaden their entitlements, by changing existing power relations which contribute to their experiences of poverty. Technical solutions which provide better access to services, more marketable skills and the like may be easier to offer, but they will be insufficient to address the root causes of poverty if social relations of power are considered central to processes of becoming and staying poor.

This difference in understanding and approach has clear gender implications. Not considering the structural factors shaping women's and men's experiences of poverty will be unlikely to appreciate or address the underlying inequalities in gender relations which create and perpetuate differential experiences of poverty within and between gender groups. As Kabeer states, 'Poverty is "gendered" because men and women experience poverty differently – and unequally – and become poor through different, though related, processes' (2003: 1). Some examples of the mechanisms through which poverty is gendered include:

- the gender division of labour, which gives women primary responsibility for reproductive tasks which often must be combined with productive work, leading to a time-squeeze;
- unequal entitlements to land and other productive resources which would enable more profitable engagement with markets;
- ideas about masculinity and femininity which lead to gender differences in the valuation of skills and returns to labour (Kabeer 2003: 1).

Multiple deprivation and alternative poverty measures

Poverty measured by an absolute head count and monetary value, as above, paints a stark picture of a north–south global divide. Yet, as Diane Perrons reminds us, 'while a north–south divide is still very evident there are important divisions within the north and south, giving rise to intricate patterns of uneven development' (2004: 33). Another way of measuring multiple deprivations is to apply a relative poverty threshold set at a proportion of national median household income. Key to this measure is the choice of poverty line – a choice that is contingent upon the culturally and politically constructed conventions of government and governance outlined in Chapter 8. For instance, while most countries in the European Union accept a threshold of 60 per cent of the median, the US 'official' poverty line is set at around 40 per cent, so that, even by this harsher definition, one in four US household lives in relative poverty.

Other measures of poverty include indices of deprivation. In Britain, for instance, the Index of Multiple Deprivation (IMD) combines a number of social, economic and housing indicators into a single deprivation score applied to small areas to produce a relative ranking of neighbourhoods and communities according to their level of deprivation. Supplementary indices measure income deprivation among children and older people.

Evidence that both the proportion of households living in poverty and the degree of inequality between rich and poor households have increased in recent decades has stimulated talk of a 'new' urban poverty in otherwise affluent countries. For instance, poverty levels recorded in 2004 in the UK were nearly double those witnessed in the late 1970s, and in both the UK and the USA a significant divide has opened up between the income and benefits associated with 'good' jobs and the poor pay and conditions of 'bad' jobs (Jarvis 2005a: 146–147; Nelson and Smith 1999). Accompanying this polarisation are concentrations of feminised poverty associated with lone parenthood, families with young children, those managing on welfare benefits and in retirement and old age (Gordon 1994; Nelson 1990; Leitner 2001). According to Enzo Mingione (2006) a new urban poor can be identified in a heterogeneous population of unemployed, underemployed, the elderly and disabled and those in jobs which fail to pay a living wage. The common features are social isolation and multiple deprivations, notably poor education, poor health, limited access to affordable healthcare, fuel poverty, exposure to crime and insecure housing (see also Small and Newman 2001; Curley 2005; Wilson 1996). This trend looks set to continue through the snowballing effects of the 'credit crunch' which began in 2007. In this context, multiple deprivations are increasingly associated with a lack of affordable housing and poor-quality, temporary and unprotected employment. The variables involved are undeniably gendered. We return to consider the way structural divisions of gender, generation, class and ethnicity are implicated in these trends of urban poverty in otherwise affluent cities later in the chapter (pp. 273–8) (see J. Anderson 2007).

Characteristics of urban poverty

At this point it is constructive to consider how the issues introduced above relate to urban poverty as distinct from rural poverty. Clearly rural and urban poverty do not differ in regards to the importance of structural factors in understanding why different people are poor. In a seminal piece on conceptualising urban poverty, Ellen Wratten (1995) contends that debates about rural and urban poverty and which is 'worse' only serve to distract efforts to mitigate the effects of the structural drivers of poverty within mainstream development agendas. She holds that what is 'urban' is itself contested and the division between urban and rural is not clear cut, given the global diversity in criteria for labelling a place 'urban' (Wratten 1995). Moreover, the everyday realities of making a living typically reside in both of these spaces and, as such, the circuits of livelihood in urban and rural populations are intimately linked. There are nevertheless a number of characteristics to poverty in urban environments which set urban poverty apart in terms of locally lived experience. These include:

- Greater environmental and health risks due to high densities of living, proximity of residential areas to industrial sites, and even location of hazardous production processes in settlement areas, poor access to clean water and sewerage services and traffic congestion leading to air pollution and risks of accidents (Wratten 1995). Women may be more affected by these risks due to the greater time spent in or near the home relative to men. (This is illustrated in case 9.3.)

- Higher degree of commercialisation of exchange and hence reliance on cash for basic needs such as food, shelter, wood for fuel and basic services (Satterthwaite 1997; Rakodi 1999; Rakodi and Lloyd-Jones 2002; Moser and McIlwaine 1997; Moser 1998). This makes poor urban households dependent on exchanging their own labour for survival, often on adverse terms. Women, due to reproductive work burdens which may limit work time or choices of work locations, and actual or perceived differences in their skills, often enter labour markets on more adverse terms than men (Kabeer 2003), as Case 9.1 shows.
- More social diversity, which can lead to less social support, greater fragmentation and higher crime rates, with distinct gender implications (Wratten 1995). The greater social heterogeneity can have positive implications for women, whose mobility may be more constrained in rural areas, where they are more easily 'known' and linked back to the extended family.
- Higher likelihood of contact with the state and police, which may be positive if policies are in place to alleviate poverty, but can also be oppressive, particularly around access to land and space to vend, as Case 9.1 illustrates (Wratten 1995).

In terms of current development practice, concern with the distinct manifestations of urban relative to rural poverty has far greater influence on policy and project responses than does awareness of the underlying causes perpetuating poverty, for men or women. In short, technical solutions are more common than political ones. For example, among multi- and bilateral development institutions such as the World Bank, USAID and JICA, urban development interventions often take the form of large-scale physical infrastructure investments to upgrade slum settlements (where land tenure is legal) and improve access to adequate and affordable services, or of projects to improve access to markets and capital such as through microfinance. These may take a gendered approach, but often under a gender-neutral rubric, where different gender roles and responsibilities are acknowledged but interventions are designed to accommodate rather than challenge these structures of constraint (Kabeer and Subrahmanian 2000). For example, women's reproductive roles make them key actors to consider in the location of basic services such as water taps in slums; this recognises women's responsibility for collecting water but it does not challenge it. Developing structures to have women manage water distribution in the settlement and building skills in maintaining the local infrastructure would be more transformatory. Some NGOs and coalitions of agencies working in urban areas are more willing to take a political stance and to work for the rights of poor urban residents, with housing being one core advocacy issue. Some of these groups take a gendered perspective and integrate women into processes of negotiation with city officials and all responsibilities within the organisation, building their voices and capacities. The local benefits of this advocacy work are illustrated in Case 9.2 in relation to improvements in partnership established between NGOs and slum dwellers in Mumbai.

Case 9.1: Adverse labour market inclusion in Kabul

Nasreen is the wife of Ahmad and mother of their six children aged 3 to 13. Ahmad vends shoes in a market near a cinema in Kabul. His income is highly variable due both to changing demand for shoes and to being forced to stop selling owing to a strong and at times violent police presence:

> [The police] don't understand our problems and living conditions, therefore they always beat and bother us to leave this place and not work. Last year they attacked me and spilled my goods in the street. Police beat me by a stick, whereas I told them to allow us to work, since I explained for them that if I don't do this work, what else could I do, robbery, stealing, kidnapping or what should we do! Police troops are harassing us 5-6 times in a day and they make us flee and clear the place, although they are receiving from each vender 20 Afs as bribe.

This variable income is insufficient to meet household needs, meaning the older boys also do vending work, combined with schooling, while Nasreen, following norms of female seclusion, works from home shelling almonds for a trader…and pays money to do so. Nasreen earns no income in cash for the work she does. Instead she pays for the nuts' transport to her home, all so that the family can reduce expenditure on fuel wood, and burn the shells during Kabul's cold and snowy winters. Thus, extreme need creates the context in which women such as Nasreen are integrated into productive work, but on highly negative, or adverse, terms – in this case for no cash pay from the trader though he can now sell his nuts at a higher price, and footing the bill herself for the almonds' transport.

This case highlights the lack of power and choice available to the urban poor, men and women alike, as they strive to piece together livelihoods which adequately provide for their families. This is not easy when local governance structures are actively working against you and when you have so little voice to negotiate employment terms that you work for cast offs, not willing to risk your current access to an important fuel source and the winter security it provides for the possibility of better conditions in the future.

Source: Kantor and Hozyainova (2008).

Multiple economies of livelihood and vulnerability

The multidimensional nature of poverty and the diverse ways through which most poor households in low- to middle-income countries make a living are captured in the concept of livelihood. Reference to livelihoods further reflects a shift in

Case 9.2: SPARC, NSDF and Mahila Milan: working with slum dwellers in Mumbai

The Society for the Promotion of Area Resource Centers (SPARC) is one of the largest Indian NGOs working on housing and infrastructure issues for the urban poor. In 1984, when SPARC was formed, it began working with the most vulnerable and invisible of Mumbai's urban poor – the pavement dwellers. SPARC's philosophy is that if we can develop solutions that work for the poorest and most marginalised in the city, then these solutions can be scaled up to work for other groups of the urban poor across the country and internationally.

Since 1986, SPARC has been working in partnership with two community-based organisations, the National Slum Dwellers Federation (NSDF) and Mahila Milan (MM). Today, these agencies work in about 70 cities in the country and have networks in about 20 countries internationally.

NSDF and MM organise hundreds of thousands of slum dwellers and pavement dwellers to address issues related to urban poverty, and collectively produce solutions for affordable housing and sanitation. The NSDF was founded in the mid-1970s and is a national organisation of community groups and leaders who live in slums/informal settlements across India. Its main aim is to mobilise the urban poor to come together, articulate their concerns and find solutions to the problems they face. Today the NSDF works with about half a million households in the country.

Mahila Milan means 'Women Together' in Hindi and is a decentralised network of poor women's collectives that manage credit and savings activities in their communities, with the funds used for a range of reasons, including ill health, housing purchases and other productive investments. Mahila Milan aims to provide a space for women to take on important decision-making roles and be recognised for their critical contributions towards improving the lives of their communities. Mahila Milan was initiated in 1986 when 500 women who lived on Mumbai's pavements organised themselves to successfully prevent the demolition of their homes. Today, Mahila Milan has given out tens of thousands of loans to poor women all across the country and has collected savings worth several scores of rupees.

SPARC, in its alliance with the NSDF and MM, is an example of bottom-up action and advocacy to address existing inequalities in the distribution of housing, infrastructure and financial services which affect the large population in India's cities living in insecure settlements, including on pavements. Its work crosses scale, working at the local level to secure men's, women's and their families' rights to land, basic services and finance as well as advocating for the diverse interests of slum dwellers at the national and international level. Addressing basic needs initially led to higher-profile activities through which previously disempowered women constructed a platform for negotiating with male authority at different scales: family, neighbourhood, local authority, police, national government.

Raising visibility is a vital aspect of the strategies by which Mahila Milan (in tandem with different slum dwellers' organisations) keeps urban poverty firmly in the focus of the authorities and powers-that-be in Mumbai; an inversion of the norm for poor groups, which tend to try and survive by being passive and making themselves invisible to authority (Patel and Mitlin 2001). Conducting censuses of pavement dwellers, for instance, brings people with no representation into a visible realm of governance. In recent years these organisations have built model houses and put on a housing display for people, government and media, each model arising from discussions with groups of women to try and reconcile their ability to pay with available materials and possible sites. This way, a once marginalised group has become a major innovator in reaching decisions on how to resolve communal problems (Patel and Mitlin 2001: 132).

Source: Patel and Mitlin (2002); Appadurai (2001b); SPARC India (2008).

thinking away from narrow definitions of poverty based on income and consumption to recognise the 'multiple economies' we introduced as a bridging concept in Chapter 1. By way of reminder, a holistic understanding of multiple economies, as compared with the partial view offered by narrow monetary definitions, encompasses a spectrum of formal, informal, paid, unpaid, exchange-based, subsistence and self-provided (sweat equity) activities, goods and services. This concept has recently been embraced by scholars analysing poverty and poor neighbourhood conditions in European and American cities inasmuch as it is closely associated with development studies scholarship. Later in this chapter we go on to show how an understanding of multiple economies (or what Owusu (2001) refers to as 'multiple modes of livelihoods') sheds light on the vital functions of non-material social networks of reciprocity and diverse informal economic activities to household livelihoods in a wide range of urban contexts.

> *Livelihood*: livelihood is defined as the means of making a living or, more specifically, 'the capabilities, assets (stores, resources, claims and access) and activities required for a means of living'.
>
> (Chambers and Conway 1992: 7)

By shifting the focus of analysis from material deprivation, as poverty does, to what the poor have – what assets the poor have to work with to achieve various desired outcomes – greater credit is given to the survival value of feminised social reproduction work (Rakodi 1999; Rakodi and Lloyd-Jones 2002; Moser and McIlwaine 1997; Moser 1998). On the one hand, this emphasis on multiple and diverse livelihood assets suggests there is positive scope (in what the poor have to work with) for creative strategies to cope in a crisis. On the other hand, livelihoods are found to be vulnerable to changes which may increase the risk of a decline in livelihood outcomes. For example, Ahmad's case (in Case 9.1, p. 253) shows how vendors are vulnerable to police intervention, leading to reduced daily income, and hence a reduced ability to buy the day's food requirements, as many live a hand to mouth existence. If this negative intervention continued and the household did not have other income sources to rely upon or to which it could switch, the well-being of its members, or some of them, could significantly decline.

> *Vunerability*: vulnerability is defined as exposure to risk, shocks and stresses, resulting from some sort of unplanned environmental change (Chambers 1989; Wratten 1995; Moser and McIlwaine 1997; Moser 1998). There are two dimensions of vulnerability, external and internal. Risks, shocks and stresses are common terms illustrating external vulnerability, while internal vulnerability refers to a household's inability to cope with these external changes without experiencing damaging loss. Vulnerability is a more complex concept than poverty, in that it has a dynamic quality, making it better suited to capturing changes in socio-economic status in contrast to poverty's focus on current socio-economic status. It captures processes of poverty and the ups and downs experienced by poor households versus just taking a static snapshot at one point in time.

The level and diversification of the assets that a household has available can protect the household's livelihood from outside risks and enable it to adapt to unavoidable risks. Assets are therefore the means by which households avoid and/or adapt to crisis, and access to them is mediated by social relations, by who you are in a given context and how that affects what you are entitled to own and control (Chambers 1989; Swift 1989; Moser and McIlwaine 1997; Moser 1998; Rakodi 1999; Rakodi and Lloyd-Jones 2002; Sebstad and Cohen 2001; Beall 2002). Assets are categorised in diverse ways by different authors, but the main categories include physical or productive assets, financial assets, human capital, social capital and political capital (Swift 1989; Chambers 1989; Sherraden 1991; Rakodi 1999; Rakodi and Lloyd-Jones 2002; Meikle 2002) (see Box 9.1). Human capital includes labour and the characteristics of that labour (education, gender, caste), both of which are central to the security of urban livelihoods. Political capital is a more recent addition to livelihood analyses and represents the resources a household or individual has to influence decision making in political systems and achieve outcomes in its interest. Levels of political capital reflect the power different groups hold (Rakodi 1999); its addition reflects a response to criticisms of livelihoods analysis and related frameworks for their lack of attention to power relations and their influence on different groups' ability to create chosen livelihoods and achieve desired outcomes (Wood and Salway 2000).

Box 9.1: Examples of different types of assets

- *Productive assets*: work equipment, secure tenure, good-quality housing, basic services.
- *Financial assets*: savings in money form, jewellery, gold. Debt is a negative financial asset.
- *Human capital assets*: spending on education and preventative healthcare, levels of education, experiences of illness, training and skills.
- *Labour*: part of human capital but central to urban poverty and vulnerability due to the great dependence on income in urban areas. Dependency ratios, occupation and employment status, work days and regularity of work, worker characteristics (age, sex).
- *Social capital assets*: networks through which people access jobs, credit, help in times of need; organisations or institutions with which one interacts. One needs to examine net claims as these relationships are two-way and can be a drain. They can also be negative or exploitative, for example, patron–client relations.
- *Natural assets*: access to natural resources such as clean air, water, land. In urban areas this access is mediated by other interests, since access to basic services is not guaranteed and is often politicised, and tenable legal land for housing is limited.
- *Political assets*: links with those with access to decision-making processes: media, politicians, international advocacy groups.

Source: Rakodi and Lloyd-Jones (2002); Carney (1998).

While there is agreement on the important role of asset diversity in mitigating risk by helping households avoid risks or adapt better to external change, it must be noted that access to assets alone is not sufficient to reduce vulnerability to livelihood decline. The level and diversification of assets a household has is determined in part by status and entitlements in a particular social context. Households and their members will differ in their abilities to access assets, and in their abilities to utilise those assets to which they have access, to transform them into the necessities of survival, such as food and income (Moser and McIlwaine 1997; Chambers and Conway 1992; Rakodi and Lloyd-Jones 2002). This brings back into the analysis the structural factors which may limit individual or group capabilities, agency or level of choice. Arguably, just as we stressed in previous chapters, this points to the benefits of an everyday life approach, whereby livelihoods are analysed in relation to the cultural norms and expectations embedded within institutions like the market, state and family which relate to gender, caste, race or class and how they mediate individuals' or households' abilities to access a range of assets and utilise them effectively in times of crisis. This also links back to our understanding of the dynamic gender systems and multi-layered governing institutions discussed in Chapter 8.

The livelihoods concept is often applied to a household. However, relying solely on the household as a unit of analysis to understand processes of poverty, livelihoods and vulnerability is problematic, as this hides inequalities in access to and control over assets, and therefore in levels of well-being and security, within the household (Cagatay 1998; Kabeer 1989, 1994; Baden and Milward 1995; Razavi 1999). Individuals within households may differ in their asset holdings based on their entitlements and status within a society, leading to differences in vulnerability and livelihood outcomes both across and within households. In many contexts it is women and girls who may have considerably different entitlements to household assets, meaning they may suffer more acutely the downturns experienced by households (Razavi 1999; Nussbaum 2000). Therefore, intra-household or individual-level analyses that delve into the outcomes of decision-making and resource allocation processes of households experiencing crisis are vital to an understanding of the influence of risks on well-being and security.

The impact that local cultural norms and expectations have on individual and household livelihoods is readily illustrated in two different examples. In the first example, Figure 9.1 illustrates the way women in Bangkok openly participate in street vending, selling the produce they have grown or made at home as activities entirely outside the formal economy. The scope these women have to exchange goods for cash outside the home contrasts with the cultural restrictions placed on Nasreen's home-based informal economic activities in Case 9.1 (p. 253). In the second example, Davis *et al.* (2006) report on the intersections of multiple capitals (assets and modes of livelihood) in their survey of 400 couples in four Chinese cities. The authors look at changes to household gender divisions in what is an increasingly marketised but still communist political economy. In addition to the capitals associated with paid employment, other income, human resources and social networks, they identify positions of political authority and public sector employment as having a significant bearing on livelihood security. Political assets (status and authority beyond simple membership of the Communist Party) were

Figure 9.1 ***Street vendors in Bangkok***
Source: Helen Jarvis

found to be highly gendered within the household but also responsible for a major distinction between the standard and stability of polarised household livelihood. In this context, new trends of poverty are being driven by market-oriented reform and a socially regressive transformation of welfare.

Responding to risk

Understanding what risk means is also important to understanding livelihoods and vulnerability. Risk is the probability that a loss will result; for example, households dependent on one livelihood activity, even if a secure and regular source of income, are more at risk, have a higher chance of facing a loss of income, than those relying on different income flows which can make up for a loss in one of them. 'Risk' is a broad term: there are many different types of risks, arising from different sources, and characterised by differing frequency and impacts (Sebstad and Cohen 2001; Lund and Srinivas 2000). Households respond to events that may cause risk or the threat of them in different ways. Three different sources of risk include:

- *structural factors*: inflation, seasonality and weather patterns that may result in floods or droughts;

- *unanticipated shocks and crises*: unemployment, serious illnesses and hospitalisation, the death of a main earner, fire, harassment by city officials and the bulldozing of settlements;
- *life cycle events, trends*: marriages, births, festivals and rituals – these are expected events but still cause a large drain on household finances.

The source of risk can determine its frequency and nature (Sebstad and Cohen 2001; Lund and Srinivas 2000). Often the risks associated with structural factors are covariate, meaning they affect all households at the same time. Such risks are generally infrequent, but may be repeated. Shocks and crises tend to be low frequency, though some illnesses may be a high-frequency risk. Shocks tend to be idiosyncratic, affecting individual households versus entire communities, and occur randomly. Lifecycle events affect individual households (marriages, births) or communities (rituals and festivals) but are not random in that they are expected to occur as households move through their lifecycles.

Responses to risk vary according to the frequency, source and nature of the risk in question. Households may develop proactive strategies to guard against risk as well as reactive strategies to adapt to events after they have occurred (Sebstad and Cohen 2001; Lund and Srinivas 2000; Siegel and Alwang 1999). Proactive strategies may include building up a diverse asset base; investing in insurance programmes to guard against income losses due to illness, death or maternity; and saving money and managing household finances well (such as avoiding extraneous expenditure, especially on leisure activities). Strategies to adapt to external changes after they have occurred may include the selling or pawning of physical assets, taking loans, reducing consumption or putting more household members into the labour force.

One noted absence in the above analysis of risk is structural risk, which may be chronic as opposed to linked to bounded, identifiable events. Specifically, ascribed characteristics like gender, class or race, which limit access to assets and the ability to transform them into valued outcomes, have been conceptualised as chronic risks (Wood and Salway 2000; Unni and Rani 2003) which limit the agency certain groups have in creating livelihood strategies. Use of the term 'strategy' can be problematic in relation to livelihoods because it implies an ability to make choices from a wide set of options, with few constraints. This purposeful sense of agency is not a reality for many poor households or their members in low- and middle-income countries. As Unni and Rani (2003) point out, who people are in a society influences how they enter the labour market, the skills available to them and how the skills they have are valued – all framing the livelihood options available and outcomes achievable. Hence strategising may be possible, but from a limited range of choices. Wood and Salway (2000) link the lack of attention to chronic risks to a tendency for livelihoods analysis to overemphasise the agency open to poor households and their members. While livelihoods as a concept was introduced to draw attention to the agency of the poor as opposed to only examining their deprivations and constraints, the analysis cannot shift too far in this direction. A balanced assessment of choices available to groups within a given socio-cultural context is needed in order to inform realistic programming and policy interventions which move beyond technical solutions to also address some of the overarching factors maintaining some people's vulnerable status. Again, as we have previously argued, the 'everyday' of micro-level interactions and actions will reflect the structures and power relations framing it.

Livelihoods in practice

Livelihoods analysis has not just been an academic exercise: the design of livelihoods frameworks has entered into development practice through the application of these concepts to grassroots practice. Here we focus on the sustainable livelihoods framework (SLF) developed by the UK's Department for International Development (DFID). At the same time we note that other international NGOs have adapted frameworks to fit their needs (e.g. CARE and Oxfam) (Carney 1998; Carney *et al.* 1999).

As Figure 9.2 shows, the SLF focuses on the assets held by households and/or individuals within households, and places this micro-level focus within the wider environment of the vulnerability context and the policies, institutions and processes affecting the ability to access assets and transform them into strategies to achieve desired livelihood outcomes. Of course, any framework is a simplification of the complexity of everyday life and this one is no exception. It is a guide to analysis which is as good as those people who apply it. It assists in structuring the collection of information on the various components of the framework and their interrelationships, with the arrows in the figure showing these relationships and not implying direct causality (Rakodi and Lloyd-Jones 2002). Critically assessing the information, informed by knowledge of the context in question, is left to the analyst. The framework makes it possible to consider the local implication of adverse trends and shocks. For instance, Lyons and Snoxell observe that 'all urban residents suffer congestion, land disputes and safety and health hazards', and a dramatic increase in informal street trade, a decline in alternative means of employment and increased taxation mean that 'new traders experience exploitation and uneven access to resources' (2005: 1303; see also Rogerson 1996; Portes *et al.* 1996). The framework helps to make the connections between these discrete aspects of livelihood.

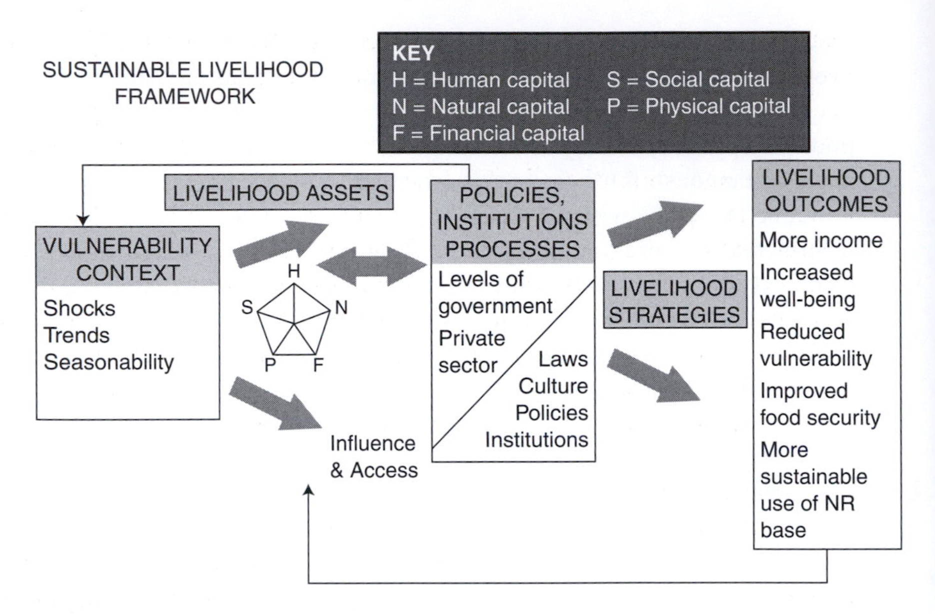

Figure 9.2 ***Sustainable livelihoods framework***
Source: DFID Sustainable Livelihoods Guidance Sheets, Section 2, available at www.livelihoods.org.

In Figure 9.2, assets are represented via a pentagon – capturing the initial five assets integrated into livelihoods thinking, but not political assets. However, the pentagon metaphor can be extended to a larger number of asset categories if appropriate to a context, by incorporating the appropriate shape – in the case of six assets, a hexagon. Representing the assets via a shape, with one side representing one asset, means that the shape and corresponding lines from the centre can be used to show the level and diversity of the different assets held by an individual or household. In turn, the short lines from the centre to the angles each represent financial, natural, political and social assets, and their short length implies low levels of these assets, and an overall concentration of asset holdings on human capital (perhaps labour power) and productive assets.

Looking for patterns across subgroups in their level and diversity of assets can inform thinking on interventions which may aid in increasing the level and diversity of assets this group can obtain. It also may highlight different interactions between assets and the need to think of some if not all of them jointly as opposed to as individual entities. Bebbington (1999) considers social assets key to enhancing the ability to access other assets. For example, relatives in many contexts are an important source of inexpensive, if not free, credit. So maintaining family contacts (building social assets) may be important to accessing financial assets. We go on to argue that social assets (or 'social capital') also feature significantly in the coping strategies of the 'new' working poor living in post-industrial as well as post-socialist Eastern European cities.

Change is important to livelihoods analysis and in many respects this is what differentiates it from the relatively static picture offered by poverty analysis (Rakodi and Lloyd-Jones 2002). The vulnerability context brings in the wider environment and the risks that exist in the macro context. Individuals and households may be able to obtain assets but the crucial question is how they can use them and in the face of what risks. As shown in Figure 9.2, the DFID framework integrates three types of risk into its assessment of the vulnerability context: trends, shocks and seasonality. These are slightly different categories from those described previously, but they cover the same broad types of risks, with the categories trends and seasonality including the life cycle events and structural risks mentioned earlier. What is missing from the framework and many applications of it is the idea of chronic risks, or those associated with identity and social relations, and how they limit opportunities and outcomes for different groups, and for different members of groups in various ways based on their intersecting identities (e.g. a poor dalit (low-caste) woman in India may experience risk and vulnerability differently than a poor woman from a different caste group). However, some aspects of chronic risk can enter the framework through attention to institutions, policies and processes, and how they frame access and influence.

The vulnerability context brings a dynamic sense to livelihoods analysis as it represents the external changes to which poor people must respond in order to preserve or improve the status quo even among chronic risks which in the long term can change in form and influence. Hence, changes in the global economy, new opportunities for organising to resist exploitative employment conditions, or altering weather patterns may influence existing livelihood patterns either positively or negatively, and different groups in different ways. Examining what in the larger context has changed and how different groups can respond is central in applying the livelihoods framework.

The 'processes' section of the framework looks at the way that different policies and institutions – or rules of the game – influence livelihoods. Institutions are public or private structures which set the rules of the game – set how life works in a particular context (North 1990; Kabeer 1994). The policies, rules or norms may be practised at a range of levels from the household through the national and international, and hence add considerable complexity to livelihoods analysis – but complexity mirroring reality. It is here that power more directly enters the analysis, including an analysis of gender and other social relations and how they may influence assets and the ability to transform them into outcomes differently for different groups (Rakodi and Lloyd-Jones 2002). Social norms, 'culture', regulations and government policies influence interactions within and across institutions, influencing access to and the ability to utilise assets (Rakodi and Lloyd-Jones 2002). For example, laws may exist providing women with inheritance rights; however, social norms may intervene in the enforcement of these laws in practice, influencing women's access to assets within the household, as well as their access to markets. Thus, laws and social norms influence how the family and market interact, and how women may enter the market (land used as collateral may allow entry into financial markets), their position in the family and, in the end, their livelihood outcomes.

Institutions and processes operate to create and maintain poverty and vulnerability. However, these structures are not immovable – dynamism exists in this part of the framework as well (Rakodi and Lloyd-Jones 2002). In fact, social and political assets may provide opportunities through which to challenge existing institutions and structures, creating changes which open new opportunities for previously marginalised groups. Looking for opportunities to facilitate such action is one aim of livelihoods analysis.

Livelihoods strategies are something of a black box in the framework – they are the intermediary between the previous interactions between context and assets, and livelihood outcomes. It is here that issues of constrained choices, of needing to seek security today, but at the cost of independence and investment in the future, may be revealed, with all their implications for how power relations provide immediate welfare in some cases but limit opportunities for future improvement (Wood 2001, 2003; Gough and Wood 2004; Wood and Gough 2006).

For example, female home-based carpet weavers, lacking mobility and hence detailed knowledge of the local carpet market, may find more security through labour contracting relations. The contractor provides carpet materials, a loom and designs, and sells the finished carpet in the open market. He pays weavers a set price per carpet according to size and complexity, meaning the women have a known income and are not open to being shortchanged in the market due to lack of knowledge, assuming they could directly access the market. While they are dependent on the contractor to provide more work (and hence must not be 'difficult' and must produce good-quality carpets on time, balancing reproductive roles with paid work), they can also ask for advances or credit in times of need, extending their social networks and access to financial assets. While the women may seem to lose out, they in fact have a known and regular income, though perhaps lower than if they sold the carpet themselves, and a patron to provide credit as needed to help with daily consumption needs or major illness events – a very valuable asset in contexts with limited credit markets and few state options for social protection. Hence, choice in livelihood strategy is limited by circumstances, leading to security today but not necessarily supporting autonomy and improvement in future.

Risky urban environments and livelihood practices

As we noted in Chapter 5, poor access to basic services such as water, toilets and electricity is common in urban areas in the global south. Urban life is often associated with high-density living environments and poor service provision – including solid waste collection. This combination increases health risks due to the presence of solid waste in communities and has also spawned the high-risk livelihood activity of collecting and recycling waste products, including household waste, medical waste and industrial waste. Scavengers, as those involved in this activity are called, can be men, women, boys or girls. Some wander the streets, collecting any 'valuable' waste found, others may buy marketable waste (glass, tin cans, metal) from households or businesses for resale, while others work or live around large dump sites, where private contractors dump large quantities of waste. The scavenging operation often involves the whole family in sorting waste in the home environment. The gendered local effects of vulnerability in this livelihood practice are explored in Case 9.3.

Case 9.3 shows the risks involved for women in taking a WID approach (see Chapter 3), where this delivers development programmes to women without integrating a gendered analysis of the social relations framing men's and women's interactions in society. It indicates that aid agencies in some circumstances can be a form of risk to their intended beneficiaries, effectively raising expectations for change which the NGOs are not able to bring about, at least in the short term. Ideally, bringing men of all ages into a broader programme, as well as engaging in more community-wide initiatives around social norms and practices – presenting alternatives in participatory and transformatory ways – offers a more socially progressive way of challenging the structures intervening in the young women's efforts to change their lives.

Gendered labour markets in the global economy

Labour markets, influenced by the norms and functioning of other institutions such as the family, community and state, shape women's range of choices of livelihood activities and the outcomes of the activities in which women and men engage. How these institutions intersect and influence the choices actually available differs across regions and changes over time; intersections are nevertheless always apparent through the gender division of reproductive work. The vast majority of reproductive work is naturalised as 'women's work' and this

Case 9.3: Challenging gender norms in Cairo to improve young women's lives

In the settlement of Muqattam in Cairo, Egypt, livelihoods revolve around the garbage trade (Assaad and Bruce 1997). Muqattam was settled in the 1970s by rural migrants and many of the social conventions of rural Egypt apply to this primarily Coptic (Christian) community – it has resisted Cairo's moves towards modernisation. This is particularly apparent in the gender division of household tasks, with these falling to women and even more so to low-status women in the household – younger adolescent girls and new daughters-in-law; in the restrictions on women's mobility upon menarche; and in the typically young age at marriage and often complete lack of choice for young women in their marriages. Muqattam is an asset- and income-poor community where many household members must work to contribute to survival; this is illustrated in high rates of child labour among both boys and girls, and less investment in education for both. Human capital levels are not high in the community due to low education investments and high health risks associated with its main economic activity – scavenging.

Adult male members of families engaged in garbage collecting go out overnight, in illegal donkey carts in some cases, or leased or owned trucks in others, to streets and byways middlemen allocate to them. Young boys or girls from the family may join in, generally to guard the collected rubbish. On returning home at daybreak, women's and adolescent girls' main roles in the activity begin – the sorting of the piles of rubbish dumped within the household compound into that which can be fed to the pigs and that which can be sold on as recyclable. This is done generally without benefit of shoes or gloves – the latter considered bulky and inefficient – and raises considerable health risks from direct contact with the variety of waste products collected. These arduous sorting tasks, taking up four to six hours per day, are added to existing reproductive work tasks for most young women (fetching water, cooking, sewing), meaning their work days are long.

The Association for the Protection of the Environment (APE) is a secular NGO which, under the auspices of its Health and Development Committee, carried out a range of projects within Muqattam, largely focusing on young women due to their perceived low status in the settlement. It started working in the community on activities related to scavenging, seeking to increase the safety and efficiency of the work. One of its early activities involved introducing waste sorting at the point of collection, where households giving waste sort it prior to collection, reducing the arduousness of and risks associated with the further sorting women and girls do. From there, APE branched out into literacy training, income-generation activities for young women (rug weaving, embroidery), and training young women to be paid community health workers, disseminating information about child and maternal health initially, and later integrating messages about the health of young women their own age. This role had the young women moving about the community, challenging norms of female seclusion, and it built their knowledge of good health practices, including reproductive health, which would have an influence on them after marriage when they began having children.

A self-assessment of the NGO's work highlighted changes in expectations among the young women – who now sought marriage after age 18, the opportunity to get to know a prospective groom, to have a voice in decisions about family size, to expect husbands to value both girl and boy children and to space child births. Young men, not targeted specifically in the projects, shared some of these interests in social change, including valuing girls' education and women's right to work outside the home. However, on visiting some programme participants after marriage, disturbing outcomes were found, largely related to a reversal of progress since the changes in attitudes identified above did not go beyond the younger generation, nor did they reach beyond changes in individual opinions to the level of social change. While programme staff did negotiate with families about young women's work burdens to free time for programme activities, and did offer monetary incentives for delayed marriage, all of these approaches were individual in scope, raising individual expectations of a different future, but not working to set the wider stage for this to happen.

This example highlights the everyday realities of young women from families engaged in a very common and very risky urban livelihood activity. The NGO's activities supported the young women in building up assets, including self-esteem, and in reducing some of the risk factors they faced. It assisted them in building human capital via literacy and other skill training, in accessing income and in making social connections with other young women in the community while concurrently reducing risks of ill health. While the young women before marriage seemed to report a change in expectations, this sadly was not maintained after marriage. Thus the programme did not adequately assess the influence of the central life cycle event in these young women's lives – marriage – and how it might intervene in the changes the programme sought to bring about; nor did it take a more collective approach to challenging the social norms shaping the young women's lives. The NGO in time recognised the problem of not including men – both peers and fathers of the young women – in a similar programme, and planned to change this. However, any resulting effects of this more inclusive approach to addressing gender norms may come too late to significantly improve the lives and livelihoods of these young women.

Source: Adapted from Assaad and Bruce (1997).

effectively limits the time and energy women have for paid work, and through norms regarding appropriate activities for women.

A common finding across many geographic contexts is that women are integrated into labour markets through a disproportionately high rate of dependence on informal employment. Informal employment is employment occurring without benefit of formal contracts, not covered in any existing labour legislation and where workers have no benefits or social protection. It includes self-employment, unpaid family workers, and wage workers in small unregistered units or in larger registered units but without contracts, domestic workers and industrial home-workers. Self-employment is more common than wage work within the informal economy, making up between 57 and 70 per cent of informal employment across regions (ILO 2002). Informal wage employment comprises between 30 and 40 per cent of informal employment worldwide. Within these groups two of the largest sub-sectors of informal workers are street vendors and home-based workers (ILO 2002). Home-working, with its feminised, vulnerable conditions, especially in piece-work for the garment industry, accounts for the lion's share of unregulated and unprotected economic activity across the globe but it is by nature 'invisible' and rarely properly counted (Boris and Prügl 1996). What form women's informal employment takes varies across regions, especially with respect to activities that take place outside the home, such as street vending. These patterns reflect norms of female seclusion, with Tunisia, India and Turkey and parts of North Africa having considerably lower numbers of women engaged in street vending (as illustrated in Figure 3.1 on p. 76) (see Duneier (1999) on illegal sidewalk vending as a male practice in New York City). Thus, social norms around women's mobility influence the type of work women can do, as do gendered divisions of household caring labour, concentrating women in the informal economy, in activities considered flexible enough to accommodate care work with the need to work to contribute to household income (Kantor 2002c).

Gendered constraints on women's labour market access have not gone unnoticed. The conditions and positions of women informal workers are of concern to a range of actors around the world, some of whom have joined together to form a global network to organise and mobilise around these workers' interests. The network is called WIEGO, Women in Informal Employment Globalising and

Organising. One major partner called HomeNet works specifically with home-based workers. It, along with a wider coalition of partners, had considerable success in advocating for greater recognition and protection of home-based informal workers, showing the use of processes related to globalisation in international advocacy to improve women's everyday lives and livelihoods.

Networking among organisations from the global north and south working with home-based workers, facilitated by meetings such as the World Conferences on Women, led to a major success in the international arena, with the International Labour Organisation's (ILO) 1996 adoption of a convention on home-workers. Home-workers are among the most exploited in the world, hidden from view and in an in-between employment status between employed and self-employed (Jhabvala and Tate 1996; Carr *et al.* 2000). This has raised questions of who is responsible for guaranteeing the rights of home-workers, as their employment status is so unclear. The ILO Convention sought clarity on this matter in order to provide a means to begin to guarantee the rights of home-workers to minimum protections as workers. Having it adopted was a major challenge, as the ILO works on a tripartite system, in which its Annual Conferences, where conventions are voted on, are attended by representatives of government, employers and trade unions from UN member states. Not all of these actors were in support of a Convention on Home-work, for varied reasons related to profitability, enforcement challenges and concerns that recognising home-workers might threaten the status of formal workers. The international network HomeNet lobbied hard to see it passed, starting in 1990 and targeting international unions and the ILO directly. Only through persistent efforts over many years did it achieve one step towards its goal of an international standard on home-working (Jhabvala and Tate 1996).

While the HomeNet example shows how the enhanced communication and networking capacities associated with globalisation processes can be put to good use to advance the rights of informal workers, the same processes can also be a grave source of vulnerability to workers, men and women alike. Globalisation has also brought greater movement of capital, leading manufacturers to invest in countries where labour costs are lower and where concessions are available within export processing zones. These zones often hire many women, whose 'nimble fingers' have marked them as highly productive labourers (Elson and Pearson 1981; Pearson 1998). However, global capital can be fickle, moving where the costs are lowest with little concern for the damage wrought. Bangladesh is a case in point, where the growing export garment industry has had a huge effect on the employment opportunities available to women – outside the home – in Dhaka. Some women have migrated to the capital from rural areas for these opportunities, living in hostels or other arrangements, which has given many a first taste of independence. The statements in case 9.4, drawn from a

Case 9.4: Empowerment and employment in Dhaka

Positive change

Afifa, a single young woman, started working due to her family's economic decline. Their changed economic situation also meant that the cost of a dowry which would be enough for her to marry a man from her social class was too much for her family to afford. Her experiences outside the home, working, led her to the decision to not marry until she found someone who would not demand a dowry:

> They say, if you change your house and move to a pucca building, we can get better proposals for your daughter. I told my father, if we have to change our house for me to get married, I will not marry that person. He is not marrying the house, what does the house matter? I will not marry anyone who demands a dowry even if it means not getting married at all. He will be marrying my dowry then. When the dowry finishes, my value finishes.

Shanu shows the effect of working and earning on intra-household relations between herself and her husband:

> We couldn't say a word before we started working. If I had not been working, my husband would have ordered me to look after his children and see to their needs…If one works, one has different rights. If you are at home and do not earn, then the man is more powerful.

Little effect

Asma Rahela left her husband when he came home with another woman, with whom he was having an affair. She kept custody of her children but in divorcing her husband had to give up claim to alimony in exchange for this. She went back to live with her natal family and started working in garment factories to be able to afford things for herself and her children. However, she sees little value in what she does and in her income, because she is ashamed of her garment work:

> I don't think my value has increased since I started working. The only difference is that now I can buy a few things for my children independently. But nothing has changed for me within the family.

Judgement

> Men and women sit in the same working place face to face. Whatever liberal arguments are put forward in favour of this arrangement, in reality the close

> proximity of opposite sexes arouses lust and love for each other which on many occasions lead to immoral and scandalous affairs between them.
>
> (Hossain 1980: 270, quoted in Kabeer 2000:83)

Source: Kabeer (2000).

Comment

This study illustrates the ability of some women to use work outside of the home, even work at low levels of pay and in questionable conditions, to challenge the structures limiting their choices and freedom, and to achieve a better standard of welfare for themselves and their families – all valued outcomes. This was done in the face of opposition to the 'garment girls' and the threat of change they have brought (Kabeer 2000). Such opposition meant that some women working in the factories gained less socially, being less comfortable identifying as a 'garment girl' and working solely due to need.

However, the opportunities for economic and social change offered via garment sector employment may not be long lasting. The global economy intervened in the form of the phasing out in January 2005 of the quota system that had been in place as part of the multi-fibre arrangement (MFA). This arrangement provided garments from Bangladesh with ready and known markets in the West. With the phasing out of the agreement these assured markets are at risk of disappearing since now Bangladesh will face more competition from garments produced in China and elsewhere. The fear is that Bangladesh's garment industry will not compete, and that large numbers of jobs will be lost (ILO 2006; Ernst *et al.* 2005), with women being most vulnerable to the changes given their concentration in the sector. This illustrates Bangladesh's exposure to risk from the external global economy and the need for diversification not only at the household level but also at the national level, in order to reduce vulnerability. Seventy-six per cent of Bangladesh's exports were from the clothing sector in 2003, a high concentration in the face of the upcoming shift in trade practice (Ernst *et al.* 2005). Women are the ones who may suffer, losing the opportunities to challenge social expectations which came for some from the experiences of working outside the home. While evidence of a sharp decline in employment had not surfaced in early 2005 (ILO 2005; Haider 2006), the need for economic diversification in the country, and diversification into sectors which continue to offer women opportunities for economic engagement, is a grave need.

study by Nuala Kabeer (2000) on the empowerment effects of wage work in garment factories on Bengali women, show the tensions brought by these new opportunities, grasped by some and resisted by others.

In short, the sustainable livelihoods framework can provide a useful means of conducting a gendered analysis of women's economic situation and their gendered labour market experiences. A focus on women's entitlements to assets – what they can and do claim and the institutions and processes framing these outcomes – will bring out how gender relations intervene in the market, family, community and state to constrain or open economic opportunities for women, highlighting issues around different access to or control over skills, productive tools, market information, financing and the like. Through the human capital asset, attention is specifically brought to labour power and how women are able to transform their labour in the market, and under what constraints, including their reproductive work responsibilities and cultural norms of seclusion. Attention to how social norms function will raise issues of female mobility in the context of interest, and of intra-household relations and how they may affect how women's income enters the home, and what valued outcomes are brought about through the addition of this earning power. Figure 9.3 conveys some sense of these intersecting aspects of poverty and vulnerability in the precarious infrastructure and powerful norms circumscribing women's movements in public space on the streets of Kabul. Finally, it is again important to point out that while the sustainable livelihoods framework can work well in facilitating a gendered analysis of women's labour market engagement and outcomes,

Figure 9.3 ***Poverty and vulnerability in Kabul***
Source: Paula Kantor

with the aim of identifying points of intervention to decrease risks and increase resilience, it is a tool that is only as good as the analysts who use it.

Beyond wages: diverse livelihoods in the global north

We noted above that the relevance and application of a livelihoods framework and an understanding of vulnerability are not confined to a southern perspective. In this final section we uncover evidence of social exclusion and poor-quality employment, which together point to a 'new' urban poverty in the global north. We consider the debates associated with each of these phenomena and also point to evidence of how this new urban poverty is manifest in European and US cities.

Recent academic debate has focused on the emergence of a 'truly disadvantaged' US inner-city population whose fragile assets, limited opportunities and widespread experience of discrimination leave particular subgroups, notably young black men, highly exposed to the dual calamity of reduced state welfare and the loss of 'good' blue-collar jobs resulting from de-industrialisation and a globalised economy (Wilson 1996). To some extent it can be argued that the everyday realities of this new poverty have been hijacked by prejudicial media representations of a sexed and racialised 'underclass' – where the tendency is to both blame and fear the victims of multiple deprivations. We saw evidence of this in Chapter 1 in the way journalists tend to 'frame' incidences of violent crime within a popular narrative (of absent fathers, teenage mothers, wayward youth and ethnic conflict) which views the symptoms of multiple deprivation and social exclusion narrowly, without discussing the structural causes.

This is not to say that scholars writing on gender, race and class have neglected the complex processes of inequality underpinning poverty, vulnerability and social exclusion. Indeed, the definition of social exclusion provided on p. 274 clearly demonstrates the constructive influence that academic debate, including a livelihoods framework, can have on urban planning and policy discourse. For instance, economists at the UK Centre for the Analysis of Social Exclusion (CASE) readily acknowledge a pressing need to interrogate persistent structural divides such as gender, disability, ethnicity, age, social class and location to explain why British and Irish working-class boys have the lowest levels of educational achievement and why women who work part time are seeing no progress with respect to the gender pay gap (Hills 2005). Moreover, there is growing awareness that exclusionary environments and practices are closely bound up with both 'outsider' perceptions and local constructions and intersections of gender, race and class. This can be demonstrated in the gendered territorial solidarity evident in case 9.5 in the French suburb or *banlieue*.

> *Social exclusion*: social exclusion is a complex and multidimensional process. It involves the lack or denial of resources, rights, goods and services, and the inability to participate in the normal relationships and activities available to the majority of people in a society, whether in economic, social, cultural or political arenas. It affects both the quality of life of individuals and the equity and cohesion of society as a whole.
>
> (Social Exclusion Unit, www.idea.gov.uk)

Unlike the American conception of a middle-class suburb, the *banlieue* (literal translation 'outskirts') has come to symbolise the very heart of social crisis in France. It vividly brings to mind the sensational media coverage in 2005 of violent clashes with police and flame-torched cars on the streets of ethnically diverse, working-class social housing projects, civil unrest which began in eastern Paris and rapidly spread to other cities across France over the course of a month-long state of emergency. The hardships common to residents of these communities – high unemployment, discrimination in the workplace, poorly constructed and stigmatised social housing and a typically oppressive police presence – intersect with a complex web of identities rooted in territory, ethnicity, gender, religion and social stratification. In Case 9.5 Quenet *et al.* (2006) identify a gendered 'territorial identity' whereby 'the crushing feeling of social oppression' gives rise to manifestations of solidarity which in turn reproduce highly segregated and repressive norms of masculinity and femininity. The authors attribute features of aggressive machismo to a fragile social environment where the socioeconomic hardships which give rise to territorial identity also serve to reinforce gender paternalism (see Nayak 2006 for a UK context). In short, conditions of poverty, vulnerability and social isolation in the *banlieue* are profoundly gendered with respect to both livelihoods (the operation of an informal 'grey market' economy) and everyday social interactions.

The derogatory labelling and experience of isolation and despair in run-down social housing projects are not the only manifestations of urban poverty in otherwise affluent countries. Poverty and vulnerability arise not only from absolute unemployment but also in situations where the available jobs are poorly paid relative to high costs of housing, food, fuel and transport. The paradox of this phenomenon is that the weekly cost of living for those trapped in casual and degraded employment is much higher than it is for those in better paying jobs: if living arrangements preclude access to adequate cooking facilities, then surviving on heavily processed foods which require only basic kitchen preparation is going to be more expensive and less healthy than freshly prepared alternatives (Jarvis 2005a: 142).

Another paradox is that those in poor-quality employment are less likely to benefit from the multiple assets, especially social capital, necessary for livelihood security and social mobility. This describes the 'poverty trap'. The significance of networks

Case 9.5: Through thick and thin: gender and social exclusion in French suburbs

All youth interviewees unanimously affirmed differential treatment between boys and girls, but the reasons for the differential treatment varied, depending largely on the gender of the respondent. Suburban boys responded that 'the suburbs are very dangerous and we would like to protect our sisters and girl-friends. It is not because they are girls, but because it is dangerous.' The suburban girls' answers reflected a general feeling of repression by boys their age as well as their community. The most common response was that 'the boys think that girls are inferior to boys, so we can't participate in the riots and fight. The role of the woman is to take care of children and clean the house and not to fight...but we would like to be rebellious like boys and help them against the political systems.'

Garges can be dangerous at night, based on the proliferation of the grey market, a makeshift economic system for stolen goods, and frequent criminal activity. However, suburban boys' use of solidarity to parent girls is a source of additional oppression for women. The girls, notably those of Turkish and Moroccan origin, claimed feeling the pressure of the social distinction between the 'tasspé' (the 'bad' girls who drift) and the 'good' girls who confined their responsibilities to the domestic arena. Because of these labels, even forthright, confident girls face negative social consequences for their freedom.

Although the view that women should be judged as 'good' and 'bad' and that their roles should be confined to the home are not held exclusively by the suburbs, it is in the suburbs that women can afford to bear these standards the least. With these social standards in place, women lack the solidarity gained by men through territorial identity. In fact, the solidarity gained by men through territorial identity can reinforce social standards that judge women. The emergence of territorial identity as a response to common oppression has the potential to render women in the suburb invisible. While territorial identity can inspire hope in challenging vicious cycles of poverty and social immobility, that very identity can reinforce repression in women. The youth of Garges and Sarcelles pick up the shards of their broken identity and form a new one in territory: impermanent but unifying, inspiring hope in some while repressing it in others.

Source: Quenet *et al.* (2006).

of reciprocity and exchange to sustainable livelihoods is well established in the community studies literature. In his seminal study of the de-industrialisation on the Isle of Sheppey (on the outskirts of metropolitan London) Ray Pahl (1984) found that it was the already busy household with multiple earners in secure employment who were the most likely to engage in a host of other forms of work, including self-provisioning (DIY) and 'cash in hand' work, because alongside secure employment they also gained access to social contacts, credit, skills and tools. Similar findings have been published from ethnographic analysis of small towns in America in the 1990s (Nelson and Smith 1999) and communities living in 'socialist-era' housing districts in Central European cities (Smith *et al.* 2008). Nelson and Smith (1999) differentiate between the assets associated with 'good job' and 'bad job' households. They observe that 'good job' households are better able to adapt to economic restructuring and to accommodate growing state emphasis on private and informal welfare provision. In turn these 'work-rich' households are made busier and 'richer' (in terms of social contacts and material resources) by their extensive commitment to self-provisioning, additional paid work and exchange activities in the informal economy. Conversely, when 'bad job' families are unable to maintain a dual-earner employment structure they are in a weaker position to supplement poor waged work with 'entrepreneurial moonlighting' (Nelson and Smith 1999: 123).

Smith *et al.* document the emergence of a 'working poor' in post-socialist cities,

> [where] high levels of relatively secure employment under state socialism gave way to widespread official and/or hidden unemployment; job security has been replaced by greater job insecurity and employee representation has weakened as political settlements shifted and trade unions declined; dependency on state enterprises to provide not only a monetary wage but also social amenities in kind, was replaced by greater differentials in formal wages and…informal legal and illegal income-generating activities.
>
> (Smith *et al.* 2008: 285)

Their ethnographic analysis highlights the myriad assets and strategies the households routinely draw upon, combining informal and illegal employment with multiple jobs and diverse practices of social reproduction work, in their 'struggle to create sustainable livelihoods in the face of dramatic employment change' (Smith *et al.* 2008: 288–289).

The intersections of poor-quality employment and greater reliance on often depleted social capital are further brought to life by Barbara Ehrenreich (2001) in her gritty window onto the hard-working maids, cleaners, nursing aides and Wal-Mart sales staff found crowded into motel rooms (and the occasional car or van) across the USA. This is illustrated in Case 9.6. Ehrenreich's book highlights similarities with

Case 9.6: The struggle to get by in low-wage America

In 2001 Barbara Ehrenreich's book *Nickel and Dimed: On (Not) Getting By in America* was released to much acclaim. Through participant observation Ehrenreich set out to illustrate the other side of the American dream, the millions who work very hard, often at multiple jobs, but do not earn enough to meet basic needs for healthcare, housing and decent food. Ehrenreich became a low-wage worker in America, spending one month in each of three cities working as a waitress, a Wal-Mart employee and a motel cleaner, to show the desperate straits many Americans live in. In her book she documents the lack of living wages and affordable housing as well as the well-documented healthcare crisis in the US. As a waitress in Key West, Florida, for example, she earns $2.43 per hour plus tips; this hourly wage is less than the legal minimum wage of $5.15 at that time due to the tips; however, there is no guarantee that tips will fill the gap, and while legally the employer must monitor this and make up this difference, this is not publicised or mentioned at the workplace. But as Ehrenreich learns from her own and co-workers' experiences, even at a wage of $6–10 per hour survival is an everyday challenge, in large part due to finding secure, affordable housing. She documents among her co-workers these living arrangements, which highlight the struggles:

- living in a van, showering in co-worker's motel room;
- sharing a room in a 'flophouse' offering rooms for rent on a weekly basis ($250 per week);
- living with a spouse in a rented trailer ($170 per week);
- living with a spouse in a chain hotel for $60 per night because it is within walking distance of the restaurant.

These insecure and expensive arrangements are necessary due to many people's inability to come up with the 'start-up' funds needed to access more secure and less costly monthly rental apartments – at least $1,000 is needed for the deposit and first month's rent. These housing arrangements also mean kitchens are not available or are minimal so low-wage workers cannot save money by cooking large quantities of cheap, nourishing food, but depend on fast food – costly and unhealthy for the most part. Unexpected ill health means trade-offs in terms of meeting other needs, and some responses include non-treatment due to inability to afford doctor visits or medicine.

Source: adapted from Ehrenreich (2001: 16–59).

regard to vulnerabilities experienced across low-wage workers in the US, who are not poor enough to qualify for state support but also cannot afford their basic requirements, and poor households in developing country contexts where state support is often minimal and insecurities abound in relation to work, housing, food access and health. She observes that 'there are no secret economies that nourish the poor; on the contrary, there are a host of special costs. If you can't put up the two months' rent you need to secure an apartment, you end up paying through the nose for a room by the week' (Ehrenreich 2001: 27; see also Toynbee 2003; Abrams 2002).

In short, these examples paint a similar picture of a new urban poverty in neoliberal, post-industrial economies where social isolation and poor-quality employment impose personal risk and the need for creative strategies in the face of fragile livelihoods. Here, a diminished welfare state coincides with poor pay and conditions, relative to living costs, at the low end of the job market, and poor-quality jobs coincide with weak social capital to reduce the level and diversification of assets the working poor have available to protect them from external shocks and risks. Consequently, while the scale and depth of the problem is greater in the global south, largely because of serious environmental and health risks associated with poor access to water, toilets and electricity, many aspects of urban poverty and livelihood insecurity are common to poor and socially excluded populations across all cities of the world. Moreover, persistent gender inequalities are a common structural factor but one which is locally mediated by gender norms regarding appropriate assets and activities for women and for men.

Summary

This chapter has focused on everyday life in the cities of the global south, and to a lesser extent also in relatively affluent post-industrial economies, to illustrate the relevance of the concepts of livelihoods and vulnerability to a multi-scalar gendered analysis of daily survival. Alongside concepts of poverty, vulnerability and social exclusion we introduced a range of evidence to convey the diverse local experiences, manifestations and ramifications of poverty in different urban contexts around the globe, considering evidence from Afghanistan, India, Egypt, China, France and the USA.

In the majority of our discussion on the global south we focused on women's agency in organising together, making choices and benefiting socially and economically from those choices, as well as the constraints within which this agency operates, including labour market segmentation, economic restructuring, global economic forces that operate outside national control, and at times the misguided benevolence of NGO programmes. We made the case for a gender-aware context-informed use of livelihoods frameworks to assess women's and men's everyday lives within their macro environments, to understand the range of risks faced, including chronic risks related to social and economic inequalities, and to identify possible responses which build on and further develop existing capabilities while challenging the norms and institutions increasing insecurity.

The vital thread linking the different urban contexts we examined within and between the global south and the global north was provided by a feminist understanding of multiple economies and a livelihoods framework which registers the typically discounted significance of social capital, political representation and social reproductive assets that extend far beyond wage labour. Having used this holistic approach to bring women's taken for granted assets, capabilities and social reproductive work to the foreground, a strong case can be made to pitch a sustainable livelihoods framework alongside inclusive participatory urban design and a gender mainstreaming of urban governance as essential constituents of a non-sexist city.

Learning activity

Review the assets presented in Box 9.1 (p. 257). Discuss in what ways access to these assets and the ability to utilise them may be gendered in diverse urban contexts, and with what consequences on livelihood security.

Then use the livelihoods framework to analyse the case stories provided below. Some questions to consider include: what risks are prevalent for the households and their members? What similarities and differences in experiences of livelihood insecurity are there within and across the households? What roles do gender norms play in this context?

Board bazaar is a neighbourhood in the centre of Jalalabad city and has about 150 households residing within it. It is an established neighbourhood, but located on government land – so residents do not have legal tenure and are at risk of being evicted if the government should decide it wants to use its land. Homes are primarily made of mud and wood, as is typical in Jalalabad.

Because of the lack of legal tenure, the area has few formal services: there is no sewerage sytem and people dump human waste into the shallow drains running along the streets. Dirty water often forms pools in the roads, where children play. Some households have connected themselves to the electricity grid but about half the households have no connection. Water sources are few in this arid location; one

NGO provided a well in the local mosque and this is where most people go to fetch water. There can be long queues. There is a government primary school nearby where children can go, but no government health clinic. Residents go to two NGO-provided clinics outside the immediate neighbourhood but within walking distance. Women can go if accompanied by other women or a male relative.

Akbar is a 70-year-old head of a household of seven people including himself. He lives with his third wife (the others died), who is currently pregnant – they are hoping for a son. The household lives in one room, given on charity due to Akbar's disability from an accident and his current wife's mental illness. The household relies on the labour of the 11-year-old eldest son, who works as a helper in a car wash centre. He also goes to school and is the only child of the five living in the household to do so; the daughters are not allowed to go. The household sent one daughter away to relatives in their village, to save on expenditure. Due to low cash income, the household has to eat very cheap food and at times must skip meals. Akbar at his age and with his disability requires medical care, which the household struggles to afford. Akbar's wife, Maryam, and her daughters perform all the household tasks; the daughters do not often leave the compound, except if they need to help bring water from the local mosque. The household has no electricity. The son gives his income to Akbar, who makes decisions about how it is spent on his own – he clearly considers himself the boss in the household and can be quite despotic at times. He does not have very good neighbourhood relations and his wife is too busy to form connections with other women; she can only occasionally attend family events (weddings, funerals) of other households in the area. Due to their low income, they do not enter into reciprocal credit relations with other households.

Loqman is head of a nine-person household living on encroached private land in Board Bazaar. They pay 500 Afs per month as rent to the landowner. They get water and electricity through a better-off neighbour, paying 50 Afs per month per bulb for the electricity (100 Afs total). They engage in a range of livelihood activities to meet living costs: Loqman sells cooked food made by his wife, pulls a cart and works for daily wages – he moves between these activities regularly; three sons work – as shop boy, pulling a cart and as a daily labourer. None of these income sources is secure or provides a regular income. Nasreen, Loqman's wife, does all the household work and cooks the food Loqman sells; one young daughter helps her and another older girl is in school – she is the only child in school. They own a cow and a donkey – the cow for milk and the donkey to pull the cart, which they own a part share of. However, when times get bad they think of selling the livestock assets for needed cash. Nasreen fell ill earlier in the year and had to be hospitalised; Loqman took credit from a neighbour to afford this and they have yet to repay the debt. The household maintains reciprocal credit ties to other nearby households, to help them cope in times of need. The boys and Loqman pool their income and decisions are made jointly about how it is allocated across daily needs and debt repayments.

Further reading

Beall, J. (ed.) (1997) *A City for All: Valuing Difference and Working with Diversity*. London: Zed Books.

Gibson-Graham, J. K. (2006) *A Post-Capitalist Politics*. Minneapolis: University of Minnesota Press.

Kabeer, N. (2008) *Mainstreaming Gender in Social Protection for the Informal Economy*. London: Commonwealth Secretariat.

Newman, K. (1999) *No Shame in My Game: The Working Poor in the Inner City*. New York: Russell Sage.

Rakodi, C. and Lloyd-Jones, T. (eds) (2002) *Urban Livelihoods: A People-Centred Approach to Reducing Poverty*. London: Earthscan.

Sassen, S. (2002) 'Women's burden: counter-geographies of globalization and the feminization of survival'. *Nordic Journal of International Law*, 71: 255–274.

10 Cities and gender – politics in practice

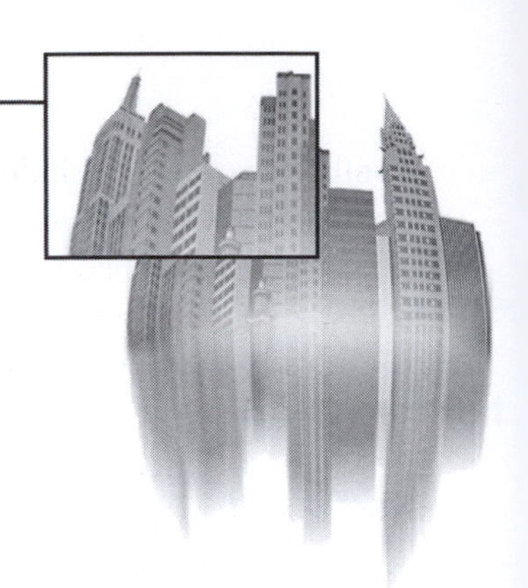

Learning objectives

- to understand how the key themes of cities and gender function as intersecting spheres
- to recognise the classroom, staff room, field trip and city as critical sites within which gender relations are performed and reproduced
- to appreciate the transformative potential of feminism by confronting normative gender stereotypes through reflective learning activities

Introduction

The purpose of this final chapter is to draw together the intersecting conceptual strands discussed in this book, both to sum up the current state of thinking on cities and gender and to reflect on outstanding questions and future directions. Those looking for clear-cut conclusions and solutions to the persistent realities of androcentrism might feel disappointed: we have little faith in the existence of a blueprint or technical fix to remove the barriers and injustices we have identified over the course of this book. Even if there were such a possibility, we do not yet believe there is sufficient capacity, in the standpoint and profile of urban scholarship and practitioner engagement, to realise the non-sexist city. This is not to say that we do not believe in the potential for mobilising socially progressive transformation. Indeed, it is in order to engage our readers in participatory action that we emphasise in this final chapter the critical role of reflexive learning and transformatory teaching practices. We conclude this book, then, in the same way that we began, by demonstrating new ways of viewing cities and gender as co-constitutive subjects – through the urban ethnographic lens of the everyday.

Back at the beginning we identified three core aims: to promote an explicitly gendered urban theory; to expose persistent inequalities in the everyday lived realities of women and men in both the global north and the global south, through the analytic lens of gender; and to influence the tone and substance of classroom debate as well as practitioner and civic engagement. This chapter summarises the journey we have taken in pursuit of these aims. We begin by reflecting on the starting premise for this book. We then summarise the key arguments raised in each of the three parts. Finally we reflect on the future direction of research and debate on cities and gender, focusing on the classroom, staff room, field trip and the city as critical sites within which gender relations are performed and reproduced

Back to the beginning

The image we chose for the front cover of this book is that of a piece of performance art called *Journey with Restraints.* In this performance, the artist Birgit Deubner walks through the streets of Liverpool, England, struggling under the weight of a pair of lead wings. We felt that the interactive urban encounter – where the city is perceived and imagined through the five senses – sight, hearing, smell, taste and touch – provided an engaging point of entry to the kind of urban ethnographic sensibility that we promote throughout this book, such as with the learning activity in Chapter 4: it also serves to animate questions of standpoint introduced in Chapter 1 (see also Cardiff 2005 for other examples of performance walks). The strength of this image lies at one and the same time in an unambiguous metaphor of the subjugated female body, doubly burdened by the practical constraints of urban daily life, coupled with an ambiguous sense of time and place whereby this silhouette could represent any city on the globe, North or South, East or West; the figure could be from any decade of the twentieth or twenty-first century.

Of course we would not wish to push these allusions too far. The title of the performance is appealing in the way it captures themes of mobile/immobile bodies and the gendered biographies and geographies which make up (and are made by) cities around the world – and yet we also stress throughout this book a more nuanced account of gender than that portrayed (almost in caricature) as the subjugated female. Indeed, we consciously made the transition early on from 'women in cities' and 'women in development', to 'gender and place' and 'gender and development'. Women are no more 'essentially' passive victims of oppression than men are always found in positions of power – by their sex alone. Constructions of femininity and masculinity – variously embodied through social interactions and performances and concrete artefacts – are continually reproduced in multiple and situated ways. At the same time, our feminist standpoint shapes the premise of this book and our ardent intention to expose persistent inequalities and injustices.

The harsh reality is that today, in the twenty-first century, the goal of women achieving empowerment on equal terms with men is far from being realised. In a study of gender equality indicators across 58 countries in 2005 the World Economic Forum acknowledged that 'achieving gender equality' across five critical areas, namely economic participation, economic opportunity, political empowerment, educational attainment, and health and well-being, 'is a grindingly slow process' (Lopez-Claros and Zahidi 2005: 1). The authors illustrate this point by observing that:

> In many parts of the world rape is not considered a crime, goes unpunished and continues to be used as a tool of war. Even in highly developed countries, violence against women of all kinds is routine, and often condoned. Female sexual slavery and forced prostitution are still terrible 'facts of life' for poor, often very young, women. Genetic testing for defects of the unborn is used in some parts of the world to determine the sex of the foetus, so that females can be aborted, while in some countries, female infants are buried alive. Forced marriage and bride-burning are still prevalent in the Asian sub-continent. A pregnant woman in Africa is 180 times more likely to die of pregnancy complications than in Western Europe. In many developed countries, where basic gender equality appears to have been achieved, the battlefront has shifted to removing the more intangible discrimination against working women – yet women still hold only 15.6 per cent of elected parliamentary seats globally.
>
> (Lopez-Claros and Zahidi 2005: 2)

Reflecting on this depressing state of affairs requires us to look beyond human rights legislation and efforts to impose stronger incentives and penalties in local and international law. In a number of case studies we have demonstrated the powerful ways that local gender contracts, norms and expectations effectively condemn or condone particular patterns of behaviour, including oppression and violence, where these cultural codes work through an effective suspension of the law – in practices ranging from 'turning a blind eye', through labelling and intolerance, to exclusion, removal and 'the state of exception' (Agamben 2005; Pratt 2005; Sanchez 2004). We are not in any way claiming a generalised suspension of the law for a category of gender alone. Instead we draw on evidence that shows how time and again the 'wicked triangle' intersecting gender with race and class in the geo-political imagination renders the most vulnerable women, live-in caregivers and cleaners, street prostitutes and mail-order brides invisible, barely human. Geraldine Pratt (2005) demonstrates the tragic consequences for a large number of migrant women on temporary visas in the Downtown Eastside in

Vancouver, displaced by gentrification, who went 'missing', to be discovered years later to have been murdered by a local man 'protecting his property'. She explores the androcentric, ethnocentric narratives which allowed the police authorities to dismiss the missing women as being merely transient. The point she makes is that for these missing women to be traced they had first to be noticed as human, as legitimate (Pratt 2005: 1053). This may be a particularly shocking case but it is sadly not exceptional (see also Viner 2005; Macklin 1992; Marston 1990).

Part I: Approaching the City

In the first part of the book we introduced a number of 'bridging' concepts intended to promote the intertwining of urban studies and gender studies and also to incorporate important lessons from development studies. First among these was 'standpoint' which recognises the existence of multiple 'lived realities' rather than a single objective truth. Second was the empirical richness of the ordinary, taken for granted, aspects of the 'everyday' – a conceptual framework which enabled us not only to bring to the surface those activities and interactions which are elsewhere considered too mundane to be entered into urban analysis, but also to promote up-close engagement with cities and gender through in-depth and participatory methods of urban ethnography. A third bridging concept introduced was that of 'multiple economies'. This allowed us to move beyond a narrow understanding of 'the' (monetary) economy to encompass the invisible but essential informal and reproductive economies, the myriad 'labours of love' undertaken by care-givers and community volunteers, who are mostly women, many of them elderly, in poor health, undertaking a 'double shift' of waged and unwaged work, propping up the formal capital economy. Alongside these bridging concepts we considered a number of cross-cutting trends, including the restructuring of the global economy, the expression of gendered power and practices across multiple scales within and between cities around the world, the conditions imposed by hegemonic neo-liberal regimes on less developed countries, via structural adjustment programmes and through geo-politically uneven development and the privileging of core 'world' cities relative to others perceived as peripheral or parochial. These deliberations on scale, placement, position and power led to the suggestion that prospects for improving gender equality depend less upon gender-conscious policies than upon a transformation of the dominant structures of representation and regulation which prioritise competitiveness over an ethic of care and concerns for inclusiveness (Jarvis 2007a; Perrons 2005a; McDowell 2004; Valentine 2005).

Part II: Gender and the Built Environment

The second part of the book explored the origins of androcentrism and ethnocentrism in the built environment. First we highlighted the skewed demographic profile and elitist standpoint of the professional practices associated with urban design, construction and management. Then we critically examined the legacy of neo-liberalism with respect to a politically motivated language of market competition, individual 'choice' and fiscal prudence. This discussion was framed by the bridging concepts previously introduced and further extended by an understanding of multiple, intersecting 'infrastructures' of daily life (Jarvis 2005b). The discussions and case material presented in this part of the book intended to promote an explicitly gendered urban theory and practice. Arguably a discourse of competitiveness dominates the literature on cities and this exerts a significant influence over all manner of practitioner engagement. It is on this moral basis that neo-liberalism is widely condemned by feminist critics as a bad model to roll out across the world (see Tronto 2004 on the loss of collective responsibility and shared endeavour). Efforts to differentiate the 'good' city for everyday life, for all, from 'the good life', for a few, between what Linda McDowell (2004) calls an ethic of care and the narrow pursuit of flagship competitiveness, could only be realised by shifting the measure of 'success' from competition to care. At present this distinction is not possible where urban quality of life is ranked on the basis of economic indicators alone, using such measures as GDP per capita, unemployment or employment rates, or stocks of vacant or derelict land (Begg 2002: 312). Economic growth is not synonymous with 'progress' as we understand it in terms of gender equality and human development.

Part III: Representation and Regulation

The final part considered the role of collective action and resistance to the prevailing gender order. Our discussion was framed by unresolved debates associated with the feminist political goals of gender mainstreaming and the corroding effects of competitiveness and growth as fundamental drivers of non-representative urban governance. We noted that it would be reasonable to expect gender mainstreaming to yield positive benefits for a more inclusive city. Gender mainstreaming is, for instance, mandatory in urban development projects financed by the EU structural funds and by the World Bank. Yet in this context Diane Perrons (2005: 389) observes that gender equality remains elusive. She suggests that gender mainstreaming remains at best a partial solution because it fails to confront the contradictions of a neo-liberal market economy, with its inherent tendency toward widening gender and class inequalities.

We pointed to a multi-layered, multi-scalar, gendered system of planning, social welfare and culturally constructed gender contracts. Within a multiple and dynamic system of planning and public policy we extended conventional notions of governance to reflect the way prevailing gender systems function through intersecting social and cultural norms and ties, household resource capabilities and livelihood practices. The vital thread linking the different urban contexts we examined within and between the global south and the global north was provided by a feminist understanding of multiple economies and a livelihoods framework which registers the typically discounted significance of social capital, political representation and social reproductive assets that extend far beyond wage labour.

Are we there yet?

There are several ways we might assess whether (and to what extent) 200 years of feminist struggle has altered the landscape of gendered social inequalities and uneven development in cities around the world. We might look for evidence of reduced inequalities through 'sameness' (inclusion, equal opportunities, equal treatment), through the celebration of equality through 'difference' (a reversal in roles, special programmes, cultural tolerance of diversity) or through 'transformation'. Notwithstanding the issue of how 'success' is to be defined in assessing the outcomes of gender mainstreaming, as explored in Chapter 8, the evidence presented above and throughout this book confounds any claim that persistent gender inequalities have been eradicated by any of these measures. Indeed, we are deeply concerned at the suggestion of a 'backlash' post-feminist movement and the idea that gender inequalities are no longer apparent or relevant as a priority for urban analysis, planning and social policy (see Box 10.1).

Earlier chapters presented case studies showing that, in affluent countries, four decades of equal opportunities policies have done little to bring about closer convergence in men's and women's rates of pay: poverty and insecurity remain notoriously feminised, especially in the stages of child-rearing and old age. A consistent gender difference is found within and between cities around the world, with women exhibiting noticeably higher rates of poverty and dependence (as witnessed in patterns of welfare receipt) than men. In an extensive study of Chicago's 'hyperghetto' *Urban Outcasts: A Comparative Sociology of Advanced Marginality*, Loïc Wacquant (2008: 106–108) observes that women are twice as likely as men to be jobless, significantly less likely to come from a household possessing a home or any other asset and far more likely to rely for survival on soup kitchens, food stamps and public aid. Yet this otherwise comprehensive study consigns the structural inequalities to the status of a footnote to malestream urban studies: the author tackles what is a distinct feminisation of urban social exclusion in a few sentences over the course of some 200 pages. This is symptomatic of the neglect of gender in urban studies texts as well as in the curricula identified in the first learning activity in Chapter 1. It also points to the urgent need for action to unsettle the attitudes, assumptions and practices underpinning the professional training of architects, designers, land use and transport planners and local government officers.

In her book *Towards Cosmopolis* Leonie Sandercock (1998) explores the idea of a progressive planning curriculum and what it should look like (see also Sandercock 2000; Thomas 2004). We welcome these developments and look forward to increased momentum in this regard. Certainly, the future prospect for inclusive, non-sexist and accessible cities will never become a reality if those trained and given authority to make key decisions are neither conscious of gender at work in our cities nor motivated to act to eradicate androcentric assumptions, barriers and infrastructures (Umemoto 2001). Yet the message of this book has been that we need new ways of approaching the city; to engage with a more holistic conceptualisation of the social and non-representational as well as the material infrastructures of the gendered city which are currently neglected within a planning-practice perspective.

Box 10.1: Anti-feminist backlash

Evidence of anti-feminism accompanies the entire history of the women's movement like a black shadow. This has escalated to form a post-feminist 'backlash' against 'militant' second-wave feminist activity from the early 1980s (Faludi 1991, 1999). Much as the term implies, post-feminism claims that the job of women's liberation is complete and that to pursue a redundant cause would have the obverse effect of provoking male wrath in a backlash against women's equality. It claims we are entering an age in which it is no longer relevant to claim gender inequality as a political issue. One of the earliest uses of the term post-feminism was in Susan Bolotin's 1982 article 'Voices of the Post-Feminist Generation', published in *New York Times Magazine*. Since this beginning, a number of magazine articles and popular surveys have drawn attention to the idea that women no longer identify with the label feminist. Natasha Walters (1998) is one of a number of feminist scholars who criticise the suggestion we are entering a 'post-feminist' age, pointing to persistent barriers to equal opportunity as well as continuing violence and threat to women around the world.

The corroding effects of neo-liberalism and post-feminism

It is not only with respect to the planning and governance of cities that competitiveness has overtaken care. In this final chapter we reflect not only on the textbook 'subject' of cities and gender but on how these are taught and represented in different walks of life. As a global phenomenon, for instance, neo-liberalism implicates numerous sites and sectors, including schools and universities, in a profound restructuring of goals and values (Hills 2005). The idealised career trajectory is one of single-minded self-sufficiency and conferences rarely seek inclusiveness: to get by you need to be relatively young, carefree, mobile and energetic (Caplan 1993). This is not simply the perennial concern with inadequate childcare (but see Ekinsmyth *et al.* 2004). Instead, Giroux (2002) points to the corrosive effect of a neo-liberal shift from collective to individual accountability in teaching and learning environments.

This is witnessed in particular in relation to an 'audit culture' permeating every aspect of university activity. Mechanisms have been introduced to measure 'teaching performance', 'research quality' and 'institutional effectiveness' (Shore and Wright 1999). While these 'audit technologies' are typically framed in terms of 'empowerment' and 'self-actualisation', the cumulative effects have been particularly disempowering for those groups and concerns (such as women's, gender and sexuality studies) which have always struggled for recognition at the margins of the curriculum. The 'rolling out' of market-like regulatory mechanisms has coincided with a wider social trend of political disengagement which is concentrated among a young, well-educated and new-media savvy cohort – the traditional student intake of higher education institutions. Disengagement can be witnessed in voter apathy and a shift from collective participation (in street protests and marches) to 'a new repertoire of non-collective political conduct' such as politicised consumption (product boycotts and the like) (Hay 2007: 25). Also associated with this disengagement is a suggested 'backlash' against explicit gender interventions. Angela McRobbie (2004: 1471) warns that contemporary popular culture serves effectively to undermine the feminist gains and struggles for gender equality of the 1970s and 1980s. While Hall and Rodriguez (2003: 879) dispute the evidence that we are living in a post-feminist era they acknowledge that a 'no, but...' version of feminism has developed alongside a de-politicised belief in the 'irrelevance' of a global women's movement. The upshot is that attacks on the value and relevance of gender and feminism, whether as taught subjects or methods of teaching, come from all directions – from university employers (van Hoven forthcoming) as well as from the student body. The result is that while university students and administrators endorse an equal opportunities framework 'as a given' they typically 'distance themselves from feminism' in practice (Rich 2005: 495).

With neo-liberalism the paradox is that widespread privatisation of once-public goods and services (such as water, education and health, as demonstrated in earlier case studies) gives rise to social exclusions across multiple scales of national, city-regional and local lived realities, at the same time as neo-liberalism in higher education tends to stifle efforts to engage meaningfully with these political consequences. This is not to say that local resistance has not previously been the focus of attention in urban studies. A literature already exists which offers a bottom-up perspective on neighbourhood struggles over public amenities (such as schools) in the American city, for instance, as well as over transport infrastructure and strategies to combat poverty (see, for instance, Cox and Jonas 1993; Gilbert 1997). Yet arguably what remains underdeveloped is adequate iteration, and inclusive conceptualisation, of the gendered and messy realities of urban daily life, as subjects of strategic policy and planning.

Towards gender neutrality?

In mainstream debate, women are *assumed* to be included and thus rendered invisible by a lack of recognition of difference. Yet when a separate space is created for women or when the language is one of achieving 'gender neutrality' women are rendered outsiders or simply made to 'disappear'. Whichever way we respond in policy or political terms to inequality, this book demonstrates that cities will always function as 'vital arenas in the embodiment, contestation, mobilisation, subversion and transformation of all aspects of gender' (Bondi 2006: 1). The gendered nature of the urban lived experience arguably transcends headline patterns of segregation and inequality to encompass gendered identities, performances and states of being in existential, emotional and unstable ways.

On one level the popular perception that we are living in a post-feminist age suggests that awareness of past injustices might be remedied by simply 'updating' cultural practices, signs, texts and language (such as with the façade of political correctness). A crude example of the way a 'post-feminist' popular culture appears to deny the relevance of gender or 'neutralise' differences can be witnessed in the adoption by young African-American students of a gender-neutral third-person pronoun: 'yo'. Some of the classroom exercises undertaken by Stotko and Troyer (2007) in two Baltimore schools (pupils age 11–13) might be usefully replicated by readers of this book – either as participatory action research in the field, or as a reflexive learning activity. It is with this practical engagement in mind that further details of this preliminary study are recounted here.

In the case of so-called 'generic' labels (such as master plan, masterpiece, 'urban man' or 'social man') previous research indicates that people do not use and interpret these terms as gender neutral (Schneider and Hacker 1973). After receiving anecdotal evidence from several Baltimore teachers of students adopting the yo form of address (as in 'Yo handin' out papers', 'She [the teacher] is handing out papers';. 'Peep yo!' 'Look at him!') Stotko and Troyer (2007: 262) designed a series of classroom tasks to try to understand the popular cultural 'rules' of gender-neutral language.

The first task involved a set of cartoons depicting two people talking to a third person: in each case the third person either looked unusual or appeared to be doing something unusual. Above the head of the two people having the conversation were empty speech balloons for the students to fill in. In two of the cartoons the third person was female; in one, the third person was male; and in the remaining picture the head of the third person was covered with a party hat, thus disguising the gender identity. The teacher distributed the cartoons to the students with the instruction to fill in the conversation bubbles using slang (defined for them as 'informal language, the way you talk to your friends') (Stotko and Troyer 2007: 266).

A second task involved a set of short conversations based on speech that had been heard at the school, variously using male, female and 'gender-neutral' pronouns. Students were asked to circle one of three choices ('I talk like this', 'Some people talk like this', 'No one talks like this') to judge the appropriateness of a particular sentence in each conversation. Further tasks involved students writing a conversation of their own to accompany a story prompt and also judging sentences by their acceptability among their peers.

Through these classroom exercises Stotko and Troyer (2007: 273) confirmed the use of a gender-neutral third-person pronoun in casual conversation among a proportion of African-American school students in Baltimore – often in ways that actually translated as 'he'. On one level this is quite a trivial and obscure example of how gender identities get buried or made over in popular 'street' culture today. Yet, on another level, and bearing in mind the usefulness of this type of exercise for learning – or to elicit an evidence base – this once again stresses the importance of uncovering the mundane and taken for granted practices of daily life.

Transformative teaching and learning

Throughout this book we have invited readers to explore new ways of seeing the world – and to reflect on their own and other people's taken for granted understanding of everyday environments, events, discourses and interactions. This message of engagement is pushed further in this chapter with explicit reference to classroom and field trip learning opportunities. We argue that transformative teaching and learning are crucial as a means to challenge normative assumptions and routine behaviour. Elsewhere Liz Bondi (2004: 175) has argued that the transformative potential of feminist teaching and practice has been neglected and connections have too rarely been made between the *subject* of gender studies (or urban studies) and questions about *how* we teach feminism (and urban theory) in the classroom.

Bound up with this recognition is the perennial challenge of drawing gender in from the margins and cultivating gender interventions in more mainstream components of urban study. In this respect a particular concern should be a persistent occupational sex segregation, both in terms of 'feminised' qualitative methods versus 'masculinist' quantitative spatial analysis and in relation to the authorship of articles by subject and journal, comparing, say, urban studies (predominantly male) with gender studies (predominantly female) (Davenport and Snyder 1995; Aitchison 2001). It is very difficult to dismantle dominant cultural narratives if transgression relies on individual expressions of dissent in a hostile environment (the lone feminist in the staff room) or if teaching and learning on gender modules remains stereotypically perceived as a 'women-only' concern. In reality, Liz Bondi reminds us that 'gender is necessarily being done all the time by all participants in teaching contexts' (Bondi 2009: forthcoming). The blending of urban studies and gender studies promoted in this book offers an exciting opportunity and challenge to demonstrate the continuing relevance of feminism by confronting normative gender stereotypes through inclusive, reflexive, collective learning activities.

'Feeling' as learning

It is widely accepted that learning involves a host of activities beyond 'thinking' and 'doing' – including 'reflexivity' and 'feeling' (Kolb 1984). While reflexivity is now well established as a pedagogic practice (which is not exclusive to feminist teaching), learning activities rooted in 'feeling' (concrete experiences dealing with immediate human situations subjectively) are far less well developed (Kolb 1984). We argue that it is entirely possible and appropriate to make gender interventions in the classroom, town hall and conference; to provoke personal and political transformation as an antidote to the instrumental pursuit of measurable learning objectives (and competitiveness) alone. One approach is to push the practice of reflexive learning beyond the most commonly mobilised form of 'intellectual oversight' (Giddens 1991). This opens up transformative potential by deploying concrete experiences, as interpreted through the medium of role play, diaries, field trips and participatory action research (see Kindon *et al.* 2008).

Similarly, we would advocate the promotion of 'feeling' as a neglected mode of learning and as a natural extension of an urban ethnographic approach to everyday life research (McDowell 2004; Rigg 2007). The close encounters advocated by reflexive learning and 'feeling' contrast with orthodox instrumental and individual instruction. Such encounters recognise that it is empowering for students to experience first hand, discover and communicate their awareness of a previously abstract concept of inequality. It is essential to this process that students are properly supported in the steps they take outside their comfort zone, to establish an ethical environment within which to provoke a state of catharsis. For example, recent decades have witnessed new debates *within* feminism *about* feminism. Rather than view this negatively as evidence that feminism is no longer relevant in the classroom, as from a post-feminist perspective, these internal debates provide the ideal canvas for teaching critical thinking – doing, reflecting and feeling – around inequalities and uneven development. The challenge is, as Low (2000: 159) observes, to consider contestation and transformation in terms of 'bridging' multiple, intersecting and hybrid gender identities (in the classroom), and in terms of realising a process of change (cultivating a more politically engaged subject). Used more particularly in relation to 'frame theory' as applied to social movements, and building on the work of Goffman (introduced in Chapter 2), the notion of 'bridging' captures the sense in which transformative feminist pedagogy might be used both to draw gender in from the margins (into mainstream urban studies, for instance) and to pave the way for a deeper understanding of intersections and interdependence – of gender, race, class, sexuality, faith and disability.

Movements of transformation?

According to Walby a vision of transformation suggests 'neither the assimilation of women into men's ways, nor the maintenance of a dualism between women and men, but rather something new, a positive form of melding, in which the outsiders, feminists, change the mainstream' (2005: 323; see also Rees 1998). We argued above that practice-based teaching and learning can enable students to reflect empathetically on the tensions, paradoxes and concrete experiences underlying gender stereotypes/sex-typing in their own and other people's lives – but what about the potential for transformative deliberations in other practice-based contexts?

In order that we do not end this book on an entirely pessimistic note concerning prospects for future gender equality and an ethic of care, we must acknowledge the valuable work underway in a number of progressive (women's) planning networks. Earlier chapters made reference to the advocacy work of EuroFEM and the promotion of inclusive design and participatory planning tools (Reeves 2005). Other networks to highlight include the progressive Planners Network UK (PNUK); the Toronto-based Women and Environments Network (see Rahder 1998; Rahder and Attilia 2004); international movements such as the Global Women's Strike and Mothers' Centers International Network for Empowerment (MINE) (46,000 women, men and children participating in community groups across 22 countries); also the high profile achieved by the role of gender in social and environmental campaigns such as waste management (Urban Waste Expertise Programme, UWEP) and public toilet provision (Greed 2003; Browne 2004; Overall 2007). In addition, we have made periodic reference in this book to the different scale of gender concerns confronting women and men in the global south as compared with the affluent north. Focusing on the issues of waste management and toilet provision, for instance, Huysman (1994) points out that in the Indian context women are found to pick up street waste as a survival strategy rather than because of attitudes towards public health and cleanliness. Similarly, Chapter 5 related the experience of women in Mumbai, where the lavatories that exist are usually filthy, broken down and generally unusable, waiting until nightfall to defecate in the street. Yet in this context too we also highlight the fledgling signs of active resistance and scope for transformation in the proliferation of grassroots women's movements (Novy and Leubolt 2005; Patel and Mitlin 2002).

We also recognise in recent urban planning scholarship and policy initiatives renewed interest in public participation, emancipatory planning, deliberative democracy and 'mainstreaming' efforts to incorporate gender in urban planning (Rahder 1998, 1999; Healey 1997), along with a 'renaissance' in the kind of 'human-scale' urban vernacular associated with the seminal writings of Jane Jacobs. At present many of the most economically 'successful' cities are hostile environments for families with young children. According to Peñalosa, if we can build a successful city for children and older people, we will have a more successful and caring city for all (Ives 2002; Developments Magazine 2008). Although not beyond criticism and not intended as a feminist approach, 'smart growth' and 'liveable cities' debates potentially signal a more receptive social and political landscape within which to reorient the focus of planning and policy from measuring growth to tackling issues of ethics, care, justice and worth.

Concluding remarks

Throughout this book we have stressed the need for greater dialogue between urban studies and gender studies as one step towards incorporating an explicitly gendered and holistic urban theory into mainstream teaching and practitioner engagement. We recognise a welcome shift in these disciplines away from binary categories, amidst growing awareness of multiple, fluid, dynamic and contested identities and identifications. To this end we have stressed a complex system of inequality, vulnerability and discrimination – irreducible to man/woman, male/female, rich/poor. Recent theoretical developments highlighting intersectionality are especially constructive in this regard (see McCall 2005), as are in-depth analyses which explore the context- and situation-specific negotiation of gendered moral rationalities whereby, at any given moment, confronted by a seemingly impossible 'choice', men and women alike are inevitably torn between affiliations and identifications. Gendered power relations intersect with other complex inequalities and capabilities, including (dis)ability, (im)mobility, sexual orientation, age, faith, ethnicity and class. Cross-cutting these embodied attributes are 'entitlements' to a web of resources, through multiple economies: of income, property, transport, 'sweat equity', gifts, inheritance, kin networks and unpaid personnel.

At the same time we have not shied away from reminding those who might need reminding that half of the urban world population still does not experience equal rights or equal treatment, that women are constrained (in different and unequal ways) by the way cities and urban infrastructures (material, cultural, institutional, moral) are designed principally by and for men. We make this provocatively 'second-wave' protest at an interesting time in the history of the feminist movement. Just as we completed the manuscript for this book the last remaining women's studies undergraduate programme in Britain closed its doors. A similar restructuring is taking place around the world. While awareness of gender as a social variable has been integrated within other disciplines this does not reflect a mainstreaming of gender on the educational agenda. Indeed, our suspicion is that the demise of women's studies is not a symptom of higher-profile gender studies but instead a worrying symptom of the 'post-feminist' backlash we pointed to earlier – the popular perception that the campaign is over, that gender equality has been achieved and that feminist concerns as such constitute a non-issue. The purpose of this book has been to illustrate through simple snapshots of everyday lived realities for men and women in very different urban contexts across the world that this is patently not the case.

Lest we appear to end on a negative note, or with a conclusion that there is no conclusion, that there are no solutions, that there is no prospect for gender democracy in the future, we end by saying that there *is* a role for progressive planning and transformative social movements. A first step is transformation in the classroom and in practitioner training and public participation. Transformation will come from a reintegration of a culture of care in public life where ruthlessly competitive structures are dismantled and replaced with infrastructures which support and enable the social reproduction work that ultimately maintains societies, livelihoods and an inclusive, gender-democratic urban environment. As an immediate priority this means striving to guarantee secure access to safe water, clean air and sanitation, green spaces and secure, affordable housing. It also means recognising social as well as material infrastructures and reorganising these around the principles of social justice and an ethic of care.

Further reading

Kindon, S., Kesby, M. and Pain, R. (2008) *Participatory Action Research Approaches and Methods: Connecting People, Participation and Place*. London: Routledge.

McCall, L. (2001) *Complex Inequality: Gender, Class and Race in the New Economy*. London: Routledge.

Oakley, A. (2002) *Gender on Planet Earth*. Oxford: Polity (useful as a means of provoking classroom discussion and debate).

Rahder, B. (2004) 'Where is feminism in planning going? Appropriation or transformation?' *Planning Theory*, 3: 107–116.

Wilson, E. (1997) 'Looking backward, nostalgia and the city'. In S. Westwood and J. Williams, (eds) *Imagining Cities: Scripts, Signs, Memories*. London: Routledge, 125–137.

Web resources

Women-friendly cities – Frankfurt: http://www.bmkadinhaklari.org/Unjp/UNJPBultenSayi8/frankfurt_en.html.

Women and Equality Unit (gender mainstreaming explained for UK/EU): http://www.womenandequalityunit.gov.uk/equality/mainstreaming_explained.htm.

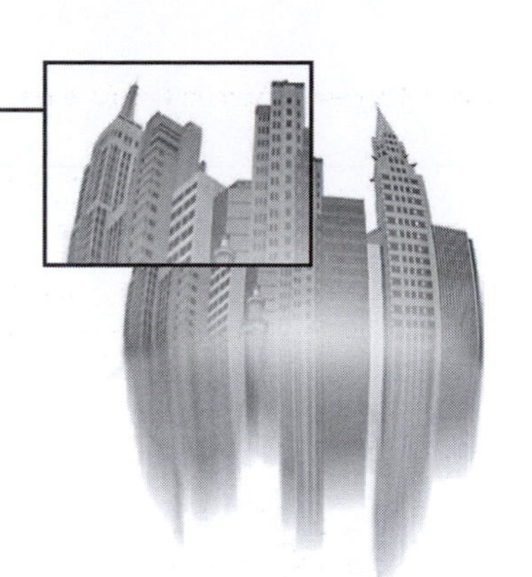

Appendix: Selected chronology to show the parallel development of urban studies and gender studies and of both in relation to external events and technological innovations

Developments in Urban Studies	Selected context of world events	Developments in Women's and Gender Studies
	1880s	
Growth of the industrial city Contrast theory: Tönnies and Simmel Concern for sanitation: Charles Booth's survey of working-class living conditions Publications by Marx and Engels Origins of the Garden City Movement of Ebenezer Howard	1851: the Great Exhibition, London, Crystal Palace, celebrated 'works of industry of all nations' and the mass production of new technology (e.g. home and transport) 1876: telephone invented 1885: first petrol-driven motor car 1895: wireless radio invented 1900: television invented/named 1914–1918: World War I	**First-wave feminism** 1902: women get the vote in Australia 1903: Women's Social and Political Union (suffragettes) formed in Britain by Emmeline Pankhurst 1915: women from Northern Europe and the USA meet to discuss peace at the first International Congress of Women 1919: Sex Disqualification Removal Act in Britain allows women to enter the professions for the first time
	1920s	
Poverty and disorder Chicago School, setting out the 'ecological approach' linking spatial form and social organisation (with reference to competition and 'natural selection') Class and race are key issues of concern Urban managerialism (and regulation theory)	1923: Rosewood Massacre, Florida, USA; Ku Klux Klan lynchings of black men accused of attacks on white women; this theme of 'protecting Southern womanhood' memorialised in Harper Lee's classic *To Kill a Mockingbird* 1928: commercial development of air travel following Charles Lindbergh's flight across the Atlantic 1929: Wall Street stock market crash, 24 October	1920: Oxford University opens its degrees to women (Cambridge did not do the same until 1948) 1922: the Law of Property Act enables both husband and wife to inherit property equally for the first time in Britain 1928: Equal Franchise Act (UK) entitles women the vote on equal terms with men; women become 'persons' in their own right in Britain by order of the Privy Council 1929: Virginia Wolf publishes *A Room of One's Own*
	1930s–1950s	
Fordism Modernist architecture and planning, especially Le Corbusier's 'cities of the sky' Greatly increased output of mass-market consumer goods (such as labour-saving 'white goods' in the home) coupled with suburbanisation **Phenomenology** 1938: Wirth publishes *Urbanism as a Way of Life*, introducing the notion of 'psychic overload' contributing to alienation and social disorder (revival of Simmel)	1932: a 'marriage bar' is introduced in the British 'establishment' (the BBC, civil service, teaching, post office) forcing women to resign from employment on marriage (not lifted until 1946–1952) 1939–1945: World War II; huge impact on women's employment prospects; 'Rosie the Riveter', etc. 1944: women get the vote in France 1950: women get the vote in India 1955: USA, Rosa Parkes refuses to give up her seat to a white man on a bus; this inspires the black civil rights movement	1949: Simone de Beauvoir publishes *The Second Sex* 1950: Harvard Law School hosts a forum on 'women's education: kitchen or career'; the 1950s is often perceived as an era of conformism (images of the aproned housewife); development of the women's liberation movement accompanied by a 'backlash' period of increased suppression of women 1950s: Women's studies debates mobilised around the creation of mass single-family housing built around the idealised nuclear family (e.g. Levittown, the paradigmatic post-war US suburb)

1960s–1970s		
Humanism (critique of Fordism) 1961: Jane Jacobs publishes *The Death and Life of Great American Cities*; Lewis Mumford publishes *The City in History* (against uncontrolled growth and 'sprawl') 1972: Oscar Newman publishes *Creating Defensible Space* **Marxist revival** 1973 David Harvey publishes *Social Justice and the City* 1977 Manuel Castells publishes *The Urban Question* and *The Rise of the Network Society* The implications of de-industrialisation and economic liberalisation are key concerns	1961: the contraceptive pill (access restricted by national law and cost) goes on sale for the first time (free in Britain since 1974) 1961: publication of *Silent Spring* by Rachel Carson marks the birth of a modern environmental movement 1961–1973: USA at war with Vietnam (associated with massive anti-war demonstrations, including self-immolation) 1964: passage of the Married Women's Property Act in Britain 1966: Indira Ghandi becomes the first woman prime minister of India; marriage bar in Australia finally removed 1967: abortion legalised in Britain 1971: Switzerland finally gives voting rights to women (some restrictions exist until 1994) 1973: women are allowed onto the floor of the London Stock Exchange for the first time 1976: severe oil/petrol supply disruption and global economic recession 1978: world's first test-tube baby	**Second-wave feminism** **Women in Cities (safe cities and mobilities discourse)** 1963: Betty Friedan publishes *The Feminine Mystique* 1966: formation of the National Organization for women (NOW) in the USA by Betty Friedan 1960: Women's studies first taught in US universities 1970: Ester Boserup publishes *Women's Role in Economic Development*; Germaine Greer publishes *The Female Eunuch*; Kate Millett publishes *Sexual Politics*; protests against Miss World contest, USA 1972: formation of the 'radical' magazines *Ms.* (USA) and *Spare Rib* (UK) 1977: International Women's Day formalised 1977: Marilyn French publishes *The Women's Room* (which for many feminists marked in fiction the origins of the international campaign Wages for Housework) 1970s: Ann Oakley first used the term 'gender' to clarify discourses about identity
1980s		
Post-industrial society Preoccupation is the transition from the industrial to the information age	1981: HIV/AIDS enters global consciousness; AIDS orphan pandemic declared by 1985 for sub-Saharan Africa 1989: fall of the Berlin Wall; end of Cold War; decline of Marxism A period of renewed optimism and economic liberalisation in the affluent north in contrast with a deepening debt crisis, structural adjustment programmes and stabilisation in the developing world Widespread expansion of telecommunications technology ICT	1981–2000: Greenham Common 'embrace the base' protest against US nuclear missile base, involving a 19-year women-only peace camp **Third-wave feminism** **Difference and diversity discourse** The splintering of feminist influence; influence of black and subaltern feminists

1990s		
Fortification and exclusion 1990: Mike Davis publishes *City of Quartz* (Fortress, LA) 1991: William Cronin publishes *Nature's Metropolis* 1991: Joel Garreau publishes *The Machine, the Garden, and Paradise* (questioning conventional notions of 'progress') **Multiculturalism** 1998: Leonie Sandercock publishes *Towards Cosmopolis: Planning for Multicultural Cities*	1992: the World Wide Web (WWW) is made available to home users; explosion of the Internet along with reduced cost of personal computers Gulf War; collapse of the Soviet Union; cloning of 'Dolly' the sheep in the UK; IVF debates around sex selection/gene therapy	1990: Irene Tinker publishes a set of essays on *Persistent Inequalities* to honour Ester Boserup's contributions to gender and development **Post-structuralism, post-feminism, queer theory**
21st century		
New urbanism Discussion around 'smart cities'; liveability; urban villages; sustainable cities; compact cities	Rapid innovations in mobile phone technology 2001: terrorist attacks on the Twin Towers (New York) and Pentagon (Washington) USA, 11 September	Intersectionality; gender mainstreaming; gender interventions
Source: author's compilation, drawing on a number of texts for reference including: Badcock, (2002: 5, 6) National Women's History Project http://www.nwhp.org/; http://en.wikipedia.org/wiki/20th_century.		

Bibliography

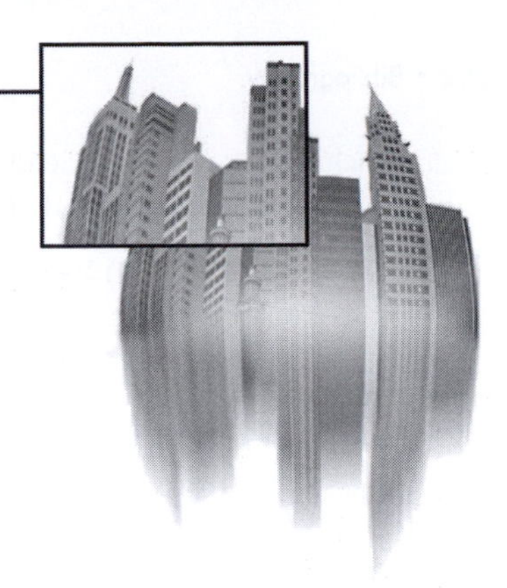

Abrams, F. (2002) *Below the Breadline: Living on the Minimum Wage*. London: Profile Books.

Acosta-Belén, E. (1986) 'Puerto Rican women in culture, history and society'. In E. Acosta-Belén (ed.) *The Puerto Rican Woman: Perspectives on Culture, History and Society*. Rio Piedras: Universidad de Puerto Rico, 1–29.

Adams, J. (1999) 'The social implications of hypermobility'. Conference proceedings of *The Economic and Social Implications of Sustainable Transportation*, Ottawa Workshop; working party on pollution prevention, working group on transport, Paris: OECD (Env/epoc/eppc/t(99)3/Final?Rev1): 99–133.

Addams, A. (1996) *Architecture in the Family Way*. Montreal and Kingston: McGill-Queens University Press, 129–162.

Agamben, G. (2005) *State of Exception* (translated by K. Attell). Chicago: University of Chicago Press.

Ahmed, S.S. and Bould, S. (2004) 'One able daughter is worth 10 illiterate sons: reframing the patriarchal family'. *Journal of Marriage and the Family*, 66.5: 1332–1341.

Ahrentzen, S. and Anthony, K. H. (1993) 'Sex, stars, and studios: a look at gendered educational practices in architecture'. *Journal of Architectural Education*, 47.1: 11–29.

AIA (2007) Membership data available on-line http://www.aia.org/siteobjects/files/cmp_ppt_membership_makeover.pdf, last accessed on-line 25/01/07.

Aitchison, C. (2001) 'Gender and leisure research: the "codification of knowledge"'. *Leisure Studies*, 23.1: 1–19.

Akkerman, A. (2006) 'Femininity and masculinity in city-form: philosophical urbanism as a history of consciousness'. *Human Studies,* 29: 229–256.

Allen, G. and Crow, G. (1989) *Home and Family: Creating the Domestic Sphere*. Basingstoke: Palgrave.

Alonso, W. (1964) *Location and Land Use*. Cambridge, MA: Harvard University Press.

American Heritage Dictionary (2000) *The American Heritage Dictionary of the English Language*. Boston: Houghton Mifflin.

Anand, N. (2005) 'Working women: issues and problems'. In S. B. Verma (ed.) *Status of Women in Modern India*. New Delhi: Deep and Deep Publications, 136–142.

Anderson, B. (2007) 'A very private business: exploring the demand for migrant domestic workers'. *European Journal of Women's Studies*, 14.3: 247–264.

Anderson, J. (2007) 'Urban poverty reborn: a gender and generational analysis'. *Journal of Developing Societies*, 23: 221–241.

Antrobus, P. (2004) *The Global Women's Movement: Origins, Issues and Strategies*. London and New York: Zed Books.

Appadurai, A. (2001a) 'Deep democracy: urban governmentality and the horizon of politics'. *Environment and Urbanization*, 13.2: 23–43.

Appadurai, A. (2001b) 'Sandas Mela…"the toilet festival"'. Mumbai: SPARC.

Assaad, M. and Bruce, J. (1997) 'Empowering the next generation: girls of the Maqattam Garbage Settlement'. *SEEDS*, 19. New York: Population Council.

Atkinson, R. and Blandy, S. (eds) (2006) *Gated Communities*. Oxford: Routledge.

Attwood, R. (2007) 'Pay bias persists as change is "too slow"'. *Times Higher Educational Supplement*, 11 May.

Avritzer, L. (2002) 'Civil society, public space and local power: a study of the participatory budget in Belo Horizonte and Porto Alegre'. *Civil Society and Democratic Governance Final Report*. Brighton: IDS.

Avritzer, L. (2006) 'New public spheres in Brazil: local democracy and deliberative politics'. *International Journal of Urban and Regional Research*, 30.3: 623–637.

Badcock, B. (2002) *Making Sense of Cities: A Geographical Survey*. London: Arnold.

Baden, S. and Milward, K. (1995) 'Gender and poverty'. *Bridge Report* No. 30. Brighton: IDS.

Badshah, A. A. (1996) *Our Urban Future: New Paradigms for Equity and Sustainability*. London and New Jersey: Zed Books.

Bailey, J. (2007) *Can Any Mother Help Me?* London: Faber and Faber.

Bailey, R. W. (2001) 'Sexual identity and urban space: economic structure and political action'. In M. Blasius (ed.) *Sexual Identity, Queer Politics*. Princeton: Princeton University Press, 231–256.

Bakker, I. and Gill, S. (2003) *Power, Production and Social Reproduction*. Basingstoke: Palgrave.

Bakker, I. and Silvey, R. (eds) (2008) *Beyond States and Markets: The Challenge of Social Reproduction*. New York: Routledge.

Bales, K. (1999) *Disposable People: New Slavery in the Global Economy*. Berkeley and London: University of California Press.

Barry, K. (1979) *Female Sexual Slavery*. New York: New York University Press.

Baxi, U. (2006) 'What may the "third world" expect from International Law', *Third World Quartlery*, 27.5: 713–725.

BBC (2008) BBC website for 10/01/08 at http: //news.bbc.co.uk/1/hi/business/7180396.stm.

Beall, J. (1996) 'Urban governance: why gender matters'. *Gender in Development Programme* (GIDP). New York: United Nations Development Programme (UNDP), available on-line http://www.ucl.ac.uk/dpu-projects/drivers_urb_change/urb_society/pdf_gender/UNDP_Beall_gender_matters.pdf.

Beall, J. (1997) *A City for All: Valuing Difference and Working in the Diversity*. London: Zed Books.

Beall, J. (2000) 'From the culture of poverty to inclusive cities: re-framing urban policy and politics'. *Journal of International Development*, 12.6: 483–856.

Beall, J. (2002) 'Living in the present, investing in the future – household security among the urban poor'. In C. Rakodi (ed.) *Urban Livelihoods*. London: Earthscan, 71–87.

Beall, J. and Kanji, N. (1999) 'Households, livelihoods and urban poverty'. *International Development Department Discussion Paper.* Birmingham, UK: University of Birmingham.

Beasley, C. (1999) *What Is Feminism? An Introduction to Feminist Theory*. London and Thousand Oaks, CA: Sage.

Beauregard, R. A. (1990) 'Trajectories of neighbourhood change: the case of gentrification'. *Environment and Planning A*, 22.7: 855–874.

Bebbington, A. (1999) 'Capitals and capabilities: a framework for analyzing peasant viability, rural livelihoods and poverty'. *World Development*, 27.12: 2021–2044.

Becker, G. (1981) *A Treatise on the Family*. London and New York: Harvard University Press.

Begg, I. (ed.) (2002) *Urban Competitiveness: Policies of Dynamic Cities*. Bristol: The Policy Press.

Belt, V., Richardson, R. and Webster, J. (1999) 'Women's work in the information economy: the case of telephone call centres'. *Information, Communication and Society*, 3.3: 366–385.

Ben-Ari, E. (1997) *Japanese Childcare: An Interpretive Study of Culture and Organization*. London/New York: Kegan Paul.

Benería, L. and Roldán, M. (1987) *The Crossroads of Class and Gender: Industrial Homework, Subcontracting, and Household Dynamics in Mexico City*. Chicago: University of Chicago Press.

Bénit, C. and Morange, M. (2006) 'Domestic workers, job access and work identity in Cape Town and Johannesburg'. In S. B. Bekker and A. Leildé (eds) *Reflections on Identity in Four African Cities*. Stellenbosch: African Minds, 69–95.

Benn, M. (2005) 'When did you last see your husband'. *Guardian* Supplement, 9 April, 14–21.

Berger, J. (1972) *Ways of Seeing*. London: Penguin.

Berube, A., Singer, A., Wilson, J. and Frey, W. (2006) 'Finding exurbia: America's fast-growing communities at the metropolitan fringe'. *Metropolitan Policy Program*, The Brookings Institution, available on-line http://www.brookings.edu/~/media/Files/rc/reports/2006/10metropolitanpolicy_berube/20061017_exurbia.pdf, last accessed 15/09/08.

Bielby, W. and Bielby, D. (1989) 'Family ties: balancing commitments to work and family in dual earner households'. *American Sociological Review*, 54: 776–789.

Bindel, J. and Atkins, H. (2007) *Streets Apart: Outdoor Prostitution in London*. London: Poppy Project/Eaves Housing for Women.

Bird, J. (1993) 'Dystopia on the Thames'. In J. Bird, B. Curtis, T. Putnam, G. Robertson and L. Tickner (eds) *Mapping the Futures: Local Cultures, Global Change*. London: Routledge, 121–138.

Black, D., Gates, G., Sanders, S. and Taylor, L. (2000) 'Demographics of the gay and lesbian population in the United States: evidence from available systematic data sources'. *Demography*, 37.2: 139–154.

Blackman, S. J. (1998) 'The school: "poxy cupid!" An ethnographic and feminist account of a resistant female youth culture: the new wave girls'. In T. Skelton and G. Valentine (eds) *Cool Places*. London: Routledge.

Blau, P. M. (1964) *Exchange and Power in Social Life*. New York: Wiley.

Blumberg, R. L. (1989) *Making the Case for the Gender Variable: Women and the Wealth and Well-Being of Nations*. Technical Reports in Gender and Development. Washington, DC: USAID–WID.

Blumberg, R. L. (1991) *Gender, Family and Economy: The Triple Overlap*. Newbury Park, CA: Sage.

Blumberg, R. L. (1995) 'Introduction: engendering wealth and well-being in an era of economic transformation'. In R. L. Blumberg, C. A. Rakowski, I. Tinker and M. Monteon (eds) *Engendering Wealth and Well-Being*. Boulder, CO: Westview Press, 1–14.

Blunt, A. and Rose, G. (1995) *Writing Women and Space: Colonial and Postcolonial Geographies*. New York and London: Guilford Press.

Böhm, S., Jones, C., Land, C. and Paterson, M. (eds) (2006) *Against Automobility*. Oxford: Blackwell.

Bohmer, S. and Briggs, J. L. (1991) Teaching privileged students about gender, race and class oppression. *Teaching Sociology*, 19.2: 154–163.

Bondi, L. (1990) 'Feminism, postmodernism and geography: space for women?' *Antipode*, 22.2: 156–167.

Bondi, L. (1991) 'Gender divisions and gentrification: a critique'. *Transactions of the Institute of British Geographers*, 16.2: 190–198.

Bondi, L. (1998) 'Gender, class, and urban space: public and private space in contemporary urban landscapes'. *Urban Geography*, 19.2: 160–185.

Bondi, L. (2004) 'Power dynamics in feminist classrooms: making the most of inequalities?' *Women and Geography Study Group, Geography and Gender Reconsidered*. WGSG: e-publication (CD): 175–182.

Bondi, L. (2006) 'Gender and the reality of cities: embodied identities, social relations and performativities'. *Feminist Studies*, January/June 2006, available on-line http://www.unb.br/ih/his/gefem/labrys9/libre/liz.htm, last accessed 15/09/08.

Bondi, L. (2009) 'Teaching Reflexivity: Undoing or reinscribing habits of gender?: *Journal of Geography in Higher Education*.

Bonney, N. and Love, J. (1991) 'Gender and migration: geographical mobility and the wife's sacrifice'. *Sociological Review*, 39: 335–348.

Booth, C. and Gilroy, R. (1999) 'Building an infrastructure for everyday life'. *European Planning Studies*, 7.3: 307–324.

Booth, C., Darke, J. and Yeandle, S. (1996) *Changing Places: Women's Lives in the City*. London: Paul Chapman Publishing.

Bordo, S. (1990) 'Feminism, postmodernism, and gender scepticism'. In L. Nicholson (ed.) *Feminism/Postmodernism*. New York: Routledge, 133–156.

Boris, E. and Prügl, E. (eds) (1996) *Homeworkers in Global Perspective – Invisible No More*. New York and London: Routledge.

Borsdorf, A., Hidalgo, R. and Sanchez, R. (2007) 'A new model of urban development in Latin America: the gated communities and fenced cities in the metropolitan area of Santiago de Chile and Valparaiso'. *Cities*, 24.5: 365–378.

Boserup, E. (1970) *Women's Role in Economic Development*. New York: St Martins Press.

Bourdieu, P. (1983) 'Forms of capital'. In J. C. Richards (ed.) *Handbook of Theory and Research for the Sociology of Education*. New York: Greenwood.

Bourdieu, P. and Wacquant, L. J. (1992) *An Invitation to Reflexive Sociology*. Cambridge: Polity Press.

Bowman, C. G. (1993) 'Street harassment and the informal ghettoization of women'. *Harvard Law Review*, 106.3: 517–580.

Boyd, M. and Grieco, E. (2003) 'Women and migration: incorporating gender into international migration theory', Migrant Policy Institute, available on-line http://www.migrationinformation.org/Feature/display.cfm?id=106, last accessed 15/09/08.

BPW USA (2008) Fact sheet on pay equity, *Business and Professional Women USA*, available on-line http://www.bpwusa.org/i4a/pages/index.cfm?pageid=4419, last accessed 15/09/08.

Bradley, H. (1999) *Gender and Power in the Workplace*. London: Macmillan.

Bradley, H. (2007) *Gender. Key Concepts*. Cambridge: Polity Press.

Brenner, N. and Theodore, N. (2002) 'Cities and geographies of actually existing neoliberalism'. *Antipode*, 34.3: 349–379.

Brenner, N., Peck, J. and Theodore, N. (2005) 'Neoliberal urbanism: cities and the rule of markets', draft paper, July 2005, available on-line http://murrum.wikispaces.com/space/showimage/Brenner-Peck-Theodore_Neoliberal_urbanism.pdf, last accessed 15/09/08.

Brion, M. (1995) *Women in the Housing Service*. London: Routledge.

Brodie, J. (2003) 'Globalization, in/security, and the paradoxes of the social'. In I. Bakker and S. Gill (eds) *Power, Production and Social Reproduction*. Basingstoke: Palgrave.

Bronson, P. (2000) *The Nudist on the Late Shift and Other True Tales of Silicon Valley*. New York: Vintage.

Brookings Institution (2006) *Finding Exurbia: America's Fast-Growing Communities at the Metropolitan Fringe*. Washington: Brookings Institution, http://www.brookings.edu/reports/2006/10metropolitanpolicy_berube.aspx.

Brown, M. and Knopp, L. (2006) 'Places or polygons? Governance, scale and the census in the Gay and Lesbian Atlas'. *Population, Space and Place*, 12.4: 223–242.

Browne, K. (2004) 'Genderism and the bathroom problem: (re)materialising sexed sites, (re)creating sexed bodies'. *Gender, Place and Culture*, 11.3: 331–346.

Bruegel, I. (1996) 'The trailing wife: a declining breed? – Careers, geographical mobility and household conflict in Britain 1970–89'. In R. Crompton, D. Gallie and K. Purcell (eds) *Changing Forms of Employment: Organisations, Skills and Gender*. London: Routledge.

Buchanan, C. (2003) *Case for London Technical Report* 1. London: Colin Buchanan and Partners for GLA Economics.

Buck, N., Gordon, I., Hall, P., Harloe, M. and Kleinman, M. (2002) *Working Capital: Life and Labour in Contemporary London*. London: Routledge.

Bunting, M. (2004) *Willing Slaves: How the Overwork Culture Is Ruling Our Lives*. London and New York: HarperCollins.

Burgess, J., Harrison, C. M. and Limb, M. (1988) 'People, parks and the urban green: a study of popular meanings and values for open spaces in the city'. *Urban Studies*, 25.6: 455–473.

Burton, L. and Snyder, A. (1998) 'The invisible man revisited: comments on the life course, history, and men's roles in American families'. In A. Booth (ed.) *Men in Families: When Do They Get Involved? What Difference Does It Make*? Hillsdale, NJ: Lawrence Erlbaum Associates, 31–39.

Butler, J. (1993) *Bodies that Matter. On the Discursive Limits of 'Sex'*. London: Routledge.

Butler, T. and Robson, G. (2003) 'Plotting the middle classes: gentrification and circuits of education in London'. *Housing Studies*, 18: 5–28.

Buzar, S., Ogden, P. E. and Hall, R. (2005) 'Households matter: the quiet demography of urban transformation'. *Progress in Human Geography*, 29.4: 413–436.

Cadstadt, J. (2006) *Influence and Invisibility: Tenants in Housing Provision in Mwanza City, Tanzania*. Stockholm: Stockholm Studies in Human Geography, 107–109.

Cagatay, N. (1998) *Gender and Poverty*. UNDP Working Paper Series, WP 5. New York: UNDP.

Cairns, S. (2004) *Drifting: Architecture and Migrancy*. London: Routledge.

Callimachi, R. (2007) 'When the lights go out, students take off to the airport'. *Guardian*, 21 July.

Campbell, B. (1993) *Goliath: Britain's Dangerous Places*. London: Methuen.

Campbell, B. (1999) 'Boys will be boys: social insecurity and crime', in J. Vail, J. Wheelock and M. Hill (eds) *Insecure Times – Living with Insecurity in Contemporary Society*. London and New York: Routledge.

Campbell, K. E. and Lee, B. A. (1998) 'Neighbour networks in blacks and whites'. In B. Wellman (ed.) *Networks in the Global Village*. Boulder, CO: Westview Press.

Campbell, S. (1996) 'Green cities, growing cities, just cities? Urban planning and the contradictions of sustainable development'. *Journal of the American Planning Association*, 62.3: 296–312.

Campkin, B. and Cox, R. (2007) *Dirt: New Geographies of Cleanliness and Contamination*. London: I. B. Taurus.

Caplan, P. J. (1993) *Lifting a Ton of Feathers: A Woman's Guide to Surviving in the Academic World*. Toronto: University of Toronto Press.

Carby, H. V. (1987) *Reconstructing Womanhood: The Emergence of the Afro-American Woman Novelist*. Oxford: Oxford University Press.

Cardiff, J. (2005) *The Walk Book*. London: Cornerhouse.

Carlstein, T., Parkes, D. and Thrift, N. (1978) *Timing Space and Spacing Time, Vol. II: Human Activity and Time Geography*. London: Arnold.

Carney, D. (1998) 'Implementing the sustainable rural livelihoods approach'. In D. Carney (ed.) *Sustainable Rural Livelihoods: What Contribution Can We Make?* London: DFID, 3–23.

Carney, D., Drinkwater, M., Rusinow, T., Neefjes, K., Wanmali, S. and Singh, N. (1999) *Livelihoods Approaches Compared*. London: DFID.

Carpentier, N. and Ducharme, F. (2003) 'Care-giver network transformations: the need for an integrated perspective'. *Ageing & Society*, 23: 507–525.

Carr, M., Chen, M. and Jhabvala, R. (1996) *Speaking Out: Women's Economic Empowerment in South Asia*. London: IT Publications.

Carr, M., Chen, M. and Tate, J. (2000) 'Globalization and homebased workers'. *Feminist Economics*, 6.3: 123–142.

Casey, B., Metcalf, H. and Milward, N. (1997) *Employers' Use of Flexible Labour*. London: Policy Studies Institute.

Castells, M. (1993) 'European cities, the information society, and the global economy'. In R. T. LeGates and F. Stout (eds) (2007) *The City Reader* (fourth edition). London and New York: Routledge, 478–488.

Castells, M. (1996) *The Rise of The Network Society*. New York: Blackwell.

Castells, M. and Portes, A. (1989) 'World underneath: the origins, dynamics, and effects of the informal economy'. In A. Portes, M. Castells and L. A. Benton (eds) *The Informal Economy: Studies in Advanced and Less Developed Countries*. Baltimore, MD: The Johns Hopkins University Press, 11–37.

Cavell, S. (1996) 'The ordinary as the uneventful'. In S. Mulhall (ed.) *The Cavell Reader*. Oxford: Blackwell, 253–259.

CEBE (2003) *Architectural Education: Studio Culture.* London: The Centre for Education in the Built Environment (CEBE), available on-line http://www.cebe.heacademy.ac.uk/news/past_events/concrete/leonie.pdf, last accessed 02/08/07.

CEBE (2006) *Guide to Supporting Student Diversity in UK Schools of Architecture*. London: UCL Eprints, available on-line http://eprints.ucl.ac.uk/2314/.

Cerrutti, M. and Bertoncello, R. (2003) 'Urbanization and internal migration patterns in Latin America'. Centro de Estudios de Población, Argentina, paper prepared for Conference on African Migration in Comparative Perspective, Johannesburg, South Africa, 4–7 June, 2003, available on-line http://pum.princeton.edu/pumconference/papers/1-Cerrutti.pdf, last accessed 15/09/08.

Chambers, D., Steiner, L. and Fleming, C. (2004) *Women and Journalism*. London: Routledge.

Chambers, R. (1989) 'Editorial introduction: vulnerability, coping and policy'. *IDS Bulletin*, 20.2: 1–7.

Chambers, R. (1997) *Whose Reality Counts?* London: ITDG Press.

Chambers, R. and Conway, G. R. (1992) *Sustainable Rural Livelihoods: Practical Concepts for the 21st Century* (IDS 296). Brighton: IDS.

Champion, T. (2001) 'Urbanization, suburbanization, counterurbanization and reurbanization'. In R. Paddison (ed.) *Handbook of Urban Studies*. London and New York: Sage, 143–162.

Chan, M. and McGuire, S. (2000) 'The NY–London life'. *Newsweek*, 10 November.

Chant, S. (1991) 'Gender, migration and urban development in Costa Rica: the case of Guanacaste'. *Geoforum*, 22.3: 237–253.

Chant, S. (2002) '"Men-streaming" gender? Questions for gender and development policy in the twenty-first century'. *Progress in Development Studies*, 2.4: 269–282.

Chant, S. (2007a) 'Gender, cities and the Millennium Development Goals in the global South'. *LSE Gender Institute New Working Paper Series*. London, available on-line http://www.lse.ac.uk/collections/genderInstitute/pdf/CHANT%20GIWP.pdf, last accessed 15/09/08.

Chant, S. (2007b) 'Children in female-headed households: interrogating the concept of inter-generational transmission of disadvantage with particular reference to the Gambia, Philippines and Costa Rica'. *LSE Gender Institute New Working Paper Series*. London, available on-line http: //www.lse.ac.uk/collections/genderInstitute/pdf/children.pdf, last accessed 15/09/08.

Cheal, D. (2002) *Sociology of Family Life*. Palgrave: London.

Chen, M., Vanek, J. and Carr. M. (2004) *Mainstreaming Informal Employment and Gender in Poverty Reduction*. London: Commonwealth Secretariat.

Chen, S. and Ravallion, M. (2007) 'The changing profile of poverty in the world'. *2020 Focus Brief on the World's Poor and Hungry People*, October, available on-line http://www.ifpri.org/2020Chinaconference/pdf/beijingbrief_ravallion2.pdf.

Chenery, H. B., Aluwalia, M. S., Bell, C. L. G., Duloy, J. H. and Jolly, R. (1974) *Redistribution with Growth*. London: Oxford University Press.

Child and Women Abuse Study Unit (2003) *Critical Examination of Responses to Prostitution in Four Countries*. London: CWASU.

Chouinard, V. and Grant, A. (1995) 'On not being anywhere near "the project": ways of putting ourselves in the picture'. *Antipode*, 27.2: 137–166.

Churchill, A. (1985) 'Forward'. In J. Lea and J. Courtney (eds) *Cities in Conflict: Planning and Management of Asian Cities*. Washington, DC: World Bank, v–vi.

CIA (2008) 'The world factbook – Honduras'. Central Intelligence Agency, available on-line http://www.cia.gov/library/publications/the-workd-factbook/geos/ho.html, last accessed 10/02/08.

CIDADE (2003) *Quem é o público do Orçamento Participativo 2002*. Porto Alegre: Cidade.

CIDADE and PMPA (2002) *Quem é o publico do Orcamento Participativo 2000*. Porto Alegre: Governo do Estado do Rio Grande do Sul/Cidade.

Citizens Advice Bureau (CAB) (2003) *In Too Deep: CAB Clients' Experience of Debt*. London: Citizens Advice, http://www.citizensadvice.org.uk.

Clark, J., McChesney, R., Munroe, D. and Irwin, E. (2006) 'Exurban settlement pattern and the exurban condition: a typology of U.S. Metropolitan areas'. Paper presented at the 53rd *Annual North American Meetings of the Regional Science Association International*. Toronto: Canada.

Clarke, A. J. (2007) 'Consuming children and making mothers: birthday parties, gifts and the pursuit of sameness'. *Horizontes Antropológicos, Porto Alegre*, 13.28: 263–287.

Cleaver, F. (2006) *Masculinities Matter! Men, Gender and Development*. London: Zed Books.

Clegg, S. R. (1989) *Frameworks of Power*. London: Sage.

Clegg, S. and Mayfield, W. (1999) 'Gendered by design: how women's place in design is still defined by gender'. *Design Issues*, 15.3: 3–16.

Cloke, J. (2002) 'The Political Economy of Micro-finance in Nicaragua', unpublished PhD, Loughborough University.

Cockburn, C. (1985) 'The material of male power'. In D. MacKenzie and J. Wajcman (eds) *The Social Shaping of Technology*. Milton Keynes: Open University Press.

Coleman, A. (1985) *Utopia on Trial: Visions and Realities in Planned Housing*. London: Hilary Shipman.

Coleman, J. C. (1988) 'Social capital in the creation of human capital'. *American Journal of Sociology*, 94: 95–120.

Collaborative Economics (2001) *The Creative Community: Leveraging Creative and Cultural Participation for Silicon Valley's Economic and Civic Future*. San Jose: Collaborative Economics.

Colls, R. (2002) 'Bodies out of bounds: fatness and transgression'. *Gender, Place and Culture*, 9.2: 218–220.

Colomina, B. (ed.) (1992) *Sexuality and Space*. Princeton: Princeton Architectural Press.

Coltrane, S. and Adams, M. (2003) 'The social construction of the divorce "problem": morality, child victims and the politics of gender'. *Family Relations*, 52.4: 363–372.

Commission on the Advancement of Women and Minorities in Science, Engineering and Technology Development (2000) *Land of Plenty: Diversity as America's Competing Edge in Science, Engineering and Technology Development*. Arlington, VA: United States Congress.

Connell, R. W. (1995) *Masculinities*. Berkeley: University of California Press.

Connell, R. W. (2000) *The Men and the Boys*. Cambridge: University of California Press.

Cooke, T. (2003) 'Family migration and the relative earnings of husband and wives'. *Annals of the Association of American Geographers*, 93.2: 338–349.

Cooke, T. J. and Bailey, A. J. (1996) 'Family migration and the employment of married women and men'. *Economic Geography*, 72.1: 38–48.

Corbett, J. (2006) *Torsten Hägerstrand: Time Geography*. Center for Spatially Integrated Social Science. Santa Barbara: UC Santa Barbara, available on-line http://www.csiss.org/classics/content/29.

Cornia, G., Jolly, R. and Stewart, F. (eds) (1987) *Adjustment with a Human Face*. Oxford: Clarendon Press.

Cornwall, A. (2000) 'Missing men? Reflections on men, masculinities and gender in GAD'. *IDS Bulletin*, 31.2: 18–27.

Council of Europe (1998) *Gender Mainstreaming: Conceptual Framework, Methodology and Presentation of Good Practices*, EG-S-MS (98) 2, Strasburg.

Cowan, J. (1998) *On Becoming an Innovative University Teacher: Reflection in Action*. Milton Keynes: Open University Press.

Cox, K. and Jonas, A. E. G. (1993) 'Urban development, collective consumption and the politics of metropolitan fragmentation'. *Political Geography*, 12.1: 8–37.

Cox, R. (2006) *The Servant Problem: Domestic Employment in a Global Economy*. London: I. B. Taurus.

Craig, S. (1992) *Men, Masculinity and the Media: Research on Men and Masculinity*. Thousand Oaks, CA: Sage.

Cresswell, T. (1999) 'Embodiment, power and the politics of mobility: the case of female tramps and hobos'. *Transactions of the Institute of British Geographers*, 24.2: 175–192.

Crozier, G. and Reay, D. (2004) *Parent Participation: Activating Partnership in School*. London: Trentham Books.

Crush, J. (ed.) (1995) *Power of Development*. London and New York: Routledge.

Curley, A. M. (2005) 'Theories of urban poverty and implications for public housing policy'. *Journal of Sociology and Social Welfare*, 32.22: 97–119.

Danziger, S. and Radin, N. (1990) 'Absent does not equal uninvolved: predictors of fathering in teen mother families'. *Journal of Marriage and the Family*, 52.3: 636–642.

Darke, J., Gilroy, R. and Woods, R. (eds) (1994) *Housing Women*. New York: Routledge.

Darke, J., Ledwith, S., Woods, R., Kennedy, H. and Campling, J. (2000) *Women and the City: Visibility and Voice in Urban Space*. Basingstoke: Palgrave.

Darrah, C. (2002) *Blurring the Edges of Work and Family: Tales from Silicon Valley*, 27 February. Atlanta, GA: The Emory Center for Myth and Ritual in American Life (MARIAL).

Davenport, E. and Snyder, H. (1995) 'Who cites women? Whom do women cite?: An exploration of gender and scholarly citation in sociology'. *Journal of Documentation*, 51.4: 404–410.

Davis, D., Bian, Y. and Wang, S. (2006) 'Material rewards to multiple capitals under market socialism'. In Y. Bian, K. Chan and T. Cheung (eds) *Social Transformations in Chinese Societies*. Hong Kong: BRILL.

de Certeau, M. (1984) *The Practice of Everyday Life*. Berkeley: University of California Press.

Dear, M. J. (1986) 'Postmodernism and planning'. *Environment and Planning D: Society and Space*, 4.3: 367–384.

Decker, A. (2006) 'Stuck at home: when driving isn't a choice'. UC Berkeley, *Access*, 29 (fall).

Deegan, M. J. (1986) *Jane Addams and the Men of the Chicago School, 1882–1918*. New Brunswick: Transaction Books.

Demos (1998) *The Good Life*. London: Demos.

Dennis, R. (2008) *Cities in Modernity: Representations and Productions of Metropolitan Space, 1840–1930*. Cambridge: Cambridge University Press.

Developments Magazine (2008) *The Inclusive City*?, available on-line http://www.developments.org/uk/, last accessed 07/03/08.

di Leonardi, M. (1987) 'The female world of cards and holidays: women, family, and the work of kinship'. *Signs*, 12.3: 440–453.

Dixon, J., Levine, M. and McCauley, R. (2006) 'Locating impropriety: street drinking, moral order, and the ideological dilemma of public space'. *Political Psychology*, 27.2: 187–206.

Dobson, R., Boztas, S. and Bray, E. (2007) 'Official: the French do less housework than anyone else…and wherever you look, it's still women who do most of the chores'. *The Independent*, Sunday, 14 October, available on-line http://www.independent.co.uk/news/uk/home-news/official-the-french-do-less-housework-than-anyone-else-396847.html, last accessed 15/09/08.

Dobson, T. and Stillwell, J. (2000) 'Changing home, changing school: toward a research agenda on children's migration'. *Area*, 32.4: 395–401.

Domosh, M. (2000) 'Unintentional transgressions and other reflections on the job search process'. *The Professional Geographer*, 52.4: 703–708.

Domosh, M. and Seager, J. (2001) *Putting Women in Place: Feminist Geographers Make Sense of the World*. New York: Guilford.

Dowling, R. (2000) 'Cultures of mothering and car use in suburban Sydney: a preliminary investigation'. *Geoforum*, 31: 345–353.

Downing, J. (1990) *Gender and the Growth and Dynamics of Microenterprises*. Gemini Working Paper No. 5. Washington, DC: USAID.

Doyle, J. and Nathan, M. (2001) *Wherever Next: Work in a Mobile World*. London: The Industrial Society.

Dreyfus, H. L. and Rabinow, P. (1982) *Michel Foucault: Beyond Structuralism and Hermeneutics* (with an afterword by Michel Foucault). Chicago: University of Chicago Press.

Duncan, S. S. (2005) 'Mothering, class and rationality'. *Sociological Review*, 53.1: 50–76.

Duncan, S. S. and Edwards, R. (1999) *Lone Mothers, Paid Work and Gendered Moral Rationalities*. Basingstoke: Macmillan.

Duncan, S. S. and Smith, D. (2001) 'Geographies of family formations: spatial differences and gender cultures in Britain'. *Transactions of the Institute of British Geographers*, NS 27.4: 471–494.

Duneier, M. (1999) *Sidewalk*. New York: Farrar, Strauss & Giroux.

Dunne, G. A. (1997) *Lesbian Lifestyles: Women's Work and the Politics of Sexuality*. Toronto: University of Toronto Press; London: Macmillan.

Dwyer, D. and Bruce, J. (eds) (1988) *A Home Divided: Women and Income in the Third World*. Palo Alto, CA: Stanford University Press.

Dyck, I. (1996) 'Mother or worker? Women's support networks, local knowledge and informal child care strategies'. In K. England (ed.) *Who Will Mind the Baby?* London: Routledge, 123–140.

Dyck, I. (2005) 'Feminist geography, the "everyday", and local–global relations'. *The Canadian Geographer,*. 49.3: 233–245.

Dyck, I., Kontos, P., Angus, J. and McKeever, P. (2005) 'The home as a site for long-term care: meanings and management of bodies and spaces'. *Health & Place*, 11: 173–185.

ECLAC (2000) 'Synthesis of national reports on the implementation of the HABITAT agenda in Latin America and the Caribbean Region'. United Nations, available on-line http://www0.un.org/ga/Istanbul+5/eclac.pdf, last accessed 15/09/08.

Edley, N. and Wetherell, M. (1995) *Men in Perspective: Practice, Power and Identity*. Hemel Hempstead: Simon & Schuster.

Ehrenreich, B. (2001) *Nickel and Dimed: On (Not) Getting By in America*. New York: Metropolitan Books.

Ehrenreich, B. (2003) 'A grubby business'. *Guardian*, 12 July: 16, also available on-line http://www.guardian.co.uk/world/2003/jul/12/gender.bookextract.

Ehrenreich, B. and Hochschild, A. R. (eds) (2003) *Global Woman: Nannies, Maids and Sex Workers in the New Economy*. London: Granta Books.

Ekinsmyth, C., Elmshirst, R., Holloway, S. and Jarvis, H. (2004) 'Love changes all: making some noise by "coming out" as mothers'. In WGSG, *Geography and Gender Reconsidered*. Dundee: WGSG, 95–107.

Elders Council (2006) *Report from a Working Group of the Elders Council of Newcastle upon Tyne*, available on-line http://www.elderscouncil.org.uk/newsletters/downloads/Report.pdf.

Electronic Times (2001) 'How do you use yours'. *Electronic Times*, 10 December.

Ellin, N. (2006) *Integral Urbanism*. London: Routledge.

Ellis, M., Wright, R. and Parks, V. (2004) 'Work together, live apart? Geographies of racial and ethnic segregation at home and at work'. *Annals of the Association of American Geographers*, 94.3: 620–637.

Elson, D. (ed.) (1991) *Male Bias in the Development Process*. Manchester: Manchester University Press.

Elson, D. and Pearson, R. (1981) 'Nimble fingers make cheap workers'. *Feminist Review*, 7 (spring): 87–107.

England, K. V. L. (1993) 'Suburban pink collar ghettos: the spatial entrapment of women?' *Annals of the American Association of Geographers*, 83.2: 225–242.

England, K. (1994) 'Getting personal: reflexivity, positionality and feminist research'. *The Professional Geographer*, 46.1: 80–89.

England, P. (1992) *Comparable Worth: Theories and Evidence*. New York: Aldine de Gruyter.

English-Lueck, J.A. (2002) *Cultures@Siliconvalley*. Stanford: Stanford University Press.

Equipo Nizkor (1999) *Informe sobre la situación de Los Derechos Humanos en Nicaragua en el año 1999*. Managua: CENIDH.

Ernst, C., Ferrer, A. H. and Zult, D. (2005) *The End of the Multi-Fibre Arrangement and Its Implications for Trade and Employment*. Employment Strategy Paper 16. Geneva: ILO.

Escobar, A. (1995) *Encountering Development: The Making and Unmaking of the Third World*. Princeton: Princeton University Press.

Escobar, A. and Harcourt, W. (2005) *Women and the Politics of Place*. Bloomfield, CT: Kumerian Press.

European Parliament (2004) *Committee on Women's Rights and Equal Opportunities. Draft Report on the Consequences of the Sex Industry in the European Union*, provisional, 2003/2107 (INI) 9 January, Rapporteur M. Eriksson.

European Women's Lobby (2000) 'Gender, trade liberalisation and services in the European Union'. EWL-LEF, Brussels, available on-line http://trade.ec.europa.eu/doclib/docs/2005/april/tradoc_122230.pdf, last accessed 15/09/08.

Evans, L. (1992) 'The impact of demographic trends in the United Kingdom on women's employment prospects in the 1990s'. In N. Folbre, B. Bergman, B. Agarwi and M. Floro (eds) *Issues in Contemporary Economics 4*. Athens, Greece: International Economics Association.

Fainstein, S. (1994) *The City Builders: Property, Politics and Planning in London and New York*. Oxford: Blackwell.

Faludi, S. (1991) *Backlash: The Undeclared War Against American Women*. New York: Crown Publishers.

Faludi, S. (1999) *Stiffed: The Betrayal of the American Man*. New York: HarperCollins.

Faune, M. (1995) 'Familias: violencia y sobreviviencia'. *Envio*, 162, August. Managua: UCA.

Ferber, M. and Nelson, J. (eds) (1993) *Beyond Economic Man: Feminist Theory and Economics*. Chicago: University of Chicago Press.

Fernandes, F. (2003) 'A response to Erica Burman'. *European Journal of Psychotherapy, Counselling and Health,* 6.4: 309–316.

Field, J. (2003) *Social Capital*. London: Routledge.

Fincher, R. and Jacobs, J. M. (eds) (1998) *Cities of Difference*. New York and London: Guilford Press.

Firestone, S. (1972) *The Dialectic of Sex*. London: Paladin.

Fisher, K. (2003) 'Demystifying critical reflection: defining criteria for assessment'. *Higher Education Research and Development*, 22.3: 318–335.

Fishman, R. (1990) 'America's new city: megalopolis unbound'. *Wilson Quarterly*, 14.1: 24–55.

Fletcher, J. K. (2001) *Disappearing Acts: Gender, Power and Relational Practices at Work*. Cambridge, MA: MIT Press.

Florida, R. (2002) *The Rise of the Creative Class – and How It's Transforming Work, Leisure, Community and Everyday Life*. New York: Basic Books.

Folbre, N. (1994) *Who Pays for the Kids: Gender and the Structure of Constraint*. New York: Routledge.

Folbre, N. (1995) '"Holding hands at midnight": the paradox of caring labour'. *Feminist Economics*, 1.1: 73–92.

Folbre, N. (2001) *The Invisible Heart: Economics and Family Values*. New York: The New Press.

Forrest, R. and Murie, A. (eds) (1995) *Housing and Family Wealth: Comparative International Perspectives*. London: Routledge.

Forskargruppen (1987) *Veier till det nye verdagslivet* (Ways to the new everyday life). Oslo: Nord.

Foucault, M. (1978) *The History of Sexuality, an Introduction. Volume 1*. Reprinted in the 1990 Vintage Books edition. New York: Random House.

Foucault, M. (1980) *Power/Knowledge – Selected Interviews and Other Writings 1972–1977* (edited by Colin Gordon). Brighton: Harvester Press.

Fraser, E. and Lacey, N. (1993) *The Politics of Community: A Feminist Critique of the Liberal Communitarian Debate*. Toronto: University of Toronto Press.

Fraser, N. (1997) *Justice Interruptus: Critical Reflections on the 'Postsocialist' Condition.* London and New York: Routledge.

Fraser, N. (2000) 'After the family wage: a post-industrial thought experiment'. In B. Hobson (ed.) *Gender and Citizenship in Transition*. New York: Routledge, 1–33.

Fraser, N. (2007) 'Mapping the feminist imagination: from redistribution to recognition to representation'. In J. Browne (ed.) *The Future of Gender*. Cambridge: Cambridge University Press.

Freedman, E. B. (2002) *No Turning Back: The History of Feminism and the Future of Women*. New York: Ballantine.

French, M. (1997) *The Women's Room* (30th Anniversary Edition). London: Virago (original published 1977).

Freund, P. (1993) *The Ecology of the Automobile*. Montreal and New York: Black Rose Books.

Frey, W. H., Liaw, K-L. and Lin, G. (2000) 'State magnets for different elderly migrant types in the United States'. *International Journal of Population Geography*, 6.1: 21–44.

Friedan, B. (1976) *It Changed my Life: Writings on the Women's Movement*. New York: Random House.

Friedmann, J. (1986) 'The world city hypothesis'. *Development and Change*, 17.1: 69–84.

Furstenberg, F. F. and Nord, C. W. (1985) 'Parenting apart: patterns of childrearing after marital disruption'. *Journal of Marriage and the Family*, 47.4: 893–904.

Fuss, D. (1989) *Essentially Speaking: Feminism, Nature and Difference*. New York: Routledge.

Fyfe, N. (2004) 'Zero tolerance, maximum surveillance? Deviance, difference and crime control in the late modern city'. In L. Lees (ed.) *The Emancipatory City? Paradoxes and Possibilities*. London: Sage, 40–57.

Galinsky, K. (1992) *Classical and Modern Interactions: Postmodern Architecture, Multiculturalism, Decline and Other Issues*. Austin: University of Texas Press.

Gans, H. J. (1961) 'The balanced community: homogeneity or heterogeneity in residential areas?' *American Institute of Planners Journal*, 27.3: 176–184.

Gardiner, M. E. (2000) *Critiques of Everyday Life*. London: Routledge.

Geddes, P. (1915) *Cities in Evolution: An Introduction to the Town Planning Movement and to the Study of Civics* (1968 edition). London: Benn.

Germain, A. and Rose, D. (2000) *Montréal: The Quest for a Metropolis*. Chichester, UK: John Wiley & Sons.

Gibson-Graham, J. K. (1996) *The End of Capitalism (As We Knew It): A Feminist Critique of Political Economy*. Oxford: Blackwell.

Gibson-Graham, J. K. (2000) 'Introduction'. In J. K. Gibson-Graham, S. A. Resnick and R. D. Wolff (eds) *Class and Its Others*. Minnesota: University of Minnesota Press.

Gibson-Graham, J. K. (2006) *The End of Capitalism (as We Knew It): A Feminist Critique of Political Economy* (second edition). Minneapolis and London: University of Minneapolis Press.

Giddens, A. (1984) *The Constitution of Society*. Oxford: Polity Press.

Giddens, A. (1991) *Modernity and Self-identity: Self and Society in the Late Modern Age*. Cambridge: Polity Press.

Giddens, A. (1993) *The Transformation of Intimacy: Sexuality, Love and Eroticism in Modern Societies*. Stanford, CA: Stanford University Press.

Gilbert, M. (1997) 'Identity, space and politics: a critique of the poverty debates'. In J. P. Jones III, H. Nast and S. M. Roberts (eds) *Thresholds in Feminist Geography: Difference, Methods and Representation*. New York: Rowman & Littlefield.

Gill, R. (2002) 'Cool, creative and egalitarian? Exploring gender in project-based new media work in Europe'. *Information Communication and Society*, 5.1: 70–89.

Gilligan, C. (1982) *In a Different Voice*. Cambridge, MA: Harvard University Press.

Gilroy, R. and Booth, C. (1999) 'Building an infrastructure for everyday lives'. *European Planning Studies*, 7.3: 307–324.

Ginsburg, R. (2004) 'Native daughter: home, segregation, and mental maps'. *Home Cultures*, 1.2: 127–145.

Giroux, H. A. (2002) 'Neoliberalism, corporate culture, and the promise of higher education: the university as a democratic public sphere'. *Harvard Educational Review*, 72.4: 425–463.

Glassman, J. (2001) 'From Seattle (and Ubon) to Bangkok: the scales of resistance to corporate globalization'. *Environment and Planing D: Society and Space*, 20.5: 513–534.

Glucksmann, M. (1995) 'Why "work"? Gender and the "total social organization of labour"'. *Gender Work & Organization*, 2.2: 63–75.

Goffman, E. (1959) *The Presentation of Self in Everyday Life*. Edinburgh: Anchor Books.

GOI (2001) *Good Urban Governance Campaign – India Launch Learning from One Another*. UNCHS (Habitat), New Delhi, Government of India, 4–6 September.

Gold, J. R. (2008) 'Modernity and utopia'. In T. Hall, P. Hubbard and J. R. Short (eds) *The Sage Companion to the City*. London: Sage, 67–87.

Goleman, D. (1995) *Emotional Intelligence.* New York: Bantam Press.

Gordon, L. (1994) *Pitied but Not Entitled. Single Mothers and the History of Welfare 1890–1935*. New York: The Free Press.

Gornick, J. C. and Meyers, M. K. (2003) *Families that Work: Policies for Reconciling Parenthood and Employment*. New York: Russell Sage Foundation.

Gortz, A. (1980) *Adieu au Prolétariat. Au delà du Socialisme*. Paris: Galilée.

Gottdiener, M. and Budd, L. (2005) *Key Concepts in Urban Studies*. London: Sage.

Gough, I. and Wood, G. (eds) (2004) *Insecurity and Welfare Regimes in Asia, Africa and Latin America: Social Policy in Development Contexts*. Cambridge: Cambridge University Press.

Gould, P. and White, R. (1992) *Mental Maps* (2nd edn). London: Routledge.

Graham, S. (1996) 'Imagining the real-time city: telecommunications, urban paradigms, and the future of cities'. In S. Westwood and J. Williams (eds) *Imagining Cities: Scripts, Signs and Memories*. London: Routledge, 31–49.

Graham, S. and Marvin, S. (2001) *Splintering Urbanism: Networked Infrastructures, Technological Mobilities and the Urban Condition*. London: Routledge.

Grais-Targow, R. (2004) 'Femicide in Guatemala', occasional paper available on-line http://marxsite.com/Femicide%20in%20Guatemala.pdf, last accessed 15/09/08.

Granovetter, M. (1973) 'The strength of weak ties'. *American Journal of Sociology*, 78.6: 1360–1380.

Greed, C. (1991) *Surveying Sisters: Women in a Traditional Male Profession*. London: Routledge.

Greed, C. (1994) *Women and Planning: Creating Gendered Realities*. London: Routledge.

Greed, C. (1999) *Social Town Planning: Planning and Social Policy*. London: Taylor & Francis.

Greed, C. (2003) *Inclusive Urban Design: Public Toilets*. London: Architectural Press.

Greed, C. (2005) 'Overcoming the factors inhibiting the mainstreaming of gender into spatial planning policy in the United Kingdom'. *Urban Studies*, 42.4: 719–749.

Greed, C. and Roberts, M. (2001) *Approaching Urban Design: The Design Process*. London: Longman.

Green, A. (2002) *Geographic Mobility and Desirability: Costs and Benefits*. Warwick: Institute of Employment Research, available on-line http://www.cabinetoffice.gov.uk/~/media/assets/www.cabinetoffice.gov.uk/strategy/gm_green%20pdf.ashx.

Green, A. E., Hogarth, T. and Shackleton, R. (1999) *Long-distance Living: Dual Location Households*. Bristol: Policy Press.

Gregson, N. and Rose, G. (2000) 'Taking Butler elsewhere: performativities, spatialities and subjectivities'. *Environment and Planning A: Society and Space*, 15: 433–452.

Griffin, J. M., Fuhrer, R., Stansfeld, S. A. and Marmot, M. (2002) 'The importance of low control at work and home on depression and anxiety: do these effects vary by gender and social class?' *Social Science & Medicine*, 54: 738–798.

Griffin, W. A. and Morgan, A. R. (1988) 'Conflict in maritally distressed military couples'. *American Journal of Family Therapy*, 16.1: 14–22.

Grimshaw, J. (1986) *Feminist Philosophers: Women's Perspectives on Philosophical Traditions*. Brighton: Wheatsheaf Books.

Grosz, E. (1994) *Volatile Bodies: Towards a Corporeal Feminism*. Bloomington, IN: Indiana University Press.

Grosz, E. (2001) *Architecture from the Outside: Essays on Virtual and Real Space*. Massachusetts: MIT Press.

Grünell, M. and Saharso, S. (1999) 'bell hooks and Nira Yuval-Davis on race, ethnicity, class and gender'. *European Journal of Women's Studies*, 6.2: 203–218.

Gullestad, M. (1991) 'The transformation of the Norwegian notion of everyday life'. *American Ethnologist*, 18.3: 480–499.

Gurley Brown, H. (1962) *Sex and the Single Girl*. New York: Harper.

Guy, S. and Shove, E. (2000) *A Sociology of Energy, Buildings and the Environment*. London and New York: Routledge.

Haider, M. (2006) 'Defying predictions, Bangladesh's garment factories thrive'. *Christian Science Monitor*, 7 February.

Haigh, M. (2001) 'Constructing Gaia: using journals to foster reflective learning'. *Journal of Geography in Higher Education*, 25.2: 167–189.

Hakim, C. (2000) *Work–Lifestyle Choices in the 21st Century: Preference Theory*. Oxford: Oxford University Press.

Halford, S. and Leonard, P. (2006) *Negotiating Gendered Identities at Work*. Basingstoke: Palgrave.

Hall, E. J. and Rodriguez, M. S. (2003) 'The myth of postfeminism'. *Gender and Society*, 17.6: 878–902.

Hall, P. and Pfeiffer, U. (2000) *Urban Future 21: A Global Agenda for Twenty-first Century Cities*. London: E. & F. Spon.

Hall, R., Ogden, P. E. and Hill, C. (1997) 'The pattern and structure of one-person households in England and Wales and France'. *International Journal of Population Geography*, 3: 161–181.

Hall, T., Hubbard, P. and Short, J. R. (eds) (2008) *The Sage Companion to the City*. London and New York: Sage.

Hammar, T., Brochmann, G., Tamas, K. and Faist, T. (eds) (1997) *International Migration, Immobility and Development: Multidisciplinary Perspectives*. Oxford: Berg.

Hanochi, S. (2003) 'Constitutionalism in a modern patriarchal state: Japan, the sex sector and social reproduction'. In I. Bakker and S. Gill (eds) *Power, Reproduction and Social Reproduction*. Basingstoke: Palgrave.

Hanson, S. and Pratt, G. (1994) 'Commentary – suburban pink collar ghettos: the spatial entrapment of women? By Kim England'. *Annals of the Association of American Geographers*, 84.3: 500–502.

Hanson, S. and Pratt, G. (1995) *Gender, Work and Space*. London: Routledge.

Haraway, D. (1991) '"Gender" for a Marxist dictionary: the sexual politics of a word'. In L. McDowell and J. P. Sharp (1997) *Space, Gender, Knowledge: Feminist Readings.* London: Edward Arnold.

Hardill, I. (2002) *Gender, Migration and the Dual Career Household*. London and New York: Routledge.

Hartley, S. (2001) *Mrs P's Journey: The Remarkable Story of the Woman Who Created the A–Z Map*. London: Simon & Schuster.

Hartmann, H. (1981) 'The unhappy marriage of Marxism and feminism: towards a more progressive union'. In L. Sargent (ed.) *The Unhappy Marriage of Marxism and Feminism: A Debate on Class and Patriarchy*. London: Pluto.

Hartsock, N. C. M. (1983) 'The feminist standpoint: developing the ground for a specifically feminist historical materialism'. In S. Harding and M. B. Hintikka (eds) *Discovering Reality.* Dordrecht and London: Reidel.

Harvey, D. (1975) 'The geography of capital accumulation: a reconstruction of the Marxian theory'. *Antipode*, 7.2: 9–21.

Hay, C. (2007) *Why We Hate Politics*. Cambridge: Polity Press.

Hay, C. and Marsh, D. (2000) 'Introduction: demystifying globalisation'. In C. Hay and D. Marsh (eds) *Demystifying Globalisation*. Basingstoke: Palgrave, 1–17.

Hayden, D. (1976) *Seven American Utopias: The Architecture of Communitarian Socialism 1790–1975*. Cambridge, MA: MIT Press.

Hayden, D. (1980) 'What would a non-sexist city be like? Speculation on housing, urban design, and human work'. *Signs*, 5.3: 170–187.

Hayden, D. (1981) *The Grand Domestic Revolution*. Cambridge, MA: MIT Press.

Hayden, D. (1984) *Redesigning the American Dream: The Future of Housing, Work and Family Life*. New York and London: W. W. Norton.

Hayden, D. (2003) *Building Suburbia: Green Fields and Urban Growth, 1820–2000*. New York: Pantheon.

Haylett, C. (2003) 'Remaking labour imaginaries: social reproduction and the international project of welfare reform'. *Political Geography*, 22.7: 765–788.

HBF (2003) *Water Privatization from a Gender Perspective*. Thailand: Heinrich Böll Foundation.

Healey, P. (1997) *Collaborative Planning: Shaping Places in Fragmented Societies*. Basingstoke and London: Macmillan.

Held, D. and McGrew, A. (2007) *Globalization/Anti-globalization: Beyond the Great Divide*. Oxford: Polity Press.

Henderson, H. (1982) *The Politics of the Solar Age: Alternatives to Economics*. Indianapolis: Knowledge Systems.

Henderson, H. (1990) *Paradigms in Progress: Life Beyond Economics*. Indianapolis: Knowledge Systems.

Hertz, R. (2006) *Single by Chance, Mothers by Choice*. Oxford: Oxford University Press.

Hetherington, E. M. and Kelly, J. (2002) *For Better or for Worse: Divorce Reconsidered*. London: W. W. Norton.

Higate, P. R. (2000) 'Tough bodies and rough sleepers: embodying homelessness amongst ex-servicemen'. *Housing, Theory and Society*, 17.3: 97–108.

Higgins, M. J. and Coen, T. L. (2000) *Streets, Bedrooms and Patios: The Ordinariness of Diversity in Urban Oazaca*. Austin, TX: University of Texas Press.

Hiller, D. V. (1984) 'Power, dependence and division of family work'. *Sex Roles*, 10.1: 1003–1019.

Hills, J. (2005) *A More Equal Society: New Labour, Poverty, Inequality and Exclusion*. Bristol: Policy Press.

Hirdman, Y. (1991) 'The gender system'. In T. Andreasen (ed.) *Moving On: New Perspectives on the Women's Movement*. Aarhus: Aarhus University Press.

Hochschild, A. R. (1979) 'Emotion work, feeling rules and social structure'. *American Journal of Sociology*, 85.3: 551–560.

Hochschild, A. R. (1984) *The Managed Heart: Commercialization of Human Feeling*. Berkeley: University of California Press.

Hochschild, A. R. (1997) *The Time Bind: When Work Becomes Home and Home Becomes Work*. New York: Metropolitan Books.

Holloway, S. (1998) 'Local childcare cultures: moral geographies of mothering and the social organization of pre-school children'. *Gender, Place and Culture*, 5.1: 29–53.

Holton, R. (2005) *Making Globalization*. Basingstoke: Palgrave.

Home Office (1994) *Paying the Price: A Consultation Paper on Prostitution*. London: Home Office.

hooks, b. (1982) *Ain't I a Woman? Black Women and Feminism*. London: Pluto Press.

hooks, b. (1984) *Feminist Theory: From Margin to Centr*e. Boston: South End Press.

Horne, S. and Maddrell, A. (2002) *Charity Shops: Retailing, Consumption and Society*. London: Routledge.

Horrelli, L. (1998) 'Creating child-friendly environments – case studies on children's participation in three European countries'. *Children*, 5.2: 225–239.

Horrelli, L. (2000) 'Mainstreaming local and regional development in the Finnish context'. In the *Council of European Conference Proceedings* of 'Gender Mainstreaming: A Step into the 21st Century'. Strasbourg: Council of Europe, 60–63.

Horrelli, L. (2002) 'A methodology of participatory planning'. In R. Bechtel and A. Churchman (eds) *Handbook of Enviromental Psychology*. New York: Wiley.

Hosek, J. R., Asch, B., Fair, C. C., Martin, C. and Mattock, M. (2002) *Married to the Military*. New York: Rand.

Hossain, M. M. (1980) *The Employment of Women, Proceedings from Conference: Thoughts on Islamic Economics*. Dhaka: Islamic Economic Research Bureau.

Howitt, R. (1998) 'Scale as relation: musical metaphors of geographical scale'. *Area*, 30: 49–58.

Hubbard, P. (1999) *Sex and the City: Geographies of Prostitution in the Urban West*. Chichester: Ashgate.

Hubbard, P. (2004) 'Revenge and injustice in the revanchist city: uncovering masculinist agendas'. *Antipode*, 36.4: 665–686.

Hubbard, P., Campbell, R., O'Neill, M., Pitcher, J. and Scoular, J. (2007) 'Prostitution, gentrification, and the limits of neighbourhood space'. In R. Atkinson and G. Helms (eds) *Securing an Urban Renaissance: Crime, Community, and British Urban Policy*. Bristol: Policy Press, 203–219.

Huysman, M. (1994) 'Waste picking as a survival strategy for women in Indian cities'. *Environment and Urbanization*, 6.2: 155–174.

Illouz, E. (2007) *Cold Intimacies: The Making of Emotional Capitalism*. Cambridge: Polity Press.

ILO (1976) *Employment, Growth and Basic Needs: A One-world Problem*. Geneva: ILO.

ILO (1998) 'Migration – the facts'. *New Internationalist* 305, available on-line http://www.newint.org/issue305/facts.html, last accessed 15/09/08.

ILO (1999) *Gender Issues in the World of Work: Gender Issues in Micro-enterprise Development*. Geneva: ILO.

ILO (2002) *Women and Men in the Informal Economy: A Statistical Picture*. Geneva: ILO.

ILO (2005) *Promoting Fair Globalization in Textiles and Clothing in a Post-MFA Environment*. Geneva: ILO.

ILO (2006) *Global Employment Trends Brief*, January. Geneva: ILO.

IMF (2004) 'United States: selected issues', IMF Country Report No. 04/228, July, available on-line http://www.imf.org/external/pubs/ft/scr/2004/cr04228.pdf, last accessed 15/09/08.

Imrie, R. and Hall, P. (2001) *Inclusive Design: Designing and Developing Accessible Environments*. London: Routledge.

INPYME/INEC/PROMICRO-OIT (2001) 'La mujer microempresaria en Nicaragua: situación y caracteristicas hacia finales de los anos noventa'. *Cuadernos de Trabajo* No. 9. San José, Costa Rica.

Ives, S. (2002) 'The politics of happiness: a conversation with Enrique Peñalosa'. *Land and People* (fall). Bogotá: The Trust for Public Land.

Jackson, C. and Palmer Jones, R. (1999) 'Rethinking gendered poverty and work'. *Development and Change*, 30.1: 557–584.

Jacobs, J. (1961) *The Death and Life of Great American Cities*. Harmondsworth: Penguin.

Jain, A. (2003) 'Viewpoint: actioning new partnerships for Indian cities'. *Cities*, 20.5: 353–359.

Jaising, I. (1995) 'Violence against women: the Indian perspective'. In J. Peters and A. Wolper (eds) *Women's Rights, Human Rights: International Feminist Perspectives*. London and New York: Routledge, 51–57.

Jamal, V. and Weeks, J. (1994) *Africa Misunderstood: Or Whatever Happened to the Rural–Urban Gap*. Basingstoke: Macmillan.

James, N. (1989) 'Emotional labour: skill and work in the social regulation of feeling'. *Sociological Review*, 37: 15–42.

Jarvis, H. (1997) 'Housing, labour markets and household structure: questioning the role of secondary data analysis in sustaining the polarization debate.' *Regional Studies*, 31.5: 521–531.

Jarvis, H. (1999a) 'Identifying the relative mobility prospects of a variety of household employment structures, 1981–1991'. *Environment and Planning A*, 31.4: 1031–1046.

Jarvis, H. (1999b) 'The tangled webs we weave: co-ordinating home, work and family life'. *Work, Employment and Society*, 13.2: 223–245.

Jarvis, H. (2005a) *Work–Life City Limits: UK–USA Comparative Household Perspectives*. Basingstoke: Palgrave.

Jarvis, H. (2005b) 'Moving to London time: household co-ordination and the infrastructure of everyday life'. *Time and Society*, 14.1: 133–154.

Jarvis, H. (2006) 'Households'. In I. Douglas, R. Hugget and C. Perkins (eds) *Companion Encyclopaedia of Geography*. London: Taylor & Francis, 351–364.

Jarvis, H. (2007a) 'Home truths about careless competitiveness'. *International Journal of Urban and Regional Research*, 31.1: 207–214.

Jarvis, H. (2007b) 'Time-geography'. In R. Hutchison (ed.) *Encyclopedia of Urban Studies*. London and New York: Sage.

Jarvis, H. (2008) '"Doing deals on the house" in a "post-welfare society": evidence of micro-market practices from Britain and the USA'. *Housing Studies*, 23.2: 213–233.

Jarvis, H. (2009) *Commentary: Gender Interventions in an Age of Disengagement. Journal of Geography in Higher Education* (special issue edited by Deborah Thien and Joyce Davison).

Jarvis, H. (forthcoming) 'Everyday life'. In B. Warf (ed.) *Encyclopaedia of Geography*. London and New York: Sage.

Jarvis, H. and Alvanides, S. (2008) 'Inequalities of school "choice": household resource strategies from a North England case study'. *Community, Work and Family*, 11.4: 385–403.

Jarvis, H. and Pratt, A. C. (2006) 'Bringing it all back home: the extensification and "overflowing" of work. the case of San Francisco's new media households'. *Geoforum,* 37.3: 331–339.

Jarvis, H., Pratt, A. C. and Wu, P. (2001) *The Secret Life of Cities: The Social Reproduction of Everyday Life*. Harlow: Prentice Hall.

Jefferys, K. and Rytina, N. (2006) *Annual Flow Report: April 2006, Department of Homeland Security: 5*. Washington, DC: Department of Homeland Security.

Jhabvala, R. and Tate, J. (1996) 'Out of the shadows: homebased workers organize for international recognition'. *SEEDS*, 18. New York: Population Council.

Johnson, R. (2006) 'Working mums: as bad as junk food?' *Daily Telegraph*, 20 September, available on-line http://www.telegraph.co.uk/health/main.jhtml?xml=/health/2006/09/20/ftmums20.xml, last accessed 15/09/08.

Johnston, L. (2005) 'Man: woman'. In P. Cloke and R. Johnston (eds) *Spaces of Geographical Thought*. London: Sage, 119–142.

Johnston, R. J., Gregory, D. and Smith, D. M. (1994) *The Dictionary of Human Geography* (3rd edn). Oxford: Blackwell.

Kabeer, N. (1989) *Monitoring Poverty as if Gender Mattered: A Methodology for Rural Bangladesh*, DP255. Brighton: IDS.

Kabeer, N. (1992) 'Feminist perspectives in development: a critical review'. In H. Hinds, A. Phoenix and J. Stacey (eds) *Working Out: New Directions for Women's Studies*. London: The Falmer Press, 101–111.

Kabeer, N. (1994) *Reversed Realities: Gender Hierarchies in Development Thought*. London: Verso.

Kabeer, N. (1999) 'Resources, agency, achievement: reflections on the measurement of women's empowerment'. *Development and Change*, 30.3: 435–464.

Kabeer, N. (2000) *The Power to Choose*. London: Verso.

Kabeer, N. (2003) *Gender Mainstreaming in Poverty Eradication and the Millennium Development Goals: A Handbook for Policy-makers and other Stakeholders*. London: Commonwealth Secretariat.

Kabeer, N. (2008) *Mainstreaming Gender in Social Protection for the Informal Economy*. London: Commonwealth Secretariat.

Kabeer, N. and Subrahmanian, R (2000) *Institutions, Relations and Outcomes: Framework and Case Studies for Gender-aware Planning*. London: Zed Books.

Kaika, M. (2004) 'Interrogating the geographies of the familiar: domesticating nature and constructing the autonomy of the modern home'. *International Journal of Urban and Regional Development*, 28.2: 265–286.

Kalabamu, F. (2005) 'Changing gender contracts in self-help housing construction in Botswana: the case of Lobatse'. *Habitat International*, 29.2: 254–268.

Kalabamu, F., Mapetla, E. and Schlyter, A. (eds) (2005) *Gender, Generation and Urban Living Conditions in Southern Africa*. Lesotho: Institute of Southern African Studies, National University of Lesotho.

Kantor, P. (2002a) 'A sectoral approach to the study of gender constraints on economic opportunities in the informal sector in India'. *Gender & Society*, 16.3: 285–302.

Kantor, P. (2002b) 'Gender, microenterprise success and cultural context: the case of South Asia'. *Entrepreneurship Theory and Practice*, 26.4: 131–144.

Kantor, P. (2002c) 'Female mobility in India: the influence of culture on economic outcomes'. *International Development Planning Review*, 24.2: 145–159.

Kantor, P. (2005) 'Determinants of women's microenterprise success in India: empowerment and economics'. *Feminist Economics*, 11.3: 63–83.

Kantor, P. and Andersen, E. (2007) *Microcredit, Informal Credit and Rural Livelihoods:* a Village Case Study in Kabul Province. AREU Case Study. Kabul: AREU.

Kantor, P. and Hozyainova, A. (2008) *Factors Influencing Decisions to Use Child Labour: A Case Study of Poor Households in Kabul*. AREU Case Study. Kabul: AREU.

Kaplan, C. (1994) 'The politics of location as transnational feminist critical practice'. In I. Grewel and C. Kaplan (eds) *Scattered Hegemonies*. Minnesota: University of Minnesota Press.

Kaplan, T. (1982) 'Female consciousness and collective action: the case of Barcelona, 1910–1919'. *Signs: Journal of Women in Culture and Society*, 7.3: 545–566.

Karpf, A. (2005) 'A time geographer's map of the world'. *Guardian*, 26 November, available on-line http://www.guardian.co.uk/print/0,,5341638–117420,00.html.

Karsten, L. (2003) 'Family gentrifiers: challenging the city as a place simultaneously to build a career and to raise children'. *Urban Studies*, 40.12: 2573–2584.

Karsten, L. (2005) 'It all used to be better? Different generations on continuity and change in urban children's daily use of space'. *Children's Geographies*. 3.3: 275–290.

Katz, C. (1996) 'The expeditions of conjurers: ethnography, power and pretense'. In D. Wolf (ed.) *Feminist Dilemmas in Fieldwork*. Boulder, CO: Westview Press, 170–184.

Katz, C. (2001) 'Vagabond capitalism and the necessity of social reproduction'. *Antipode*, 33.4: 709–728.

Katz, C. (2006) 'Messing with "the project"'. In N. Castree and D. Gregory (eds) *David Harvey: A Critical Reader*. Oxford: Blackwell, 234–246.

Katz, C. and Monk, J. (eds) (1993) *Full Circles: Geographies of Women Over the Life Course*. London and New York: Routledge.

Kearns, A. and Parkes, A. (2005) 'Living in and leaving poor neighbourhood conditions in England'. In J. Friedrichs, G. Galster and S. Musterd (eds) *Life in Poverty Neighbourhoods*. London: Routledge, 31–57.

Keating, M. (2007) 'Real work: if your face fits, it's a pretty good-looking job'. *The Guardian*, Work, 24 March: 4.

Keith, M. (2000) 'Walter Benjamin, urban studies and the narratives of city life'. In G. Bridge and S. Watson (eds) *A Companion to the City*. Oxford: Blackwell.

Kenney, M. (2001) *Mapping Gay LA: The Intersection of Place and Politics*. Los Angeles: Temple University Press.

Kindon, S., Kesby, M. and Pain, R. (2008) *Participatory Action Research Approaches and Methods: Connecting People, Participation and Place*. London: Routledge.

Knopp, L. (1995) 'If you're going to get all hyped up, you'd better go somewhere!' *Gender, Place and Culture*, 2: 85–88.

Knopp, L. (1998) 'Sexuality and urban space: gay male identities, communities and cultures in the U.S., U.K. and Australia'. In R. Fincher and J. Jacobs (eds) *Cities of Difference*. New York: Guilford, 149–176.

Knopp, L. and Brown, M. P. (2003) 'Queer diffusions'. *Environment and Planning D: Society and Space*, 21: 409–424.

Knox, P. and McCarthy, L. (2005) *Urbanization* (2nd edn). Englewood Cliffs, NJ: Prentice Hall.

Knox, P. and Pinch, S. (2006) *Urban Social Geography: An Introduction* (5th edn). Harlow: Prentice Hall.

Kolb, D. A. (1984) *Experiential Learning: Experiences as the Source of Learning and Development*. Englewood Cliffs, NJ: Prentice-Hall.

Komter, A. E. (1989) 'Hidden power in marriage'. *Gender and Society*, 2: 187–216.

Koskeh, H. and Pain, R. (2000) 'Revisiting fear and place: women's fear of attack and the built environment'. *Geoforum*, 31.2: 269–280.

Kwan, M. P. (1999) 'Gender and individual access to urban opportunities: a study using space–time measures'. *The Professional Geographer*, 51.2: 210–227.

Kwan, M. P. (2000) 'Gender differences in space–time constraints'. *Area*, 32.2: 145–156.

Landman, K. and Schönteich, M. (2002) 'Urban fortresses: gated communities as a response to crime'. *African Security Review*, 11.4, available on-line http://www.iss.co.za/Pubs/ASR/11No4/Landman.html, last accessed 15/09/08.

Latour, B. (1993) *We Have Never Been Modern*. Brighton: Harvester Wheatsheaf.

Laurie, N., Dwyer, C., Holloway, S. and Smith, F. (1999) *Geographies of New Femininities*. Harlow: Pearson.

Law, R. (1999) 'Beyond "women and transport": towards new geographies of gender and daily mobility'. *Progress in Human Geography*, 23.4: 567–588.

Leccese, M. and McCormick, K. (1999) *Charter of the New Urbanism*. New York: McGraw-Hill.

Lees, L. (2000) 'A reappraisal of gentrification: towards a "geography of gentrification"' *Progress in Human Geography,* 24.3: 389–408.

Lees, L. (ed.) (2004) *The Emancipatory City: Paradoxes and Possibilities*. London: Sage.

Lefebvre, H. (1971) *Everyday Life in the Modern World*. New York: Harper & Row.

Lefebvre, H. (1984) *Everyday Life in the Modern World*, S. Rabinovitch (trans.). New Brunswick, NJ: Transaction Publishers.

Lefebvre, H. (1991) *The Production of Space*. Oxford: Wiley Blackwell.

LeGates, R. T. and Stout, F. (eds) (2007) *The City Reader* (4th edn). London and New York: Routledge.

Leisch, H. (2002) 'Gated communities in Indonesia'. *Cities*, 19.5: 341–350.

Leitner, S. (2001) 'Sex and gender discrimination within EU pension systems'. *Journal of European Social Policy*, 11.2: 99–115.

Levine, J. A. and Pittinsky, T. L. (1998) *Working Fathers: New Strategies for Balancing Work and Family*. San Francisco: Harvest.

Lewis, A. W. (1954) 'Economic development with unlimited supplies of labour'. *Manchester School of Economic and Social Studies*, 22: 139–91.

Lewis, J. (2002) 'Individualisation, assumptions about the existence of an adult worker model and the shift towards contractualism'. In A. Carline, S. Duncan and R. Edwards (eds) *Analysing Families: Morality and Rationality in Policy and Practice*. London: Routledge.

Ley, D. (1974) *The Black Inner City as Frontier Post*. Washington, DC: Association of American Geographers.

Li, W. (1998) 'Anatomy of a new ethnic settlement: the Chinese ethnoburb in Los Angeles'. *Urban Studies*, 35: 479–501.

Lingam, L. (2005) *Structural Adjustment, Gender and Household Survival Strategies: Review of Evidences and Concerns*. Center for the Education of Women, University of Michigan, available on-line http://www.umich.edu/~cew/PDFs/pubs/lingamrept.pdf, last accessed 15/09/08.

Lipsky, M. (1980) *Street-level Bureaucracy: Dilemmas of the Individual in Public Services*. New York: Russell Sage.

Lipton, M. (1977) *Why Poor People Stay Poor: A Study of the Urban Bias in World Development*. London: Temple Smith.

Lloyds (2008) *India 2010: A Lloyds View*, available on-line http://www.lloyds.com/NR/rdonlyres/3B9FD18C-7D3A-41FE-AB0F-7AABD77FA359/0/ReportIndiaJun07.pdf, last accessed 15/09/08.

Logan, J. R. and Swanstrom, R. (eds) (1990) *Beyond the City Limits: Urban Policy and Economic Restructuring in Comparative Perspective*. Philadelphia: Temple University Press.

Longhurst, R. (1995) 'The body and geography'. *Gender, Place and Culture*, 2: 97–105.

Longhurst, R. (2001) *Bodies: Exploring Fluid Boundaries*. London: Routledge.

Lopez-Claros, A. and Zahidi, S. (2005) 'Women's empowerment: measuring the global gender gap'. World Economic Forum, available on-lines http://www.weforum.org/pdf/Global_Competitiveness_Report/Report/gender_gap.pdf.

Low, G. Ching-Liang (2000) 'Rethinking/teaching identity'. In S. Ahmed, J. Kilby, C. Lury and B. Skeggs (eds) *Transformations: Thinking Through Feminism*. London: Routledge, 159–60.

Lowe, D. M. (1995) *The Body in Late-capitalist USA*. Durham, NC: Duke University Press.

Ludwig, A. (2006) 'Differences between women? Intersecting voices in a female narrative'. *European Journal of Women's Studies*, 13.3: 245–258.

Lund, F. and Srinivas, S. (2000) *Learning from Experience: A Gendered Approach to Social Protection for Workers in the Informal Economy*. Geneva: ILO.

Lynch, K. (1960) *The Image of the City*. Cambridge, MA: MIT Press.

Lyon, D. (2007) *Surveillance Studies: An Overview*. Oxford: Polity Press.

Lyons, M. (1996) 'Employment, feminisation, and gentrification in London, 1981–93'. *Environment and Planning A*, 28.2: 341–356.

Lyons, M. and Snoxell, S. (2005) 'Sustainable urban livelihoods and marketplace social capital: crisis and strategy in petty trade'. *Urban Studies*, 42.8: 1301–1320.

Mac an Ghaill, M. (1994) *The Making of Men: Masculinities, Sexualities and Schooling*. Milton Keynes: Open University Press.

Mac an Ghaill, M. (ed.) (1996) *Understanding Masculinities: Social Relations and Cultural Arenas*. Milton Keynes: Open University Press.

McCall, L. (2001) *Complex Inequality: Gender, Class and Race in the New Economy*. London: Routledge.

McCall, L. (2005) 'The complexity of intersectionality'. *Signs Journal of Women in Culture and Society*, 30.3: 1771–1802.

McDowell, L. (1983) 'Towards an understanding of the gender division of urban space'. *Environment and Planning D: Society and Space*, 1: 15–30.

McDowell, L. (1991) 'Life without father and Ford: the new gender order of post Fordism'. *Transactions of the Institute of British Georaphers*, 16: 400–419.

McDowell, L. (1992) 'Doing gender: feminisms, feminists and research methods in human geography'. *Transactions of the Institute of British Geographers*, 17: 399–416.

McDowell, L. (1997a) *Capital Culture: Gender at Work in the City*. Oxford: Blackwell.

McDowell, L. (1997b) 'Women/gender/feminisms: doing feminist geography'. *Journal of Geography in Higher Education*, 21.3: 381–400.

McDowell, L. (1999) *Gender, Identity & Place: Understanding Feminist Geographies*. Oxford: Polity Press.

McDowell, L. (2003) *Redundant Masculinities: Employment Change and White Working Class Youth*. Oxford: Blackwell.

McDowell, L. (2004) 'Work, workfare, work/life balance and an ethic of care'. *Progress in Human Geography*, 28.2: 145–163.

McDowell, L. (2006) 'Reconfigurations of gender and class relations: class differences, class condescension and the changing place of class relations'. *Antipode*, 38.4: 825–850.

McDowell, L. and Sharp, J. P. (eds) (1997) *Space, Gender, Knowledge: Feminist Readings*. London: Arnold.

McDowell, L. and Sharp, J. P. (eds) (1999) *A Feminist Glossary of Human Geography*. Oxford: Arnold.

McDowell, L., Ward, K., Fagan, C., Perrons, D. and Ray, K. (2006) 'Connecting time and space: the signification of transformations in women's work in the city'. *International Journal of Urban and Regional Research*, 30: 159–171.

McGill, R. (1998) 'Viewpoint: urban management in developing countries'. *Cities*, 15.6: 463–471.

McGregor, D., Simon, D. and Thompson, D. (eds) (2006) *The Peri-urban Interface: Approaches to Sustainable Natural and Human Resource Use*. London: Earthscan.

McGuinness, M. (2009) 'Putting themselves in the picture? Reflecting on the use of diaries in a feminist geography module'. *Journal of Geography in Higher Education*.

McKenzie, D. J., Gibson, J. and Stillman, S. (2007) *A Land of Milk and Honey with Streets Paved with Gold: Do Emigrants Have Over-optimistic Expectations?* World Bank Policy Research Working Paper 4141, March.

Macklin, A. (1992) 'Foreign domestic workers: surrogate housewife or mail order servant?' *McGill Law Journal*, 37: 681–800.

McNeill, D. (2006) 'Globalization and the ethics of architectural design'. *City*, 10.1: 49–58.

McRae, S. (1986) *Cross-class Families: A Study of Wives' Occupational Superiority*. Oxford: Clarendon Press.

McRobbie, A. (2004) 'Notes on postfeminism and popular culture: Bridget Jones and the new gender regime'. In A. Harris (ed.) *All About the Girl: Culture, Power and Identity*. London: Routledge.

Maddrell, A. (1994) 'A scheme for the effective use of role plays for an emancipatory geography'. *Journal of Geography in Higher Education*, 18.2: 155–162.

Maddrell, A. (2007) 'The "map girls". British women geographers' war work, shifting gender boundaries and reflections on the history of geography'. *Transactions of the Institute of British Geographers*, 33.1: 127–148.

MADRE (2008) *Murder in Basra*. Madre Organization, available on-line http://www.madre.org/articles/me/womenbasra010908.html, last accessed 15/09/08.

Maguire, S. (1998) 'Gender differences in attitudes to undergraduate fieldwork'. *Area*, 30.3: 207–214.

Maher, L. (1997) *Sexed-work: Gender, Race and Resistance in a Brooklyn Drug Market*. Oxford: Oxford University Press.

Mahler, S. and Pessar, P. (2001) 'Gendered geographies of power: analysing gender across transnational spaces'. *Identities: Global Studies in Culture and Power*, 7.4: 441–459.

Mains, S. P. (2002) 'Maintaining national identity at the border: scale, masculinity and the policing of immigrants in Southern California'. In A. Herod and M. W. Wright (eds) *Geographies of Power: Placing Scale*. Oxford: Blackwell, 192–212.

Management Issues (2004) '10 million migrants needed to halt pensions crisis, *Management Issues*, 2 November, available on-line http://www.management-issues.com/display_page.asp?section=research&id=1633, last accessed 23/01/08.

Markovich, J. and Hendler, S. (2006) 'Beyond "soccer moms": feminist and new urbanist critical approaches to suburbs'. *Journal of Planning Education and Research*, 22: 410–427.

Markusen, A. R. (2005) 'City spatial structure, women's household work and national urban policy'. In S. S. Fainstein and L. J. Servon (eds) *Gender and Planning: A Reader*. Brunswick, NJ: Rutgers University Press, 67–85.

Marotta, M. (2002) 'Time, space and motherhood'. In C. L. Mui and J. S. Murphy (eds) *Gender Struggles: Practical Approaches to Contemporary Feminism*. London: Rowman & Littlefield.

Marquetti, A. (2002) 'Participacao e redistribuicao: o orcamento participativo em Porto Alegre', available on-line: http://www.democraciaparticipativa.org/files/

AdalmirMarquettiParticipa%E7%E3oeRedistribui%E7%E3ooOr%E7amento ParticipativoemPorto%20Alegre.pdf, last accessed 15/09/08.

Marston, S. (1990) 'Who are "the people"?: Gender, citizenship, and the making of the American nation'. *Environment and Planning D: Society and Space*, 8: 449–458.

Martell, L. (2007) 'The third wave in globalisation theory'. *International Studies Review*, 9.2: 173–196.

Martin, P. (2004) 'Contextualizing feminist political theory'. In L. Staeheli, E. Kofman and L. J. Peake (eds) *Mapping Women, Making Politics: Feminist Perspectives on Political Geography*. London and New York: Routledge.

Martinez, P. J. and Reboiras, F. L. (2001) *International Migration and Development in the Americas*. Santiago: ECLAC.

Massey, D. (1993) 'Power geometries and a progressive sense of place'. In J. Bird, B. Curtis, T. Putnam, G. Robertson and L. Tickner (eds). *Mapping the Futures: Local Cultures, Global Change*. London: Routledge.

Massey, D. (1995) 'Masculinity, dualisms and high technology'. *Transactions of the Institute of British Geographers*, 20.4: 487–499.

Massey, D., Allen, J. and Pile, S. (eds) (1999) *City Worlds*. London: Routledge.

Matrix (1984) *Making Space: Women and the Man-made Environmen*t. London: Pluto Press.

Mattes, J. (1994) *Single Mothers by Choice: A Guidebook For Single Women Who Are Considering or Have Chosen Motherhood.* New York: Three Rivers Press.

Matthew, G. (2003) *Silicon Valley, Women and the Californian Dream: Gender, Class and Opportunity in the Twentieth Century.* Stanford: Stanford University Press.

Mattingly, D. (2001) 'The home and the world: domestic service and international networks of caring labour'. *Annals of the Association of American Geographers*, 91.2: 370–386.

Maushart, S. (1997) *The Mask of Motherhood: How Motherhood Changes Everything and Why We Pretend It Doesn't*. Sydney: Vintage.

Mayoux, L. (1995) *From Vicious to Virtuous Circles? Gender and Micro-enterprise Development*. Occasional Paper No. 1, Fourth World Conference on Women. Geneva: UNRISD.

Meikle, S. (2002) 'The urban context and poor people'. In C. Rakodi (ed.) *Urban Livelihoods*. London: Earthscan, 37–51.

Mencher, J. (1988) 'Women's work and poverty: women's contribution to household maintenance in South India'. In D. Dwyer and J. Bruce (eds) *A Home Divided: Women and Income in the Third World*. Stanford: Stanford University Press, 99–119.

Merrett, C. D. (2004) 'Social justice: what is it? Why teach it?' *Journal of Geography*, 103.3: 93–101.

Merrifield, A. (2002) *Metromarxism: A Marxist Tale of the City.* London: Routledge.

Micklethwait, A. and Wooldridge, J. (2000) *A Future Perfect: The Challenge and Hidden Promise of Globalization*. London: William Heinemann.

Miller, D. (2001) *Car Cultures*. Oxford: Berg.

Millet, K. (1971) *Sexual Politics*. London: Rupert Hart-Davis.

Milligan, C. (2000) '"Bearing the burden": towards a restructured geography of caring'. *Area*, 32.1: 49–58.

Mills, C. (1993) 'Myths and meanings of gentrification'. In J. Duncan and D. Ley (eds) *Place/Culture/Representation*. New York: Routledge, 149–172.

Mincer, J. (1978) 'Family migration decision'. *Journal of Political Economy*, 86: 749–773.

Mingione, E. (1991) *Fragmented Societies: A Sociology of Economic Life Beyond the Market Paradigm*. Oxford: Blackwell.

Mingione, E. (2006) 'New urban poverty and the crisis in the citizenship/welfare system: the Italian experience'. *Antipode*, 25.3: 206–222.

Minton, A. (2002) *Building Balanced Communities: The US and UK Compared*. London: RICS.

Mitchell, K., Marston, S. and Katz, C. (2004) *Life's Work: Geographics of Social Reproduction*. Oxford: Blackwell.

Mohanty, C. T. (1988) 'Under Western eyes: feminist scholarship and colonial discourses'. *Feminist Review*, 30: 61–88.

Mohanty, C. T., Russo, A. and Torres, L. (1991) *Third World Women and the Politics of Feminism*. Bloomington and Indianapolis: Indiana University Press.

mollymaid.com (2008) http: //www.mollymaid.com/, last accessed 15/09/08.

Molyneux, M. (1985) 'Mobilisation without emancipation? Women's interests, state and revolution in Nicaragua'. *Feminist Studies*, 11.2: 227–54.

Molyneux, M. (1986) 'Mobilization without emancipation? Women's interests, state and revolution in Nicaragua'. In R. R. Fagen, C. D. Deere and J. L. Goraggio (eds) *Transition and Development: Problems of Third World Socialism*. New York: Monthly Review Press and Center for the Study of the Americas, 280–302.

Monzini, P. (2005) *Sex Traffic: Prostitution, Crime and Exploitation*. London and New York: Zed Books.

Moorhead, J. (2007) 'Nice week at the office darling?' *Guardian*, Family, 13 January.

Morgan, D. (1992) *Discovering Men*. London: Routledge.

Morgan, D. (1996) *Family Connections. An Introduction to Family Studies*. Cambridge: Polity Press.

Morgan, D., Brandth, B. and Kvande, E. (eds) (2005) *Gender, Bodies and Work*. Aldershot: Ashgate.

Morris, L. (1995) *Social Divisions: Economic Decline and Social Structural Change*. London: Routledge.

Morriss, P. (1987) *Power – A Philosophical Analysis*. Manchester: Manchester University Press.

Mortensen, E. (ed.) (2006) S*ex, Breath and Force: Sexual Difference in a Post-feminist Era*. New York: Lexington Books.

Moser, C. (1987) 'Women, human settlement, and housing: a conceptual framework for analysis and policy-making'. In C. Moser and L. Peake (eds) *Women, Human Settlement and Housing*. London: Tavistock, 12–32.

Moser, C. (1993) *Gender Planning and Development: Theory, Practice and Training*. London: Routledge.

Moser, C. (1998) 'The asset vulnerability framework: re-assessing urban poverty reduction strategies'. *World Development*, 26.1: 1–19.

Moser, C. and McIlwaine, C. (1997) *Household Responses to Poverty and Vulnerability, Volume 3: Confronting Crisis in Commonwealth, Metro Manila, The Philippines*. Washington, DC: World Bank.

Motavalli, J. (2000) *Forward Drive*. San Francisco: Sierra Club.

MoUD (2001) *Learning From One Another. Good Urban Governance Campaign–India Launch*. Ministry of Urban Development, Government of India, New Delhi, available on-line http://www.un.org.in/unnew/iawg/Decntz/onactivty/guccpage1.doc, last accessed 15/09/08.

Murdoch, J. (2006) *Post-structuralist Geography: A Guide to Relational Space*. London: Sage.

Murray, C. (1996) 'The emerging British underclass'. In R. Lister (ed.) *Charles Murray and the Underclass: The Developing Debate*. London: Institute for Economic Affairs.

Nagar, R. (2004) 'Mapping feminisms and difference'. In L. A. Staeheli, E. Kofman and L. J. Peake (eds) *Mapping Women, Making Politics: Feminist Perspectives on Political Geography*. New York: Routledge, 31–48.

Nagel, J. H. (1975) *The Descriptive Analysis of Power*. New Haven, CT: Yale University Press.

Naples, N. A. (1998) *Grassroots Warriors: Activist Mothering, Community Work, and the War on Poverty*. London: Routledge.

Narayan, D. with R. Patel, K. Schafft, A. Rademacher and S. Koch-Schulte (2000) *Voices of the Poor: Can Anyone Hear Us?* Oxford: Oxford University Press.

Nash, C. (1996) 'Reclaiming vision: looking at landscape and the body'. *Gender, Place and Culture*, 3: 149–169.

Nayak, A. (2006) 'Displaced masculinities: chavs, youth and class in a post-industrial city'. *Sociology*, 40.5: 813–831.

Nelson, B. (1990) 'The origins of the two-channel welfare state: Workmen's Compensation and Mother's Aid'. In L. Gordon (ed.) *Women, the State and Welfare.* Madison: University of Wisconsin Press, 123–151.

Nelson, M. K. and Smith, J. (1999) *Working Hard and Making Do: Surviving in Small Town America*. Berkeley: University of California Press.

Newman, K. (1999) *No Shame in My Game: The Working Poor in the Inner City*. New York: Russell Sage.

Nicholson, L. (ed.) (1990) *Feminism/Postfeminism*. London: Routledge, 4 -14.

Nitlápan-Envío (2004) 'Nicaragua briefs: Nicaragua gets bad grades for education'. *Revista*, 278. Managua: UCA.

Norris, P. (2001) *Digital Divide: Civic Engagement, Information Poverty and the Internet Worldwide*. Cambridge: Cambridge University Press.

North, D. (1990) *Institutions, Institutional Change and Economic Performance*. Cambridge: Cambridge University Press.

Novy, A. and Leubolt, B. (2005) 'Participatory budgeting in Porto Alegre: social innovation and the dialectical relationship of state and civil society'. *Urban Studies*, 42: 2023.

Nussbaum, M. (2000) *Women and Human Development: The Capabilities Approach*. Cambridge: Cambridge University Press.

O'Connor, J. S., Orloff, A. S. and Shaver, S. (1999) *States, Markets, Families: Gender, Liberalism and Social Policy in Australia, Canada, Great Britain and the United States*. Cambridge: Cambridge University Press.

O'Neill, B. C., MacKellar, F. L. and Lutz, W. (2000) *Population and Climate Change.* Cambridge: Cambridge University Press.

Oakley, A. (1974) *The Sociology of Housework*. Harmondsworth: Penguin.

Oakley, A. (2002) *Gender on Planet Earth*. Oxford: Polity Press.

OECD (2002) *Handbook for Measurement of the Non-observed Economy*. New York: OECD/ILO.

Olds, K. (2001) *Globalization and Urban Change: Capital, Culture and Pacific Rim Mega-projects*. Oxford: Oxford University Press, 141–161.

Orleans, P. (1973) 'Differential cognition of urban residents: effects of scale on social mapping'. In R. Downs and D. Stea (eds) *Image and Environment*. Chicago, IL: Aldine.

Overall, C. (2007) 'Public toilets: sex segregation revisited'. *Ethics and Environment*, 12.2: 71–91.

Owusu, F. (2001) 'Conceptualizing livelihood strategies in African cities'. *Journal of Planning Education and Research*, 26: 450–465.

Pacione, M. (2001) *Urban Geography: A Global Perspective.* London: Routledge.

Paglia, C. (1992) *Sex, Art and American Culture*. Harmondsworth: Penguin.

Pahl, R. (1984) *Divisions of Labour.* Oxford: Blackwell.

Pahl, R. (2000) *On Friendship*. Oxford: Wiley.

Pain, R. (1997) 'Social geographies of women's fear of crime'. *Transactions of the Institute of British Geographers*, 22.2: 231–244.

Pain, R. (1999) 'Theorising age in criminology: the case of home abuse'. *The British Criminology Conference Selected Proceedings*, vol. 2, 1–12.

Pain, R. (2001) 'Crime, space and inequality'. In R. Pain, M. Barke, J. Gough, R. Macfarlene, G. Mowl and D. Fuller (eds) *Introducing Social Geographies*. London: Arnold, 231–253.

Panelli, R. (2004) *Social Geographies: From Difference to Action.* London: Sage.

Park, R. E. (1925) 'Suggestions for investigation of human behaviour in the urban environment'. In R. Park, E. Burgess and McKenzie (eds) *The City*. Chicago: Chicago University Press, 45–46.

Park, R., Burgess, E. and McKenzie, R. (eds) (1925) *The City*. Chicago: University Press.

Parpart, J., Connelly, M. P. and Barriteau, V. E. (2000) *Theoretical Perspectives on Gender and Development*. Ottawa: IDRC.

Parreñas, R. (2001) *Servants of Globalization: Women, Migration and Domestic Work*. Stanford, CA: Stanford University Press.

Parsons, T. (1951) *The Social System*. Glencoe, IL: Free Press.

Parsons, T. and Bales, R. F. (eds) (1956) *Family, Socialization and Interaction Process*. London: Routledge and Kegan Paul.

Passel, J., Capps, R. and Fix, M. (2004) 'Undocumented immigrants: facts and figures'. Urban Studies Institute, January, p. 2, available on-line http://www.urban.org/UploadedPDF/1000587_undoc_immigrants_facts.pdf.

Patel, S. and Mitlin, D. (2001) *The Work of SPARC, the National Slum Dwellers Federation and Mahila Milan*. IIED Working Paper Series on Poverty Reduction in Urban Areas No. 5. IIED Human Settlements Programme.

Patel, S. and Mitlin, D. (2002) 'Sharing experiences and changing lives'. *Community Development Journal*, 37.2: 125–136.

Pateman, C. (1988) *The Sexual Contract*. Cambridge: Polity Press.

Pavlovskaya, M. (2004) 'Other transitions: multiple economies of Moscow households in the 1990s'. *Annals of the Association of American Geographers*, 94.2: 329–351.

Paxton, P. and Hughes, M. (2007) *Women, Politics and Power: A Global Perspective*. London: Pine Forge Press.

Peach, C. (ed.) (1975) *Urban Social Segregation*. London: Longman.

Pearsall, P. (1990) *From Bedsitter to Household Name: The Personal Story of A–Z Maps*. Sevenoaks: Geographers' A-Z Map Co.

Pearson, R. (1998) '"Nimble fingers" revisited: reflections on women and Third World industrialisation in the late twentieth century'. In C. Jackson and R. Pearson (eds) *Feminist Visions of Development*. London: Routledge.

Pemberton, J. (2001) *Global Metaphors: Modernity and the Quest for One World*. London: Pluto Press.

Perrons, D. (2003) 'The new economy and work–life balance: conceptual explorations and a case study of new media'. *Gender, Work and Organization*, 10.1: 65–93.

Perrons, D. (2004) *Globalization and Social Change: People and Places in a Divided World*. London: Routledge.

Perrons, D. (2005) 'Gender mainstreaming and gender equality in the new (market) economy: an analysis of contradictions'. *Social Politics: International Studies in Gender, State and Society*, 12.3: 389–411.

Pessar, P. (2005) *Women, Gender and International Migration Across and Beyond the Americas: Inequalities and Limited Empowerment*, UN/POP/EGM-MIG/2005/08, Expert Group Meeting on International Migration and Development in Latin America and the Caribbean, Population Division, Department of Economic and Social Affairs United Nations Secretariat, Mexico City, 30 November–2 December.

Peters, D. (2002) 'Gender and transport in less developed countries'. A background paper in preparation for CSD-9. London and Berlin: UNED Forum, available on-line http://www.cityshelter.org/13_mobil/Gender%20and%20Transport%20in%20the%20South.pdf, last accessed 15/09/08.

Pfau-Effinger, B. (1993) 'Modernisation, culture and part-time work'. *Work, Employment and Society*, 7.3: 383–410.

Pfau-Effinger, B. (2004) *Development of Culture, Welfare States and Women's Employment in Europe: Theoretical Framework and Analysis of Development Paths*. Basingstoke: Ashgate.

Phillips, A. (1995) *The Politics of Presence: Political Representation by Gender, Ethnicity and Race*. Oxford: Oxford University Press.

Phillips, A. (1999) *Which Equalities Matter?* Cambridge: Polity Press.

Phillipson, C., Bernard, M., Phillips, J. and Ogg, J. (1999) 'Older people's experience of community life: patterns of neighbourhood in three urban areas'. *Sociological Review*, 47.4: 715–743.

Pinder, D. (2002) *Visions of the City: Utopianism, Power and Politics in 20th Century Urbanism*. Edinburgh: Edinburgh University Press.

Piore, M. and Sabel, C. (1984) *The Second Industrial Divide*. New York: Basic Books.

Podmore, J. A. (2006) 'Gone "underground"? Lesbian visibility and the consolidation of queer space in Montréal'. *Social and Cultural Geography*, 7.4: 595–625.

Portes, A., Castells, M. and Benton, L. A. (eds) (1996) *The Informal Economy: Studies in Advanced and Less Developed Countries*. Baltimore, MD: The Johns Hopkins University Press.

Potts, D. (1995) 'Shall we go home? Increasing urban poverty in African cities and migration processes'. *Geographical Journal*, 161.3: 245–264.

Pratt, A. C. (2000) 'New media, the new economy and new spaces'. *Geoforum*, 31.4: 425–436.

Pratt, A. C. (2002) 'Hot jobs in cool places. The material cultures of new media product spaces: the case of South of the Market, San Francisco'. *Information, Communication and Society*, 5.1: 27–50.

Pratt, G. (1998) 'Grids of difference: place and identity formation'. In R. Fincher and J. M. Jacobs (eds) *Cities of Difference*. London: Guilford, 26–48.

Pratt, G. (1999) 'From registered nurse to registered nanny: discursive geographies of Filipina domestic workers in Vancouver, B.C.' *Economic Geography*, 75: 215–236.

Pratt, G. (2003) 'Valuing childcare: troubles in suburbia'. *Antipode*, 35: 581–602.

Pratt, G. (2004) *Working Feminism.* Edinburgh: Edinburgh University Press.

Pratt, G. (2005) 'Abandoned women and spaces of the exception'. *Antipode*, 37.5: 1052–1078.

Pratt, G. and San Juan, R. M. (2004) 'In search of the horizon: utopia in *The Truman Show* and *The Matrix*'. In L. Lees (ed.) *The Emancipatory City: Paradoxes and Possibilities*. London: Sage.

Predelli, L. N. (2004) 'Interpreting gender in Islam: a case study of immigrant Muslim women in Oslo, Norway'. *Gender and Society*, 18.4: 473–493.

Presser, H. B. (2003) *Working in a 24/7 Economy: Challenges for American Families*. New York: Russell Sage Foundation.

Putnam, R. D. (2000) *Bowling Alone: The Collapse and Revival of American Communities*. New York: Simon & Schuster.

Putnam, T. (1992) 'Beyond the modern home; shifting parameters of residence'. In J. Bird, B. Curtis, T. Putnam, T. Robertson and L. Tickner (eds) *Mapping the Futures: Local Cultures, Global Change*. London: Routledge, 151–169.

Pyle, J. L. (2006) 'Globalization and the increase in transnational care work: the flip side'. *Globalizations*, 3.3: 297–315.

Quenet, N., Duflo, C. and Patel, R. (2006) 'A grey hope: thin territorial identity among French suburban youth in Garges and Sarcelles'. Paris: Humanity in Action.

RAC Foundation (2003) 'Commuting: the facts', 22 July, available on-line http://www/racfoundation.org/index.php?option=com_content&task=view&id=182&Itemid=0, last accessed 30/01/08.

Rahder, B. (1998) 'Women plan Toronto: incorporating gender issues in urban planning'. *Planners Network*, 180, July: 32–40.

Rahder, B. (1999) 'Victims no longer: participatory planning with a diversity of women at risk of abuse'. *Journal of Planning Education and Research*, 18.3: 221–232.

Rahder, B. (2004) 'Where is feminism in planning going? Appropriation or transformation?' *Planning Theory*, 3: 107–116.

Rahder, B. and Attilia, C. (2004) 'Where is feminism in planning going? Appropriation or transformation?' *Planning Theory*, 3: 107–116.

Rahman, A. (1999) *Women and Microcredit in Rural Bangladesh: An Anthropological Study of Grameen Bank Lending*. Philadelphia, PA: Perseus.

Rajan, S. C. (2006) 'Automobility and the liberal disposition'. In S. Böhm, C. Jones, C. Land and M. Paterson (eds) *Against Automobility*. Oxford: Blackwell, 113–131.

Rakodi, C. (1995) 'Poverty lines or household strategies? A review of conceptual issues in the study of urban poverty'. *Habitat International*, 19.4: 407–426.

Rakodi, C. (1999) 'A capital assets framework for analyzing household livelihood strategies: implications for policy', *Development Policy Review*, 17: 315–342.

Rakodi, C. and Lloyd-Jones, T. (eds) (2002) *Urban Livelihoods: A People-centred Approach to Reducing Poverty*. London: Earthscan.

Randolph, W. (1991) 'Housing markets, labour markets, discontinuity theory'. In J. Allen and C. Hamnett (eds) *Housing and Labour Markets: Building the Connections*. London: Unwin Hyman, 16–47.

Rankin, K. (2003) 'Cultures of economies: gender and socio-spatial change in Nepal'. *Gender, Place and Culture*, 10: 111–130.

Razavi, S. (1999) 'Gendered poverty and well-being: introduction'. *Development and Change*, 30: 409–433.

Reed, H. C. (1981) *The Pre-eminence of International Financial Centres*. New York: Praeger.

Rees, T. (1998) *Mainstreaming Equality in the European Union*. London: Routledge.

Reeves, D. (2005) *Planning for Diversity: Policy and Planning in a World of Difference*. London: Routledge.

Reeves, R. (2001) *Happy Mondays: Putting the Pleasure Back into Work*. Harlow: Momentum.

Rich, E. (2005) 'Young women, feminist identities and neo-liberalism'. *Women's Studies International Forum*, 28.6: 495–508.

Rigg, J. (2007) *An Everyday Geography of the Global South*. London: Routledge.

Rimmer, P. J. (1991) 'The global intelligence corps and world cities: engineering consultancies on the move'. In P. W. Daniels (ed.) *Services and Metropolitan Development: International Perspectives*. London: Routledge, 66–106.

Ritzdorf, M. (1989) 'The political economy of urban service distribution'. In R. Rich (ed.) *The Politics of Urban Public Services*. Lexington, MA: Lexington Books.

Roberts, M. (1991) *Living in a Man-made World: Gender Assumptions in Modern Housing Design*. London: Routledge.

Roberts, P. and Hill, J. (2005) *A Survival Guide to the Near Future*. Brand Papers, available on-line http://www.brandchannel.com/papers_review.asp?sp_id=677.

Robeyns, I. (2003) 'Sen's capability approach and gender inequality: selecting relevant capabilities'. *Feminist Economics*, 9.2–3: 61–92.

Robins, I. (2007) 'When will society be gender just?' In J. Brown (ed.) *The Future of Gender*. Cambridge: Cambridge University Press, 54–75.

Robinson, J. (1998) 'Octavia Hill. Women housing managers in South Africa: feminism and urban government'. *Journal of Historical Geography*, 24.4: 459–481.

Robinson, J. (2006) *Ordinary Cities: Between Modernity and Development*. London: Routledge.

Robson, E. (2007) 'Listening to geographers from the global south'. *Journal of Geography in Higher Education*, 31.3: 345–352.

Roesneil, S. (2004) 'Why we should care about friends: an argument for queering the care imaginary in social policy'. *Social Policy and Society*, 3.4: 409–419.

Rogerson, C. M. (1996) 'Urban poverty and the informal economy in South Africa's economic heartland'. *Environment and Urbanization*, 8.1: 167–181.

Romaine, S. (1999) *Communicating Gender*. Michigan: L. Erlbaum Associates.

Rose, D. (1989) 'A feminist perspective of employment restructuring and gentrification: the case of Montreal'. In J. Wolch and M. Dear (eds) *The Power of Geography: How Territory Shapes Social Life*. London: Unwin Hyman.

Rose, G. (1993) *Feminism and Geography*. Cambridge: Polity Press.

Rose, M. (1995) *Cities of Light and Heat: Domesticating Gas and Electricity in Urban America*. University Park, PA: University of Pennsylvania Press.

Rosenbloom, S. (1993) 'Women's travel patterns at various stages of their lives'. In C. Katz and J. Monk (eds) *Full Circles: Geographies of Women Over the Life Course*. New York: Routledge.

Ross, A. (1997) 'Jobs in cyberspace', posted to *Nettime*, 27 June (copy available at Nettime archive, http: //www.nettime.org/nettime.w3archive/).

Rostow, W. W. (1960) *The Stages of Economic Growth: A Non-communist Manifesto*. Cambridge: Cambridge University Press.

Rowbotham, S. (1997) *A Century of Women: The History of Women in Britain and the United States*. London: Penguin.

Rowlands, J. (1997) *Questioning Empowerment*. Oxford: Oxfam.

Roy, K. (2004) 'Three-block fathers: spatial perceptions and kin-work in low-income African American neighbourhoods'. *Social Problems*, 51.4: 528–548.

Rubin, G. (1975) 'The traffic in women: notes on the "political economy" of sex'. In R. R. Reiter (ed.) *Toward an Anthropology of Women*. New York: Monthly Review Press, 157–210.

Rubin, G. (2000) 'Sites, settlements and urban sex: archaeologies and the sex of gay leathermen in San Francisco'. In R. A. Schmidt and B. L. Voss (eds) *Archaeology of Sexuality*. New York: Routledge, 62–89.

Ruddick, S. (1996) 'Constructing differences in public space: race, class and gender as interlocking system'. *Urban Geography*, 17; 132–151.

Sachs, W. (1992) *The Development Dictionary: A Guide to Knowledge as Power*. London: Zed Books.

Safa, H. (1995) 'Economic restructuring and gender subordination'. *Latin American Perspectives*, 22.2: 32–50.

Said, E. (1978) *Orientalism*. New York: Vintage Books.

Said, T. and Carter, P. (2005) 'The entwined spaces of "race", sex and gender'. *Gender, Place and Culture. A Journal of Feminist Geography*, 12: 49–52.

Sanchez, L. E. (2004) 'The global e-rotic subject, the ban, and the prostitute-free zone: sex work and the theory of differential exclusion'. *Environment and Planning D: Society and Space*, 22: 861–883.

Sandercock, L. (1998) *Towards Cosmopolis: Planning for Multicultural Cities*. Sydney: John Wiley.

Sandercock, L. (2000) 'When strangers become neighbours: managing cities of difference'. *Planning Theory and Practice*, 1.11: 13–30.

Sandercock, L. (2003) *Cosmopolis II: Mongrel Cities of the 21st Century*. Sydney: Continuum Press.

Sassen, S. (2002) 'Women's burden: counter-geographies of globalization and the feminization of survival'. *Nordic Journal of International Law*, 71.1: 255–274.

Sassen, S. (2006) 'Global cities and survival circuits'. In M. K. Zimmerman, J. S. Litt and C. E. Bose (eds) *Global Dimensions of Gender and Care-work*. Stanford, CA: Stanford University Press, 30–48.

Satterthwaite, D. (1997) 'Urban poverty: reconsidering its scale and nature'. *IDS Bulletin*, 28(2): 9–23.

Satterthwaite, D. (2002) 'Chapter 5.1 Urbanization in developing countries'. In V. Desai and R. Potter (eds) *The Companion to Development Studies*. London: Hodder Arnold.

Sayer, A. (2005) *The Moral Significance of Class*. Cambridge: Cambridge University Press.

Schmalzbauer, L. (2004) 'Searching for wages and mothering from afar: the case of Honduran transnational families'. *Journal of Marriage and Family,* 66: 1317–1331.

Schneider, J. W. and Hacker, S. L. (1973) 'Sex role imagery and use of the generic "man" in introductory texts: a case in the sociology of sociology'. *American Sociologist*, 8: 12–18.

Scott, J. (1976) *The Moral Economy of the Peasant: Rebellion and Subsistence in Southeast Asia.* New Haven, CT, and London: Yale University Press.

Sebstad, J. and Cohen, M. (2001) *Microfinance, Risk Management, and Poverty*. Washington, DC: World Bank.

Segal, L. (1990) *Slow Motion: Changing Masculinities, Changing Men*. London: Virago.

Sen, A. (1983) 'Poor, relatively speaking'. *Oxford Economic Papers*, 35.2: 153–169.

Sen, A. (1987) *On Ethics and Economics*. Oxford: Blackwell.

Sen, A. (1999) *Development as Freedom*. New York: Oxford University Press.

Sen, G. and Grown, C. (1987) *Development, Crises, and Alternative Visions: Third World Women's Perspectives.* New York: Monthly Review Press.

Sennett, R. (1996) *Flesh and Stone: The Body and the City in Western Civilization*. London: W. W. Norton & Co.

Sennett, R. (1998) *The Corrosion of Character: The Personal Consequences of Work in the New Capitalism*. New York: W. W. Norton.

Sharma, K. (2007) *Rediscovering Dharavi: Stories from Asia's Largest Slum* (Penguin India, 2000), available on-line http:// www.unep.org/ourplanet/imgversn/144/sharma.html, accessed 01/11/07.

Shelby, T. (2005) *We Who Are Dark: The Philosophical Foundations of Black Solidarity.* Cambridge, MA: Harvard University Press.

Sheldon, S. and Collier, R. (2006) *Fathers' Rights Activism and Law Reform in Comparative Perspective*. Oxford, and Portland, OR: Hart Publishing.

Sheller, M. and Urry, J. (2000) 'The city and the car'. *International Journal of Urban and Regional Research*, 24: 737–757.

Shelley, T. (2007) *Exploited: Migrant Labour in the New Global Economy*. London: Zed Books.

Shelter (2006) *Cathy Come Home 40 Years On*. London: Shelter.

Sherraden, M. (1991) *Assets and the Poor: A New American Welfare Policy*. Armonk, NY: M. E. Sharpe, Inc.

Shore, C. and Wright, S. (1999) 'Audit culture and anthropology: neo-liberalism in British higher education'. *Journal of the Royal Anthropological Institute*, 5: 557–575.

Short, J. R. (2006) *Urban Theory: A Critical Assessment*. Basingstoke: Palgrave.

Shove, E. (2003) 'Users, technologies and expectations of comfort, cleanliness and convenience'. *European Journal of Social Science Research*, 6.2: 193–206.

Sibley, D. (1990) 'Invisible women? The contribution of the Chicago School of Social Service Administration to urban analysis'. *Environment and planning A*, 22: 733–745.

Sibley, D. (1995) 'Gender, science, politics and geographies of the city'. *Gender Place and Culture*, 2.1: 37–50.

Siegel, P. and Alwang, J. (1999) 'An asset-based approach to social risk management'. SP Discussion Series 9926, *Human Development Network. Social Protection Unit.* Washington, DC: World Bank.

Silvey, R. (2004) 'Power, difference and mobility: feminist advances in migration studies'. *Progress in Human Geography*, 28.4: 490–506.

Silvey, R. and Elmhirst, R. (2003) 'Engendering social capital: women workers and rural urban networks in Indonesia's crisis'. *World Development*, 31.5: 865–879.

Simmel, G. (1905) *The Metropolis and Mental Life.* Harmondsworth: Penguin.

Simonsen, K. (2003) 'The embodied city: from bodily practice to urban life'. In J. Öhman and K. Simonsen (eds) *Voices from the North: New Trends in Nordic Human Geography*. Basingstoke: Ashgate, 157–171.

Skeggs, B. (1997) *Formation of Class and Gender: Becoming Respectable*. London: Sage.

Skelton, T. (2001) 'Girls in the club: researching working class girls' lives'. *Ethics, Place and Environment*, 4.2: 167–173.

Small, M. L. and Newman, K. (2001) 'Urban poverty after *The Truly Disadvantaged*: the rediscovery of the family, the neighbourhood and culture'. *Annual Review of Sociology*, 27: 23–45.

Smart, C. and Neale, B. (1998) *Family Fragments?* Cambridge: Polity Press.

Smith, A., Stenning, A. Rochovská, A. and Świątek, D. (2008) 'The emergence of a working poor: labour markets, neoliberalisation and diverse economies in post-socialist cities'. *Antipode*, 40.2: 283–311.

Smith, D. E. (1989) 'Feminist reflections on political economy'. *Studies in Political Economy*, 30: 37–59.

Smith, D. E. (1990) *The Conceptual Practices of Power: A Feminist Sociology of Knowledge*. Boston: Northeastern University Press.

Smith, J. and Wallerstein, I. (1992) *Creating and Transforming Households: The Constraints of the World Economy*. Cambridge: Cambridge University Press.

Smith, N. (1992) 'Homeless/global: scaling places'. In J. Bird, B. Curtis, T. Putnam, G. Robertson and L. Tickner (eds) *Mapping the Futures: Local Cultures, Global Change*. London: Routledge, 87–121.

Smith, N. (1996) *The New Urban Frontier: Gentrification and the Revanchist City*. London: Routledge.

Smith, N. and Williams, P. (1986) *Gentrification of the City*. Boston: Allen & Unwin.

Smith, R. (2003) 'World city actor-networks'. *Progress in Human Geography*, 27.1: 25–44.

Smits, J., Mulder, C. H. and Hoiimeijer, P. (2003) 'Changing gender roles, shifting power balance and long-distance migration'. *Urban Studies*, 40.3: 603–613.

Soja, E. W. (1989) *Postmodern Geographies: The Reassertion of Space in Critical Social Theory*. London: Verso.

Somerville, J. (2000) *Feminist and the Family: Politics and Society in the UK and USA*. Basingstoke: Macmillan.

Southerton, D. (2003) '"Squeezing time": allocating practices, co-ordinating networks and scheduling society'. *Time and Society*, 12.1: 5–25.

Spain, D. (1993) *Gendered Spaces*. Raleigh, NC: UNC Press.

SPARC India (2008) www.sparcindia.org, accessed 07/03/08.

Sparks, R. E. G. and Loader, I. (2001) 'Fear and everyday urban lives'. *Urban Studies*, 38,5–6: 885–898.

Sparr, P. (1994) *Mortgaging Women's Lives: Feminist Critiques of Structural Adjustment.* London: Zed Books.

Spelman, E. (1988) *Inessential Woman: Problems of Exclusion in Feminist Thought.* Boston: Beacon Press.

Spigel, L. (1992) 'The suburban home companion: television and the neighbourhood ideal in postwar America'. In B. Colmina (ed.) *Sexuality and Space*. Princeton: Princeton Architectural Press, 185–218.

Spivak, G. (1985) 'Can the subaltern speak? Speculations on widow-sacrifice'. *Wedge*, 7/8 (winter/spring): 120–130.

Squires, J. (2005) 'Is mainstreaming transformative? Theorizing mainstreaming in the context of diversity and deliberation'. *Social Politics: International Studies in Gender, State and Society*, 12.3: 366–388.

Stacey, J. (1998) *Brave New Families: Stories of Domestic Upheaval in Late Twentieth Century America*. Berkeley: University of California Press.

Stack, C. (1983) *All Our Kin: Strategies for Survival in a Black Community*. New York: Basic Books.

Staeheli, L. A., Kofman, E. and Peake, L. J. (eds) (2004) *Mapping Women, Making Politics: Feminist Perspectives on Political Geography*. London and New York: Routledge.

Stafford, W. (1995) 'Ferdinand Tönnies on gender, women and the family'. *History of Political Thought*, XVI.3: 391–415.

Stasiulis, D. K. and Bakan, A. B. (eds) (1997) *Not One of the Family: Foreign Domestic Workers in Canada*. Toronto: Toronto University Press.

Stasiulis, D. K. and Bakan, A. B. (2005) *Negotiating Citizenship: Migrant Women in Canada and the Global System*. Toronto: University of Toronto Press.

Steigerwald, B. (2001) *City Views: Urban Studies Legend Jane Jacobs on Gentrification, the New Urbanism and Her Legacy*. Reasononline, available on-line http: //www.reason.com/news/show/28053/html.

Stenson, K. (2007) 'Framing the governance of urban space'. In R. Atkinson and G. Helms (eds) *Securing an Urban Renaissance: Crime, Community, and British Urban Policy*. Bristol: Policy Press, 23–39.

Stephen, L. (1997) *Women and Social Movements in Latin America: Power from Below*. Austin: University of Texas Press.

Stephenson, C. (1982) 'Feminism, pacifism, nationalism and the United Nations Decade for Women'. *Women's Studies International Forum*, 5.3–4: 287–300.

Stichter, S. and Parpart, J. C. (eds) (1988) *Patriarchy and Class: African Women in the Home and the Workforce*. Boulder, CO: Westview Press, 1–12.

Stotko, E. M. and Troyer, M. (2007) 'A new gender-neutral pronoun in Baltimore, Maryland: a preliminary study'. *American Speech*, 82.3: 262–279.

Superson, A. M. and Cudd, A. E. (eds) (2002) *Theorizing Backlash: Philosophical Reflections on the Resistance to Feminism*. New York and Oxford: Rowman & Littlefield.

Susskind. Y. (2008) 'Who is killing the women of Basra?' MADRE organization, 9 January, available on-line http://www.madre.org/articles/me/womenbasra010908.html, last accessed 15/09/08.

Suttles, G. D. (1990) *The Man-made City: The Land-use Confidence Game in Chicago*. Chicago: University of Chicago Press.

Swift, J. (1989) 'Why are rural people vulnerable to famine?' *IDS Bulletin*, 20: 8–15.

Swyngedouw, E. (1997) 'Excluding the other: the production of scale and scaled politics'. In R. Lee and J. Wills (eds) *Geographies of Economies*. London: Arnold, 167–176.

Swyngedouw, E. (2004) *Social Power and the Urbanization of Water*. New York: Oxford University Press.

Tallon, A. (2003) 'Residential transformation and the promotion of inner city centre living'. *Town and Country Planning*, 7: 190–191.

Taylor, P. (1995) 'Beyond containers: inter-nationality, inter-stateness, inter-territoriality'. *Progress in Human Geography*, 19: 1–15.

Theile, B. (1986) 'Vanishing acts in social and political thought: tricks of the trade'. In C. Pateman and E. Gross (eds) *Feminist Challenges: Social and Political Theory.* Sydney: Allen & Unwin.

Thien, D. and Davidson, J. (2009) *Gender Interventions in Research, Teaching and Practice* (special issue). *Journal of Geography in Higher Education*.

Thomas, J. M. (2004) 'Neighbourhood planning: uses of oral history'. *Journal of Planning History*, 3.1: 50–70.

Thomas, M. E. (2005) 'Girls, consumption space and the contradictions of hanging out in the city'. *Social and Cultural Geography*, 6.4: 587– 605.

Thomas, R. (1999) 'Fathers must be more than "walking wallets": Home Office helpline aims to foster a nation of "actively involved" dads'. *Observer*, Society, 25 April, 5.

Thrift, N. (1983) 'On the determination of social action in space and time'. *Environment and Planning D: Society and Space*, 1: 23–57.

Thrift, N. (1995) 'A hyperactive world'. In R. Johnston, P. Taylor and M. Watts (eds) *Geographies of Global Change: Remapping the World in the Late Twentieth Century*. Oxford: Blackwell, 18–35.

Thrift, N. (2000) 'Performing cultures in the new economy'. *Annals of the American Association of Geographers*, 90.4: 674–692.

Tickel, A. and Peck, J. (2003) 'Making global rules: globalisation or neoliberalisation?' *GaWC Research Bulletin* 102, available on-line http://www.lboro.ac.uk/gawc/rb/rb102.html, last accessed 15/09/08.

Tinker, I. (ed.) (1990) *Persistent Inequality: Women and World Development*. Oxford: Oxford University Press.

Tivers, J. (1988) 'Women with young children: constraints on activities in the urban environment'. In J. Little, L. Peake and R. Richardson (eds) *Women in Cities: Gender and the Urban Environment*. London: Macmillan.

Tivers, J. (1999) 'The home of the British Army: the iconic construction of military defence landscapes'. *Landscape Research*, 24.3: 303–319.

Tooke, J. (2000) 'Institutional power geometries: enduring and shifting work relations in cleansing depots'. *Geoforum*, 31.4: 567–574.

Torbin, J. and Nordhaus, W. D. (1972) 'Is growth obsolete?' *Cowles Foundation Paper*, 398: 1–57.

Townley, B. (1994) *Reframing Human Resource Management: Power, Ethics and the Subject at Work.* London: Sage.

Townsend, P. (1971) *The Concept of Poverty*. London: Heinemann Educational Books.

Townsend, P. (1985) 'Sociological approach to the measurement of poverty – a rejoinder to Professor Amartya Sen'. *Oxford Economic Papers*, 37.4: 659–668.

Townshend, J., Zapata, E., Rowlands, J., Alberti, P. and Mercado, M. (1999) *Women and Power: Fighting Patriarchies and Poverty*. London and New York: Zed Books.

Townshend, T. J. (2006) 'From public neighbourhoods to multi-tier private neighbourhoods: the evolving ecology of the neighbourhood privatization in Calgary'. *GeoJournal*, 66.1–2: 103–120.

Toynbee, P. (2003) *Hard Work: Life in Low-pay Britain*. London: Bloomsbury.

Transparency International (2002) *Corruption in South Asia: Insights & Benchmarks from Citizen Feedback Surveys in Five Countries.* Berlin: Transparency International Secretariat.

Tronto, J. (2004) 'Vicious circles of privatised care'. Paper presented in the 'Rethinking Care' session of the CAVE *Rethinking Care Relations, Family Lives and Policies Symposium*, 3–4 December, University of Leeds.

Tuana, N. and Tong, R. (1995) *Feminism and Philosophy*. Boulder, CO: Westview Press.

Ullman, E. (1997) *Close to the Machine*. San Francisco: City Lights.

Umemoto, K. (2001) 'Walking in another's shoes: epistemological challenges in participatory planning'. *Journal of Planning Education and Research*, 21: 17–31.

UN (2005) *World Migration Stock: The Revised 2005 Population Database*. Department of Economic and Social Affairs, Population Division, available on-line http://esa.un.org/migration/index.asp?panel=2, last accessed 15/09/08.

UNDP (1997) *Human Development Report*. Oxford: Oxford University Press.

UNFPA (2001) *State of the World Population. Footprints and Milestones: Population and Environmental Change*. New York: United Nations Population Fund.

UN-Habitat (2000) *Habitat Agenda for Latin America*. New York: UN-Habitat.

UN-Habitat (2003) *The Challenge of Slums: Global Report on Human Settlements 2003*. London: Earthscan.

UNIFEM (2005) *Progress of the World's Women: Women, Work and Poverty*. New York: UNIFEM.

United Nations (1999) *World Urbanization Prospects: The 1999 Revision*, available on-line http://www.prb.org/Educators/TeachersGuides/HumanPopulation/Urbanization.aspx?p=1, last accessed 15/09/08.

United Nations Population Division (2001) *World Population Monitoring 2000, Gender and Development*. New York: United Nations.

United States Census Bureau (USCB) (2004) America's Families and Living Arrangements: 2003. Populations Characteristics. Issued November 2004, available on-line http://www.census.gov/prod/2004pubs/p20–553.pdf (p. 2 and Table 1, p. 3).

UNMP/Task Force on Education and Gender Equality (TFEGE) (2005) *Taking Action: Achieving Gender Equality and Empowering Women.* United Nations Millennium Project. London: Earthscan.

Unni, J. and Rani, U. (2003) 'Social protection for informal workers in India: insecurities, instruments and institutional mechanisms'. *Development & Change*, 34.1: 127–161.

Urban Ecology (2008) *Child Friendly Cities*, available on-line http://www.urbanecology.org.au/, last accessed 06/07/08.

Urry, J. (2004) 'The "system" of automobility'. *Theory, Culture and Society*, 21.4/5: 25–39.

Urry, J. (2006) 'Inhabiting the car'. In S. Böhm, C. Jones, C. Land and M. Paterson (eds) *Against Automobility*. Oxford: Blackwell, 17–32.

US Department of Housing and Urban Development (2004) *An Analysis of Mortgage Refinancing, 2001–3*. Washington, DC: U.S. Department of Housing and Urban Development Office of Policy Development and Research.

Uttal, L. (1999) 'Using kin for child care: embedment in the socioeconomic networks of extended family'. *Journal of Marriage and the Family*, 61: 845–857.

UWEP (Urban Waste Expertise Programme) (2008) http://www.gdrc.org/uem/waste/z-doc.html, last accessed 15/09/08.

Vaiou, D. and Lykogianni, R. (2006) 'Women, neighbourhoods and everyday life'. *Urban Studies*, 43.4: 731–743.

Valentine, G. (1993) 'Negotiating and managing multiple sexual identities: lesbian time–space strategies'. *Transactions of the Institute of British Geographers*, 18.2: 237–248.

Valentine, G. (2001) 'Whatever happened to the social: reflections on the "cultural turn" in British human geography'. *Norsk Geografisk Tiksskrift*, 55.3: 166–172.

Valentine, G. (2005) 'Geography and ethics: moral geographies? Ethical commitment in research and teaching'. *Progress in Human Geography*, 29.4: 483–487.

Valentine, G. (2007) 'Theorizing and researching intersectionality: a challenge for feminist geographers'. *The Professional Geographer*, 59.1: 10–21.

van Hoven, B. and Hörschelmann, K. (eds) (2005) *Spaces of Masculinities*. London: Routledge, 1–16.

van Hoven, B. (2009) 'An experimental account of teaching gender geography in the Netherlands'. *Journal of Geography in Higher Education*. 33.3 (forthcoming).

Van Leeuwen, M. S. (1993) *After Eden: Facing the Challenge of Gender Reconciliation.* New York: Wm. B. Eerdman Publishing.

Vancouver Rape Relief Society (2007) 'Lawyer for male-to-female transsexual complainant asserts that women share nothing in common with each other'. Press release, available on-line http://www.rapereliefshelter.bc.ca/issues/knixon_pr_Aug222003.html, last accessed 15/09/08.

Vasudha, V. (2005) 'Testimonies of harassment'. *India Together*, December, available on-line http://www.indiatogether.org/2005/dec/wom-blank.htm, last accessed 15/09/08.

Verloo, M. (2005) 'Displacement and empowerment: reflections on the concept and practice of the Council of Europe approach to gender mainstreaming and gender equality'. *Social Politics: International Studies in Gender, State and Society*, 12.3: 344–365.

Vincent, F. (1995) *Alternative Financing, Volume I.* Geneva: Development Innovations and Networks (IRED).

Viner, K. (2005) 'A year of killing'. Special Report. *Guardian Weekend*, 10 December.

Viswanath, V. (2001) 'Women's micro-enterprises for food security in India'. *Development*, 44.4: 90–93.

Vuchic, V. R. (1999) *Transportation for Livable Cities.* Rutgers: Centre for Urban Policy Research.

Walby, S. (1990) *Theorising Patriarchy*. Oxford: Blackwell.

Walby, S. (2001) 'Community to coalition: the politics of recognition as the handmaiden of the politics of redistribution'. *Theory, Culture and Society*, 18.2–3: 113–135.

Walby, S. (2005) 'Gender mainstreaming: productive tensions in theory and practice'. *Social Politics: International Studies in Gender, State and Society*, 12.3: 321–343.

Walkerdine, V. (2003) 'Reclassifying upward mobility: femininity and the neo-liberal subject'. *Gender and Education*, 15.3: 237–248.

Warde, A. (1991) 'Gentrification as consumption: issues of class and gender'. *Environment and Planning D: Society and Space*, 9: 223–232.

Waring, M. (1990) *Counting for Nothing: What Men Value and What Women Are Worth.* Wellington, NZ: Bridget Williams.

Warren, B. (1980) *Imperialism: Pioneer of Capitalism*. London: Verso.

Watts, J. (2007) 'Migrants suffering for China boom, says study'. *Guardian*, 2 March, available on-line http://www.guardian.co.uk/world/2007/mar/02/china.jonathanwatts.

Weber, R. (2002) 'Extracting value from the city: neoliberalism and urban redevelopment'. *Antipode*, 34.3: 519–540.

Weidemann, J. (1995) *Microenterprise and Gender in India: Issues and Options*. Gemini Technical Report No. 93. Washington, DC: USAID.

WEDO (2003) *Diverting the Flow: A Resource Guide to Gender, Rights and Water Privatization*. New York: WEDO.

Wekerle, G. R. (1984) 'A woman's place is in the city'. *Antipode*, 17.2–3: 145–153.

WGSG (Women and Geography Study Group) (1997) *Feminist Geographies: Explorations in Diversity and Difference*. Harlow: Addison Wesley Longman.

Wheelock, J. (1990) *Husbands at Home: The Domestic Economy in a Post-industrial Society*. London and New York: Routledge.

Wheelock, J. and Oughton, E. (1996) 'The household as a focus for research'. *Journal of Economic Issues*, 30.1: 143–159.

WhichFranchise (2001) 'Report on MollyMaid market share', available on-line http://www.whichfranchise.com/franchisorPage.cfm?companyId=295, last accessed 22/06/07.

White, M. (2001) 'GATS and women'. *Foreign Policy in Focus*, 6.2, January.

Whitelegg, J. (1992) *Critical Mass*. London: Pluto Press.

Whitelegg, J. (1997) *Critical Mass: Transport, Environment and Society in the Twenty-first Century*. London: Pluto Press.

Whitely, N. (1993) *Design for Society*. London: Reaktion Books.

Williams, A. M., King, R. and Warnes, T. (1997) 'A place in the sun: international retirement migration from northern to southern Europe'. *European Urban and Regional Studies*, 4.2: 115–154.

Williams, C. C. and Windebank, J. (2000) 'Reconceptualising paid informal exchange: some evidence from English cities'. *Environment and Planning A*, 31: 121–140.

Williams, F. and Roseneil, S. (2004) 'Public values of parenting and partnering: voluntary organizations and welfare politics in New Labour's Britain'. *Social Politics*, 11.2: 181–216.

Williams, M. (2002) 'IMF–World Bank–WTO coherence: the implications for water privatization'. Presentation for GERA/TWN Public Forum. Accra, Ghana, 27 November.

Willmott, P. (1987) 'Community and social structure'. *Policy Studies*, 8.1: 52–63.

Willott, S. and Griffin, C. (1996) 'Men, masculinity and the challenge of long term unemployment'. In M. Mac An Ghaill (ed.) *Understanding Masculinities*. Milton Keynes: Open University Press.

Wilson, E. (1991) *The Sphinx in the City: Urban Life, the Control of Disorder, and Women*. London: Verso.

Wilson, E. (1995) 'The invisible flâneur'. In S. Watson and K. Gibson (eds) *Postmodern Cities and Spaces*. Oxford: Blackwell, 59–80.

Wilson, E. (1997) 'Looking backward, nostalgia and the city'. In S. Westwood and J. Williams (eds) *Imagining Cities: Scripts, Signs, Memories*. London: Routledge, 125–137.

Wilson, J. Q. and Kelling, G. L. (1982) 'Broken windows: the policy and neighbourhood safety'. *Atlantic Monthly*, March.

Wilson, W. (1987) *The Truly Disadvantaged. The Inner City, the Underclass and Public Policy*. Chicago: University of Chicago Press.

Wilson, W. (1996) *When Work Disappears: The World of the New Urban Poor*. New York: Knopf.

Winnicott, D. W. (1960) 'The theory of the parent–infant relationship'. *International Journal of Psycho-Analysis*, 41: 585–595.

Witz, A. (2007) 'Georg Simmel and the masculinity of modernity'. *Journal of Classical Sociology*, 1.3: 353–370.

Wolff, J. (1995) *Resident Alien: Feminist Cultural Criticism*. New Haven, CT: Yale University Press.

Wolff, J. (2000) 'The feminine in modern art: Benjamin, Simmel and the gender of modernity'. *Theory, Culture and Society*, 17.6: 33–53.

Women and Environments Network (2008) available on-line http://www.utoronto.ca/iwsgs/we.mag/womenet.html, last accessed 15/09/08.

Wood, G. (2001) 'Desperately seeking security'. *Journal of International Development*, 13.5: 523–534.

Wood, G. (2003) 'Staying secure, staying poor: the "Faustian bargain"'. *World Development*, 31.3: 455–471.

Wood, G. and Gough, I. (2006) 'A comparative welfare regime approach to global social policy'. *World Development*, 34.10: 1696–1712.

Wood, G. and Salway, S. (2000) 'Introduction: securing livelihoods in Dhaka slums'. *Journal of International Development*, 12.5: 669–688.

Woodward, A. (2003) 'European gender mainstreaming: promises and pitfalls of transformative policy'. *Review of Policy Research*, 20.1: 65–88.

World Bank (1993) *Housing: Enabling Markets to Work with Technical Supplements*. Washington, DC: World Bank.

World Bank (2001a) *World Development Report: Attacking Poverty*. Oxford: Oxford University Press.

World Bank (2001b) *Engendering Development*. Oxford: Oxford University Press.

World Bank (2003) *World Development Report*, available on-line http://www.dynamicsustainabledevelopment.org/.

Wratten, E. (1995) 'Conceptualizing urban poverty'. *Environment and Urbanization*, 7: 11–36.

Wylie, I. (2006) 'Balancing work and life'. *Guardian*, Saturday, 14 October, also available on-line http://www.guardian.co.uk/money/2006/oct/14/careers.work1.

Young, B. (2005) 'Globalization and shifting gender governance order(s)'. *Sowi-online Journal*, available on-line http://www.sowi-online.de/journal/2005–2/globalization_young.html.

Young, I. M. (1995) 'The ideal of community and the politics of difference'. In P. A. Weiss and M. Friedman (eds) *Feminism and Community*. New York: Temple University Press, 233–259.

Young, I. M. (2005) 'Justice and the politics of difference'. In S. S. Fainstein and L. J. Servon (eds) *Gender and Planning: A Reader*. Brunswick, NJ: Rutgers University Press, 86–103.

Yunus, M. (1994) 'Redefining development'. In K. Danaher (ed.) *50 Years Is Enough: The Case against the World Bank and the International Monetary Fund*. Boston: South End Press.

Yuval-Davis, N. (1999) 'The multi-layered citizen'. *International Feminist Journal of Politics*, 1.1: 119–136.

Zanetta, C. (2001) 'The evolution of the World Bank's urban lending in Latin America: from sites and services to municipal reform and beyond'. *Habitat International*, 25: 513–533.

Zhang Ye (2002) 'Hope for China's migrant women workers'. *China Business Review*, May–June, available on-line http://www.chinabusinessreview.com/public/0205/ye.html, last accessed 15/09/08.

Zook, M. (2001) 'Old hierarchies or new networks of centrality? The global geography of the Internet content market'. *American Behavioural Scientist*, 44.10: 1679–1696.

Index

Numbers in **bold** represent figures